Reconfigurable Computing

João M.P. Cardoso • Michael Hübner
Editors

Reconfigurable Computing

From FPGAs to Hardware/Software Codesign

Springer

Editors
João M.P. Cardoso
Departamento de Engenharia Informática
Faculdade de Engenharia (FEUP)
Universidade do Porto
Rua Dr. Roberto Frias, s/n,
4200–465 Porto, Portugal
jmpc@acm.org

Michael Hübner
Institut für Technik der
Informationsverarbeitung, Fakultät für
Elektrotechnik und Informationstechnik
Karlsruher Institut für Technologie (KIT)
Kaiserstr. 12
Karlsruhe, Germany
michael.huebner@kit.edu

ISBN 978-1-4899-8859-1 ISBN 978-1-4614-0061-5 (eBook)
DOI 10.1007/978-1-4614-0061-5
Springer New York Dordrecht Heidelberg London

Printed on acid-free paper

Springer is part of Springer Science+Business Media (www.springer.com)

Preface

Dr. Panagiotis Tsarchopoulos

The objective of the European research programme in Information and Communication Technologies (ICT) is to improve the competitiveness of European industry and enable Europe to master and shape future developments in ICT. ICT is at the very core of the knowledge based society. EU research funding has as target to strengthen Europe's scientific and technology base and to ensure European leadership in ICT, help drive and stimulate product, service and process innovation and creativity through ICT use and value creation in Europe, and ensure that ICT progress is rapidly transformed into benefits for Europe's citizens, businesses, industry and governments.

Over the past years, the European Commission has constantly increased the amount of funding going to research in computing architectures and tools through the European research programme in Information and Communication Technologies. In this context, the European Commission has funded a number of European research projects in the area of reconfigurable computing. Results from these projects are being presented in this book providing a valuable reference point, which describes the efforts of several international research teams.

Reconfigurable Computing is a fascinating alternative to mainstream computing. But is it always going to remain just an alternative occupying a market niche? The potential for reconfigurable computing has not yet been fully unleashed although there have been notable successes – mostly for 'fine-grain' reconfigurability. Now there are technological developments and market opportunities that suggest breakthroughs in the future for coarse-grain reconfigurability – a field in which Europe has particular strengths. As a matter of fact, the 'coarse-grain' market is showing increasing potential. Tile-based architectures, for example, offer a balance of flexibility and ease of programming, drawing on libraries of pre-defined functionality. Europe has a considerable track-record in research into coarse-grained reconfigurability, and this offers a re-entry route for Europe into the Reconfigurable Computing market, based upon an integrated approach of hardware together with development systems for specific application domains.

In reconfigurable computing, one important observation – that is also becoming reality in mainstream computing with the advent of multicore architectures – is that parallelism is omnipresent. Most reconfigurable computing exploits the potential for parallel processing as much as possible using different "flavours" of parallelism. The exploding interest in parallelism presents another opportunity for reconfigurable computing.

If it is to be effective, any European strategy for RTD in Reconfigurable Computing *must* be set in the context of its potential use by applications developers and systems designers. It must take account of the market – the market for supply of Reconfigurable Computing technologies; the evolution of the general purpose computing market; and the markets of the users. And for embedded systems applications, it must also take account of the evolution of the methodologies ad requirements of the users. Technology is not enough.

The markets for Reconfigurable Computing may be divided into two – High Performance Computing and Embedded Systems. These markets – and the technological solutions appropriate to them – are quite different. However, they share one very important property: the fundamental obstacle to take-up of Reconfigurable Computing is the difficulty of programming. While localised solutions might be devised for specific technologies, such solutions are generally not viable, given their limited markets.

The highest priority need for RTD is therefore to enable commercially viable programmability of Reconfigurable Computing technology. This requires coherent, integrated (or "integrable") suites of processes, methods and tools spanning:

- application level support for reconfigurability that supplements existing design methodologies, including support for verification and validation of reconfigurable behaviour and reconfigurability properties of the system so as to satisfy qualification requirements;
- mapping from the output of application design to reconfigurable hardware via intermediate layer(s) of abstraction with standard libraries of functions based on open and widely accepted standards; and
- run-time support for reconfiguration, typically through OS extensions for resource allocation, scheduling, and discovery; debugging and monitoring; and fast re-layout of reconfigurable units.

Future European RTD in these topics must recognise the need for compatibility with development paradigms and processes, methods and tools in the applications sectors. Indeed, RTD in Reconfigurable Computing should be application-driven. Application sectors where Europe could gain particular advantage include embedded healthcare, (multi)physical system modeling, biomedical, cognitive radio, portable consumer devices, automotive/avionics, infotainment, and user-driven reconfigurable products.

The book that you have in your hands will give you a glimpse of the future: research results that will be coming out of labs towards market introduction; unresolved issues and new research challenges that need to be solved; relentless efforts to produce the last missing piece of magic that will make everything work.... but above all, I am sure, you will feel the enthusiasm and passion of the researchers and engineers that make all this happen.

Dr. Panagiotis Tsarchopoulos
ICT Research Programme
European Commission

Disclaimer: The views expressed are those of the author and do not necessarily represent the official view of the European Commission on the subject.

Contents

Contributors

Tapani Ahonen Tampere University of Technology, Tampere, Finland

José Carlos Alves Departamento de Engenharia Electrótecnica, Faculdade de Engenharia (FEUP), Universidade do Porto, Porto, Portugal

Michel Auguin CNRS, Orsay, France

Juergen Becker Institut fur Technik der Informationsverarbeitung, Fakultat fur Elektrotechnik und Informationstechnik, Karlsruhe Institute für Technology, Karlsruhe, Germany

Jürgen Becker Institut für Technik in der Informationsverarbeitung (ITIV), Karlsruhe Institute of Technology KIT, Karlsruhe, Germany

Koen Bertels Computer Engineering Lab, Faculty Electrical Engineering, Mathematics and Computer Science, Technische Universiteit Delft, TUD, Delft, The Netherlands

Labros Bisdounis INTRACOM, Athens, Greece

Umberto Bondi Università della Svizzera italiana, Lugano, Switzerland

Philippe Bonnot Thales Research & Technology, Paris, France

Timon D. ter Braak University of Twente, Enschede, The Netherlands

Lars Braun Institut für Technik der Informationsverarbeitung, Fakultät für Elektrotechnik und Informationstechnik, Karlsruher Institut für Technologie (KIT), Karlsruhe, Germany

Paul Brelet Thales Research & Technology, Paris, France

Stephen T. Burgess Tampere University of Technology, Tampere, Finland

Marcel van de Burgwal University of Twente, Computer Science, Enschede, The Netherlands

Joan Cabestany Universitat Politècnica de Catalunya, Catalonia, Spain

Fabio Campi STMicroelectronics SRL, Agrate Brianza, Italy

João M.P. Cardoso Departamento de Engenharia Informática, Faculdade de Engenharia (FEUP), Universidade do Porto, Porto, Portugal

Luigi Carro Universidade do Rio Grande do Sul, Passo Fundo, Brazil

Marcelo Cintra University of Edinburgh, Edinburgh, UK

George A. Constantinides Department of Electrical & Electronic Engineering, Imperial College London, London, UK

José Gabriel F. de Coutinho Department of Computing, Imperial College London, London, UK

Martin Danek UTIA AV CR, Ostrava, Czech Republic

Giuseppe Desoli ST Microelectronics, Agrate Brianza, Italy

Jean-Philippe Diguet CNRS, Orsay, France

Pedro C. Diniz Electronic Systems Design and Automation Research Group, INESC-ID, Lisboa, Portugal

Susan Eisenbach Imperial College, London, UK

Fabrizio Ferrandi Dipartimento di Elettronica e Informazione Politechnico di Milano, Milano, Italy

Alberto Ferrante Università della Svizzera italiana, Lugano, Switzerland

João Canas Ferreira Departamento de Engenharia Electrótecnica, Faculdade de Engenharia (FEUP), Universidade do Porto, Porto, Portugal

Paolo Gai Evidence, Edinburgh, Italy

Christian Gamrat CEA, LIST, Centre de Saclay - Point Courrier 94, Gif sur Yvette Cedex, France

Georgi N. Gaydadjiev Computer Engineering, TU Delft, The Netherlands

Richard Geißler Atmel Automotive GmbH, Heilbronn, Germany

Roberto Giorgi Universita' degli Studi di Siena, Siena, Italy

Fernando Gonçalves Coreworks – Projectos de Circuitos e Sistemas Electrónicos S.A., CW, Porto, Lisboa, Portugal

Kim Grüttner OFFIS – Institute for Information Technology, Oldenburg, Germany

Arnaud Grasset Thales Research & Technology Campus Polytechnique1, Palaiseau Cedex, France

Reiner Hartenstein Fachbereich Informatik, Technische Universität Kaiserslautern, Baden-Baden, Germany

Philipp A. Hartmann OFFIS – Institute for Information Technology, Oldenburg, Germany

Andreas Herrholz OFFIS – Institute for Information Technology, Oldenburg, Germany

Paul M. Heysters Recore Systems, Enschede, The Netherlands

Michael Hübner Institut für Technik der Informationsverarbeitung, Fakultät für Elektrotechnik und Informationstechnik, Karlsruher Institut für Technologie (KIT), Karlsruhe, Germany

Heikki Hurskainen Tampere University of Technology, Tampere, Finland

Chris Jesshope University of Amsterdam, Amsterdam, The Netherlands

Jiri Kadlec UTIA AV CR, Ostrava, Czech Republic

Stefanos Kaxiras Industrial Systems Institute, Patras, Greece

Hans G. Kerkhoff University of Twente, Enschede, The Netherlands

André B. J. Kokkeler University of Twente, Enschede, The Netherlands

Ralf König Institut für Technik in der Informationsverarbeitung (ITIV), Karlsruhe Institute of Technology KIT, Karlsruhe, Germany

Kamil Krátký Advanced Technology Europe, Honeywell International, Brno, Czech Republic

Matthias Kühnle Institut für Technik der Informationsverarbeitung, Fakultät für Elektrotechnik und Informationstechnik, Karlsruher Institut für Technologie (KIT), Karlsruhe, Germany

Georgi Kuzmanov Computer Engineering Lab, Faculty Electrical Engineering, Mathematics and Computer Science, Technische Universiteit Delft, TUD, Delft, The Netherlands

Eric Lenormand THALES, Paris, France

Wayne Luk Department of Computing, Imperial College London, London, UK

Giacomo Marchiori Dipartimento di Fisica Università di Ferrara, Ferrara, Italy

Debora Motta Matos Universidade do Rio Grande do Sul, Passo Fundo, Brazil

Sally A. Mckee Chalmers University, Gothenburg, Sweden

Philippe Millet Thales Research & Technology, Paris, France

Razvan Nane Computer Engineering Lab, Faculty Electrical Engineering, Mathematics and Computer Science, Technische Universiteit Delft, TUD, Delft, The Netherlands

Jari Nurmi Tampere University of Technology, Tampere, Finland

Bryan Olivier ACE Associated Compiler Experts b.v., Amsterdam, The Netherlands

Frank Oppenheimer OFFIS – Institute for Information Technology, Oldenburg, Germany

Juha Pärsinnen VTT, Espoo, Finland

Zlatko Petrov Advanced Technology Europe, Honeywell International, Brno, Czech Republic

Jean-Marc Philippe CEA, LIST, Paris, France

Wolfram Putzke-Roeming Deutsche Thomson OHG, Hanover, Germany

Nikola Puzovic Universita' degli Studi di Siena, Siena, Italy

Jussi Raasakka Tampere University of Technology, Tampere, Finland

Gerard Rauwerda Recore Systems, Enschede, The Netherlands

Jean-Luc Roux ACIES, Paris, France

Mateus Rutzig Universidade do Rio Grande do Sul, Passo Fundo, Brazil

Axel Schneider Alcatel-Lucent Deutschland AG, Stuttgart, Germany

Alex Shafarenko University of Hertfordshire, Hatfield, UK

Eberhard Schüler PACT XPP Technologies AG, Munich, Los Gatos, Germany

Vlad Mihai Sima Computer Engineering Lab, Faculty Electrical Engineering, Mathematics and Computer Science, Technische Universiteit Delft, TUD, Delft, The Netherlands

Gerard J.M. Smit University of Twente, Computer Science, Enschede, The Netherlands

Hans van Someren ACE Associated Compiler Experts b.v., Amsterdam, The Netherlands

Ioannis Sourdis Computer Engineering, Chalmers University of Technology, Gothenburg, Sweden

Kim Sunesen Recore Systems, Enschede, The Netherlands

Benoit Tain CEA, LIST, Paris, France

Florian Thoma Institut für Technik der Informationsverarbeitung, Fakultät für Elektrotechnik und Informationstechnik, Karlsruher Institut für Technologie (KIT), Karlsruhe, Germany

Raffaele Tripiccione Dipartimento di Fisica Università di Ferrara, Ferrara, Italy

Bart Vermeulen NXP Semiconductors, Eindhoven, The Netherlands

Nikolaos S. Voros Department of Telecommunication Systems & Networks, Technological Educational Institute of Mesolonghi, Mesolonghi, Greece

Stephan Wong Technische Universiteit Delft, Delft, The Netherlands

Ayal Zaks IBM, Haifa, Israel

Xiao Zhang University of Twente, Enschede, The Netherlands

Henk van Zonneveld Thales Netherlands, Hengelo, The Netherlands

Chapter 1
Introduction

João M.P. Cardoso and Michael Hübner

1.1 Introduction

Reconfigurable Computing has evolved to a mature level, with a wide range of applications, from high-performance to embedded computing domains. The research community has witnessed exciting advancements, promoted by technology and scientific findings. Early stages of reconfigurable computing were dominated by the searching for killer applications. The real applications of reconfigurable computing came, in some cases as not as one expected, and widespread over almost all computing domains. However, there are still many challenges and open avenues to do research in this area and this fact makes reconfigurable computing one of the most exciting research areas related to computing!

European industry and academia have been very dynamic on researching Reconfigurable Computing architectures, tools, and applications (see, e.g., [1, 2]). In the last 5 years, the EU (European Union) has funded various IP (Integrated Projects) and STREP (Specific Targeted Research Projects) projects through FP6 and FP7 frameworks. Projects such as MORPHEUS, hArtes, 4s, AETHER, ANDRES, REFLECT, CRISP, ERA, include or have included many of the academic groups working on reconfigurable computing and relevant companies interested in reconfigurable computing in Europe.

This book mainly consists of contributions from those projects. By including some of the most active research groups and institutions in Europe, we believe this book is representative of the R&D performed in Europe and partially funded by the EU about reconfigurable computing. There were many other compelling and outstanding research projects in Europe that would deserve to be included in this book.

J.M.P. Cardoso (✉)
Departamento de Engenharia Informática, Faculdade de Engenharia (FEUP),
Universidade do Porto, Rua Dr. Roberto Frias, s/n, 4200–465 Porto, Portugal
e-mail: jmpc@acm.org

J.M.P. Cardoso and M. Hübner (eds.), *Reconfigurable Computing: From FPGAs to Hardware/Software Codesign*, DOI 10.1007/978-1-4614-0061-5_1,

However, it has been our intention since the very beginning to edit a book about the research projects about reconfigurable computing funded by EU. We call the attention to the fact that the projects presented in this book are also unique in the sense they had industrial partners and give a real and practical image of reconfigurable computing.

There are several books about reconfigurable computing (see, e.g., [3–6]), especially ones focused on specific aspects, such as compilation techniques, design automation, architectures, applications and tools. However, there is no book showing a representative landscape of the bigger European R&D projects carried in the last 5 years. This book allows researchers to get a high quality overview to the achieved results in the reconfigurable computing area. Furthermore, it can be a guideline to prepare novel project proposals, which are aligned on the roadmap which was defined by the projects described in this book. This book is therefore more than a collection of results, it is a scientific guideline for readers and interested scientists.

We believe that this book can be used in courses about reconfigurable computing, in M.Sc. and Ph.D. programs, and as a library reference. It is a source of different perspectives, results, goals, etc. With this book, M.Sc. and Ph.D. students can learn about challenges addressed in recent EU projects. This is a mandatory soft skill for future researchers, which is definitely appreciated in this book in synergy with high level technical descriptions.

1.2 Overall Organization

This book is organized in eleven chapters. Chapters 2 and 3 present the relevance of reconfigurable computing and its main challenges, especially the most relevant ones as seen by members of the Reconfigurable Computing Cluster of HiPEAC NoE (High Performance and Embedded Architectures and Compilation Network of Excellence). The subsequent chapters present eight projects related to reconfigurable computing and partially funded by the EU. We include six projects that have already finished (Chaps. 4–9) and two projects that started in the beginning of 2010 (Chaps. 10 and 11). We now briefly describe each of the subsequent chapters.

Chapter 2 gives a clear introduction of reconfigurable computing and the main relevance and challenges. The chapter aims at identifying the main reasons why reconfigurable computing should become mainstream. The chapter was written by Reiner Hartenstein and we think that it comes in the right moment, not only because the many-core trend in general and the widespread of embedded computing in particular, but also because it reflects a personal view supported by the many years of experience of Reiner in the subject. Reiner has been pushing forward the reconfigurable computing area since many years. Reiner and his students have proposed one of the first reconfigurable computing machines and the associated tool support: the Xputer machine paradigm [7]. His visions and work has influenced many of us and we are very proud to start this book with his chapter.

Chapter 3 presents a number of issues which need to be addressed in order to make Reconfigurable Computing a widely used solution for future systems. The chapter is a result of the fruitful discussions among members of the Reconfigurable Computing Cluster of the HiPEAC NoE. The chapter starts with an introduction and then describes the technology trends according to power wall, transistor count, clock frequency, reliability, and how these trends affect reconfigurable computing. In a subsequent section are enumerated the eight most important challenges in reconfigurable computing according to the authors and as a corollary of the opinions expressed in the Reconfigurable Computing Cluster meetings.

Chapter 4 presents the MORPHEUS project. The MORPHEUS project had the objective to develop a heterogeneous chip featuring an ARM9 General Purpose Processor which manages three types of FPGA technology with different granularity. On the programming side, the project developed tools to partitioning an application on this architecture, to generate the code to synchronize the General Purpose Processor and the accelerator IPs mapped on FPGA, and to schedule the data transfers. One of the most achievements of MORPHEUS was undoubtedly the fabrication of an integrated circuit with the MORPHEUS proposed heterogeneous architecture. The chapter clearly describes the MORPHEUS project and its main achievements and is a privileged medium to start reading about it.

Chapter 5 describes the hArtes project. The hArtes project addressed the development of heterogeneous hardware platforms comprising embedded processors, digital signal processing and reconfigurable hardware. The hArtes project targets the entire design cycle for a heterogeneous computing platform and using a conventional approach in the hardware generation for its algorithm exploration and hardware-software system co-design tool chain. hArtes proposes an end-to-end design flow for developing real-time systems by mapping input applications on the target reconfigurable heterogeneous system. The input application consists of C code, manually developed or generated from other sources such as MATLAB or a graphical representation. The project included hardware/software partitioning techniques and also addressed techniques to take advantage of the dynamic nature of reconfigurable hardware.

Chapter 6 describes the 4S project (Smart Chips for Smart Surroundings). The overall goal of the 4S project was to define and develop efficient, flexible, reconfigurable core building blocks, including the supporting tools, for future Ambient System Devices. Reconfigurable hardware was used to enable an adaptation to fast changing standards, e.g., in mobile communication, and to support current and future trends in software defined/cognitive radio. 4S also included the development of a Network-on-Chip which connected the different heterogeneous IP-cores on the chip. A further step forward which was achieved with the 4S project was to develop a hardware component aware add-on for the selected operation system. For this purpose, the hardware scheduling software OSYRES was used to realize a situation and quality of service tailored task mapping on the different hardware components of 4S.

Chapter 7 presents the AETHER project. AETHER, completed at the end of 2008, had the objective to study languages, middleware, and architectures for future self-adaptive devices. The AETHER project includes hardware aspects of self-adaptive computing elements and novel programming paradigms and software techniques to efficiently use self-adaptive computing entities, and also looks at critical aspects of interfacing hardware and software. They defined the concept of a SANE, the smallest computing entity at which a local and autonomous decision process occurs in relation to its own operation. A SANE is composed of four parts: a data flow computing engine, a local control unit, a communication mechanism, and a local monitoring of computation. AETHER was able to define a standardizable language to define huge networked multiprocessor systems. With the novel description language called S-Net, it is possible to develop scalable networks with heterogeneous nodes as processing element. The achievements led to further projects with the goal to improve the language and to exploit it on further application scenarios.

Chapter 8 presents the ANDRES project (Analysis and Design or runtime Reconfigurable, heterogeneous Systems). The ANDRES project addressed a tool chain to design runtime reconfigurable and heterogeneous systems using SystemC. ANDRES includes the translation of adaptive models onto RTL (Register-Transfer Level) descriptions for being implemented with runtime reconfigurable hardware. Designing embedded hardware/software systems on a higher level of abstraction emphasizing in particular the application of run-time reconfigurable architectures was one of the main goals of ANDRES. ANDRES addressed an operating-system like approach for supporting dynamic reconfiguration.

Chapter 9 describes the CRISP project. CRISP was an EU co-funded Project that proposed a scalable reconfigurable multi-core architecture for data streaming applications. CRISP uses coarse-grained reconfigurable cores (based on MONTIUM) and a network on a chip concept. In the project, novel multiprocessor architectures were evaluated and developed. The CRISP project main topics were to create a highly scalable and dependable reconfigurable system concept for a wide range of tomorrow's streaming DSP applications. Also in this project, a network-on-chip based multicore processor approach was used to handle the streaming applications. For this purpose, a novel dependable infrastructure with a run-time resource management was developed and demonstrated the coarse-grained core-level reconfigurable scalable system. The scalability was demonstrated in terms of computational power, flexibility for different applications, and power consumption.

Chapter 10 presents a recent FP7 EU project focused on researching new methodologies in both tools and hardware design to break through power and memory walls in embedded systems. The ERA ultimate objective is to help design the next-generation embedded systems platforms. To accomplish these objectives, ERA uses adaptive hardware to provide the highest possible performance with limited power budgets. The target architecture of ERA consists of a mixture of commercially available off-the-shelf processor cores, industry-owned IP cores, and application-specific/dedicated cores. The cores used in ERA are dynamically adapted regarding their composition, organization, and even instruction-set architectures in order to provide the best possible performance/power trade-offs.

Chapter 11 presents another recent FP7-ICT EU partially funded project aiming at developing a new design flow for FPGA-based embedded computing systems. The project uses an aspect-oriented approach to complement the description of an application (in C or MATLAB) with non-functional requirements, design patterns, and user's knowledge about the input application. Such approach has the particularly to allow users to control and guide the design flow to generate FPGA-based implementations consisting of a microprocessor connected to hardware cores that best fit the requirements. REFLECT exploits the use of aspect oriented methods enabling the use of FPGA architectures using hardware templates and code generation techniques with higher abstraction levels. E.g., the REFLECT approach can generate efficient streaming processing solutions by using design patterns specially tailored to data-streaming behaviors. In fact, REFLECT ensures the capability to generate customized cores implemented by reconfigurable logic hardware. Thus, REFLECT is focused on both the programming aspects and design flow strategies for mapping computations to multi-core reconfigurable architectures.

1.3 Summary

This first chapter introduced the content of the book. As can be seen from the description about each of the chapters, this book aims at giving an overview about some of the most compelling European projects about reconfigurable computing, all partially funded by the EU under FP6 and FP7 framework programs. We believe that these projects will help to shape the face of the reconfigurable computing field.

Finally, we have a special wish to the readers: we hope you enjoy reading this book as much as we have enjoyed organizing and working on it.

Acknowledgments We are fully grateful to Springer and especially to our Springer editor, Charles Glaser, Springer USA. He gave us his fully support since the very beginning.

We gratefully acknowledge Dr. Panagiotis Tsarchopoulos (ICT Research Programme, EU Commission) by his contribution with the preface to this book.

We are extremely gratefully to the authors of the chapters by their contributions. Without them, this book would not have been a reality.

We would like to acknowledge all the work performed by the reviewers of the chapters in this book. Each chapter was reviewed by one or more co-authors of the additional chapters in the book. They have been very important to help the authors to improve their chapters.

The recent funding support given by DAAD/FCT to a collaborative project involving the Karlsruhe Institute of Technology (KIT) and the University of Porto, and a past funding support given by DAAD/CRUP to a collaborative project involving the University of Karlsruhe and the INESC-ID in Lisbon have been a great help as they allowed a stronger cooperation between us.

Finally, we are very thankful to the EU FP7 REFLECT project as it also helped us both direct and indirectly.

References

1. Nikolaos S. Voros, Alberto Rosti and Michael Hübner, *Dynamic System Reconfiguration in Heterogeneous Platforms: The MORPHEUS Approach*, Lecture Notes in Electrical Engineering, Volume 40, Springer, 2009.
2. K. Bertels, V. Sima, Y. Yankova, G. Kuzmanov, W. Luk, J. Coutinho, F. Ferrandi, C. Pilato, M. Lattuada, D. Sciuto, A. Michelotti, "HArtes: Hardware-Software Codesign for Heterogeneous Multicore Platforms," *IEEE Micro*, 30(5): 2010, pp. 88–97.
3. Christophe Bobda, *Introduction to Reconfigurable Computing: Architectures, algorithms and applications*, Springer, 1st Edition (November 9, 2007).
4. Maya B. Gokhale, and Paul S. Graham, *Reconfigurable Computing: Accelerating Computation with Field-Programmable Gate Arrays*, Springer, 1st Edition (December 14, 2005).
5. João M. P. Cardoso, and Pedro C. Diniz, *Compilation Techniques for Reconfigurable Architectures*, Springer, 1st Edition (October, 2008).
6. Scott Hauck, and André DeHon, *Reconfigurable Computing: The Theory and Practice of FPGA-Based Computation (Systems on Silicon)*, Morgan Kaufmann (November 16, 2007).
7. Reiner W. Hartenstein, Alexander G. Hirschbiel, M. Weber, "Xputers: An Open Family of Non-Von Neumann Architectures," in ARCS'1990, P. Müller-Stoy (Ed.): Architektur von Rechensystemen, Tagungsband, 11. ITG/GI-Fachtagung, 7.-9. März 1990, München, Germany. VDE Verlag 1990, ISBN 3-8007-1688-7, pp. 45–58.

Chapter 2
The Relevance of Reconfigurable Computing

Reiner Hartenstein

Abstract This chapter introduces the highly promising future role of Reconfigurable Computing (RC) and emphasizes, that it is a critical survival issue for computing-supported infrastructures worldwide and stresses the urgency of moving RC from niche to mainstream. It urges acceptance of the massive challenge of reinventing computing, away from its currently obsolete CPU-processor-centric Aristotelian CS world model, over to a twin-paradigm Copernican model including and interlacing both, software and configware. It gives a flavor of the fundamentals of RC and the massive impact on the efficiency of computing it promises. Furthermore the chapter outlines the educational barriers we have to surmount and the urgent need for major funding on a global scale to run a world-wide mass movement, of a dimension at least as far reaching as the Mead-&-Conway-style VLSI design revolution in the early 1980s. The scenarios are similar: around 1980 an urgently needed designer population has been missing. Now a properly qualified programmer population is not existing. But this time the scenario is much more complex and the problem is more difficult, requiring not only a twin-paradigm approach for programming heterogeneous systems including both: many-core processors and reconfigurable accelerators, but also to find a solution to the parallelism crisis also called the "Programming wall". The presentation of recent R&D advances in RC, especially those ones funded by the EU, are also subject of all other chapters of this book.

R. Hartenstein (✉)
Fachbereich informatik, Technische Universität Kaiserslautern,
Traubenstr. 29, 76532 Baden-Baden, Germany
e-mail: reiner@hartenstein.de

J.M.P. Cardoso and M. Hübner (eds.), *Reconfigurable Computing: From FPGAs to Hardware/Software Codesign*, DOI 10.1007/978-1-4614-0061-5_2,

2.1 Introduction

Reconfigurable computing is an important survival issue of the world-wide computing infrastructures. This chapter stresses, that all our computer-based infrastructures worldwide are extremely important to avoid a massive crisis of the global and local economy (Sects. 2.1.1 and 2.2.3). This chapter warns of the future unaffordability of the electricity consumption of the entirety of all computers worldwide, visible and embedded (Sect. 2.2), and, that green computing, although important and welcome, is by far not sufficient to guarantee affordability and not at all to support further progress for future applications of high performance computing (Sect. 2.2.2).

Reconfigurable Computing, the second RAM-based machine paradigm offers drastic reduction of the electric energy budget and speedup factors by up to several orders of magnitude – compared to using the von Neumann paradigm [1], now beginning to loose its dominance. This chapter stresses the urgency of moving Reconfigurable Computing (RC) from niche to mainstream (Sect. 2.2.3) and urges, that we need a worldwide mass movement of a larger format than that of the VLSI design revolution around 1980, where only an urgently needed designer population has been missing [2–6]. This time a properly qualified programmer population is missing. But we need to push the envelope into two different directions. The VLSI design revolution has been the most effective project in the modern history of computing. But this time we need even more. A dual rail effort is needed for simultaneously developing the scene toward parallel programming for manycore architectures and to structural programming for reconfigurable computing (RC), as well as heterogeneous systems including the cooperation of both paradigms.

Currently the dominance of the basic computing paradigm is gradually wearing off with the growth of the area of RC applications – bringing profound changes to the practice of both, scientific computing and ubiquitous embedded systems, as well as new promise of disruptive new horizons for affordable very high performance computing. Due to RC also the desk-top personal supercomputer is near. To obtain the payoff from RC we need a new understanding of computing and supercomputing, as well as of the use of accelerators. For bridging the translational gap, the software/configware chasm, we need to think outside the box.

Section 2.3 tries to introduce (almost) non-expert readers to the flavor of RC. FPLAs, FPGAs, platform FPGAs, fine grain and coarse grain. Section 2.3.1 discusses the differences of RC applications in embedded systems and in supercomputing. It introduces the fundamentals of RC and the massive impact on the efficiency of computing it promises. Furthermore the chapter outlines the educational barriers we have to surmount and the urgent need for major funding on a global scale to run a world-wide mass movement, of a dimension reaching further than the Mead-&-Conway-style microelectronics revolution in the early 1980s. Section 2.2.2 illustrates the Reconfigurable Computing Paradox, and Sect. 2.2.3 Why von Neumann is so inefficient. Section 2.3 tries to convince the reader, why we need to reinvent computing. This chapter advocates to introduce a dual paradigm trans disciplinary education by using Configware Engineering as the counterpart of Software Engineering by new curricula in CS (Computer Science) and CE (Computer Engineering) for providing an integrating dual paradigm mind set to cure severe qualification deficiencies of our graduates Sect. 2.3.3 tells us what problems we must solve, Sect. 2.3.1 the "parallel programming problem", Sect. 2.3.4 how to introduce RC and Sect. 2.3.5 describes the way toward a new world model of computing. It urges acceptance of the massive challenge to reinvent computing, away from its currently obsolete CPU-processor-centric Aristotelian CS world model, to a twin-paradigm Copernican model. For more details about the problem area see [7].

2.1.1 Why Computers Are Important

Computers are very important for all of us. By many millions of people around the world computers are used in hundreds of application areas featuring ten-thousands of programs with millions of lines of code developed by thousands of man-years by investment volumes up to billions of dollars [8]. Computers running this legacy software are indispensable in our world [9]. Completing their tasks manually would require much more time, or, would be completely impossible, especially if networking is involved. Not only to maintain our global economy we must maintain these important infrastructures. However, threatening unaffordable operation cost by excessive power consumption is a massive future survival problem for our existing cyber infrastructures, which we must not surrender.

2.1.2 Unaffordable Energy Consumption by Computing

It has been predicted that by the year 2030, if current trends continue, the worldwide electricity consumption by ICT (Information and Communication Technology) infrastructures will grow by a factor of 30 [10], reaching much more than the current total electricity consumption of the entire world for everything, not just for computing. The trends are illustrated by an expanding wireless internet, and by a growing number of internet users. as well as with tendencies toward more video on demand, HDTV over the internet, shipping electronic books, efforts toward more cloud computing and many other services. Other estimations claim, that already now the greenhouse gas emission from power plants generating the electricity needed to run the internet is higher than that of the total world-wide air traffic. For more predictions see [11]. The electricity bill is a key issue not only for Google, Microsoft, Yahoo and Amazon with their huge data farms at Columbia River [12]. That's why Google recently submitted an application asking the Federal Energy Regulatory Commission for the authority to sell electricity [13], and has a patent for water-based data centers, using the ocean to provide cooling and power (using the motion of ocean surface waves to create electricity) [14]. Already in the near future the electricity bill of most facilities will here be substantially higher than the value of their equipment [15]. Already in 2005, Google's electricity bill was about 50 millions US-$ higher than the value of its equipment. Meanwhile the cost of a data center is calculated solely by the monthly power bill, not by the cost of hardware or maintenance [16]. As Google's employee L. A. Barroso said [17]: "The possibility of computer equipment power consumption spiraling out of control could have serious consequences for the overall affordability of computing." Power consumption estimations for an exascale supercomputer (1,018 calculations/s) like expected for about 2,018 range between 200 MW and 10 GW (the double of New York 16 million people energy consumption) [18, 19].

2.1.3 Peak Oil: Dig More Coal for Computing?

Rapidly growing energy prices are predicted since the oil production has reached its peak by about the year 2009 [20–23]. Already currently 80% of crude oil is coming from decline fields (Fig. 2.1). However, the demand is growing because of developing standards of living in China, India, Brazil, Mexico and newly industrializing countries. We need at least "six more Saudi Arabias for the demand predicted for 2030" (Fatih Birol, Chief Economist IEA [20]). I believe that these predictions do not yet consider the growing electricity consumption of computers. Maybe, we will need ten more Saudi Arabias. About 50% of the shrinking oil reserves are under water [24]. In consequence of the Gulf of Mexico oil spill not all deepwater explorations will be allowed and the crude oil prices will go further up. All this will cause a massive future survival problem for our cyber infrastructures, which we must not surrender because this is an important global economy issue. Or, should we dig more coal [22]? It makes sense, to measure computing performance not just by MIPS (million instructions per second), but by MIPS/W instead.

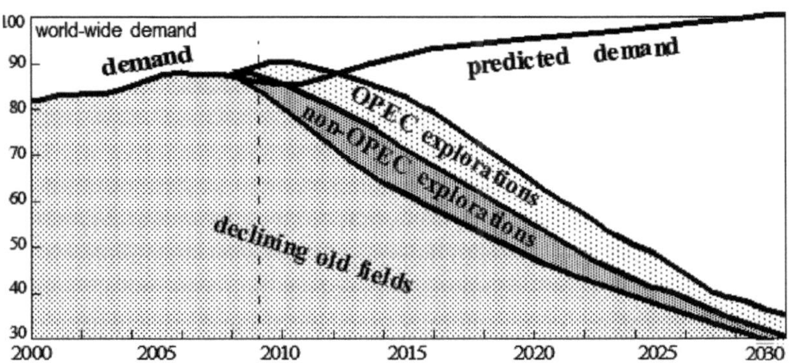

Fig. 2.1 Beyond Peak Oil: massively declining future crude oil production [22]

2.1.4 Green Computing: Important, but Limited

Green Computing tends to use conservative methods to save energy by more efficient modules and components. For example LED flat panel displays need much less power than LCD-based displays. Also much more power-efficient power supply modules are possible. The potential to save power is substantially less than an order of magnitude: maybe, a factor of about 3–5. A scene separate from Green Computing is Low Power Circuit Design, now also called Low Power System on Chip Design (LPSoCD). Its most important conference series are about 30 years old: the PATMOS (oldest) and the ISLPED conference series.

Several methods are known for LPSoCD, such as: Active Body Bias (ABB), Adaptive Voltage Scaling (AVS), Dynamic Voltage Scaling (DVS), Multiple Supply Voltages (MSV), Multi-Threshold CMOS (MTCMOS), Power Gating (PG), Power Gating with Retention (RPG), etc. [16]. However, the order of magnitude of the benefit to be expected from this subarea LPSoCD is rather low. By MSV in using 3 Vdds the power reduction ratio at best is about 0.4 [16]. LPSoCD is a matter of ASIC design, e. g. of hardwired accelerator design. Only 3% of all design starts are ASIC designs (Fig. 2.2) with a trend leading further down. But in fact, low power design is also used for developing better power-efficient FPGAs – to the benefit of Reconfigurable Computing. But we need a much higher potential of saving energy because "Energy cost may overtake IT equipment cost in the near future" [9]. "Green Computing has become an industry-wide issue: incremental improvements are on track" [23], But "we may ultimately need revolutionary new solutions." [25] Let me correct this statement by "we will ultimately also need revolutionary solutions (like reconfigurable computing), since we need much higher efficiency."

2.1.5 Massively Saving Energy by RC

The idea of saving energy by using Reconfigurable Computing is not new [26, 27]. Being very important to massively reduce the energy consumption of computing, by up to several orders of magnitude, Reconfigurable Computing is extremely

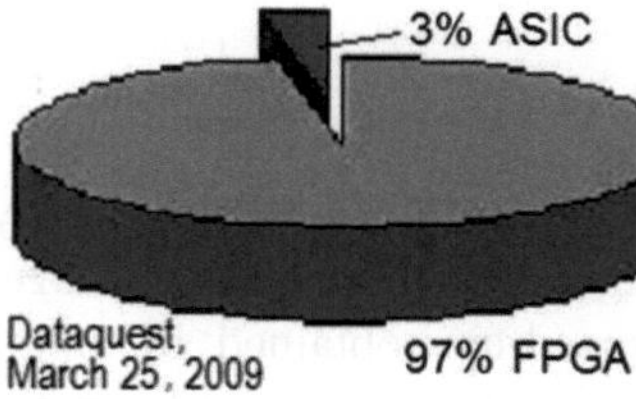

Fig. 2.2 FPGA to ASIC design start ratio

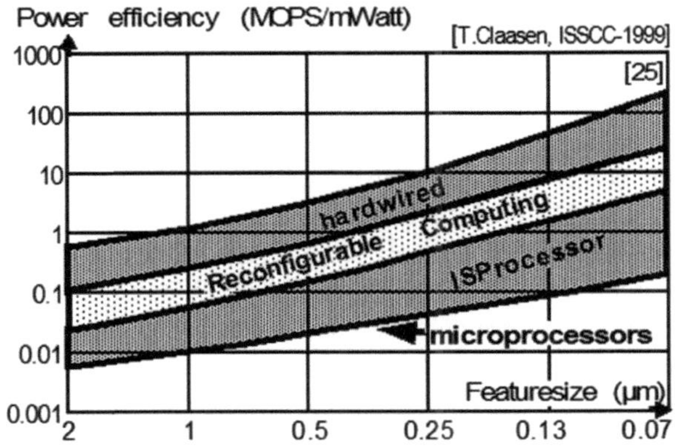

Fig. 2.3 Better power efficiency by accelerators

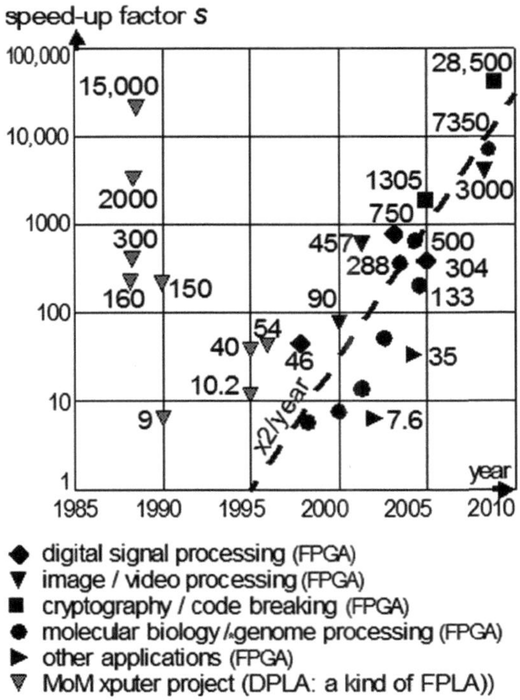

Fig. 2.4 Speed-up factors

important for the survival of the world economy. Already a partial paradigm shift promises to save electricity by orders of magnitude. Dozens of papers (ref. in [28]) have been published on speed-ups obtained by migrating applications from software running on a CPU, over to configware for programming FPGAs [28]. It has been reported already more than a decade ago, that for a given feature size, microprocessors using traditional compilers have been up to 500 times more power hungry than a pure hardware mapping of an algorithm in silicon [27] (Fig. 2.3). Speedup factors up to 4 orders of magnitude have been reported from software to FPGA migrations [26–49]. Here the energy saving factor is roughly about 10% of the speedup factor, i.e., still up to >3 orders of magnitude.

Figure 2.4 shows a few speedup factors picked up from literature, reporting a factor of 7.6 in accelerating radiosity calculations [46], a factor of 10 for FFT (fast Fourier transform), a speedup factor of 35 in traffic simulations [47]. A speedup by a factor of 304 is reported for an R/T spectrum analyzer [48]. For digital signal processing and wireless communication, as well as image processing and multimedia, speed-ups by 2 to almost 4 orders of magnitude have been reported. In the DSP area for MAC operations a speedup factor of 100 has been reported compared to the fastest DSP on the market (2004) [49]. Already in 1997, a speedup between 7 and 46 has been obtained over the fastest DSP [26]. In the multimedia area we find factors ranging from 60 to 90 in video rate stereo vision [34] and from 60 to 90 in real-time face detection [35], and of 457 for hyperspectral image compression [36]. In communication technology we find a speedup by 750 for UAV radar electronics [37]. For cryptography speed-ups by 3 to >5 orders of magnitude have been obtained. For a commercially available Lanman/NTLM Key Recovery Server [50] a speedup of 50–70 is reported. Another cryptology application reports a factor of 1,305. More recently for DES breaking a speed-up by × 28,514 has been reported [51] (Table 2.1).

For Bioinformatics applications [52] (also see [29]) speed-ups have been obtained by 2–4 orders of magnitude. Compared to software implementations sensational speed-up factors have been reported for software to FPGA migrations. A speedup of up to 30 has been shown in protein identification [30], by 133 [31] and up to 500 [32] in genome analysis. The Smith-Waterman algorithm, which is used for protein and gene sequence alignment, is basically a string-matching operation that requires a lot of computational power [52]. Here another study demonstrates speed-ups of 100x using Xilinx Virtex-4 hardware matched against a 2.2 GHz Opteron [53]. A speedup by 288 has been obtained with Smith-Waterman at the National Cancer Institute [33]. More recently a speed-up higher by more than an order of magnitude has been obtained here [45]. The CHREC project (supported by 24 industry partners [54]) reports running Smith-Waterman on a Novo-G supercomputer, a cluster of 24 Linux servers, each housing four Altera Stratix-III E260 FPGAs. According to this CHREC study, a four-FPGA node ran 2,665 times faster than a single 2.4 GHz Opteron core [55]. Another Smith-Waterman DNA sequencing application that would take 2.5 years on one 2.2 GHz Opteron is reported to take only 6 weeks for 150 Opterons running in parallel. Using 150 FPGAs on NRL's Cray XD1 (speedup by 43) is reported to further reduce this time to 24 h, which means a total speedup of 7,350X over a single Opteron [42]. These are just a few examples from a wide range of publications [29–51] reporting substantial speedups by FPGAs.

Recently not only energy saving factors have been reported, roughly one order of magnitude lower than the speed-up. More recently has been reported [51] for DES

Table 2.1 Recent speed-up/power save data from software to configware migration [51]

SGI Altix 4,700 w. RC 100 RASC vs. Beowulf cluster	Speed-up factor	Save factor Power	Cost	Size
DNA and protein sequencing	8,723	779	22	253
DES braking	28,514	3,439	96	1,116

breaking (a crypto application): 28,500 (speed-up) vs. 3,439 (saving energy) and for DNA sequencing 8,723 (speed-up) vs. 779 (saving energy) etc. (Table 2.1). This paper also reports factors for saving equipment cost (up to x96) and equipment size (up to 1,116, see Table 2.1). No hangar full of equipment is needed when FPGAs are used in Scientific Computing. The Pervasiveness of FPGAs is not limited to embedded systems, but is also spread over practically all areas of scientific computing, where high performance is required and access to a supercomputing center is not available or not affordable. The desk-top supercomputer is near.

2.1.6 A Mass Movement Needed as Soon as Possible

This subsection emphasizes that RC is a critical survival issue for computing-supported infrastructures worldwide and stresses the urgency of moving RC from niche to mainstream. It urges acceptance of the massive challenge of reinventing computing, away from its currently obsolete CPU-processor-centric Aristotelian CS world model, over to a twin-paradigm Copernican model. A massive software to configware migration campaign is needed. First this requires clever planning to optimize the effort versus its expected results. Which software packets should be migrated first. All this requires massive R&D and education efforts taking many years. Lobbying for the massive funding should be started right now. We should address politicians at all levels: community level, state level, national level, and European Union level.

To explain all this to politicians is very difficult. Since politicians always watch the sentiment of their voter population, we efficiently have to teach the public, which is a challenge. Without a strong tailwind from the media a successful lobbying seems to be almost successless. All this has to be completed as soon as possible, as long as we can still afford such a campaign. To succeed with such a challenging educational campaign the foundation of a consortium is needed for running an at least Europe-wide project.

2.2 Reconfigurable Computing

This section introduces a flavor of Reconfigurable Computing, its history, its more recent developments and, the massive impact on the efficiency of computing it promises. It is not easy to write this section such that it may be (almost) readable for non-experts – like the booklet FPGAs for Dummies which may help a little bit [56]. A Classical application for reconfigurable computing subsystems is the use as an accelerator to support the CPU ("central processing unit"). According to the state of the art in the 1990s, having been the tail wagging the dog, this typically was and is a non-von-Neumann accelerator [57]. But we have to distinguish two kinds of such accelerators: made from hardwired logic or from field-programmable logic. These two kinds are distinguished by binding time of their functionality: (1) before fabrication for fixed logic or hardwired logic devices (HWD) vs. (2) after fabrication for (field-)programmable logic devices (PLD). The term "field-programmable" indicates, that by reconfiguration the functionality can be changed also at the users site by receiving new configuration code: from some memory, or, even over the internet.

First field-programmable blocks from the early 1980s have been so-called FPLAs featuring very area-efficient layout similar as known from ePROM memory for the price of being able to compute only Boolean functions in sum-of-product form. Very high speed-up could be obtained by matching hundreds of boolean expressions within a single clock cycle instead of computing them sequentially by a microprocessor. Together with a reconfigurable address generator [58] this brought a speed-up by factor up to 15,000 [59–63] for a grid-based design rule checker – already in the early 1980s. Via the multi project chip organization of the E.I.S. project such a FPLA (which was called DPLA) has been manufactured on a multi-project chip of the multi university E.I.S. project: the German contribution to the Mead-&-Conway VLSI design revolution. This DPLA has the capacity of 256 first FPGAs (field-programmable gate array) just appearing on the market (by Xilinx in 1984). This again demonstrates the massive area-inefficiency of FPGAs contributing to the Reconfigurable Computing Paradox (see Sect. 2.3.2) and the very early high speed-ups (Fig. 2.4).

The usual acronyms (Table 2.2) FPLA and FPGA are highly confusing being really not intuitive. From the straight-forward language feeling there does not seem to be any difference between "logic" in logic array (LA) and "gate" in Gate Array (GA). What is really different with FPGAs? In fact, FPGAs feature much more flexibility by introducing CLBs and routable wiring fabrics for interconnect between CLBs (Fig. 2.5). In contrast to FPLAs, the CLB in FPGAs allows for instance to select one of 16 logic functions from simple LUTs (look-up tables, Fig. 2.6). However, PLAs [64, 65] are not routable and allow only to implement Boolean functions in sum-of-product form.

Beyond such fine grained reconfigurability the progress of Moore's law leads to higher abstraction levels with "coarse-grained reconfigurability" featuring also CFBs (configurable function blocks), which may be adders, multipliers and/or many other functions. The next step is coarse-grained "platform FPGAs", which also include one or several microprocessors, like the PowerPC in earlier platform FPGAs from Xilinx.

2.2.1 *Embedded Systems vs. Supercomputing*

A growing trend is the use of FPGAs in embedded systems: ERC (Embedded Reconfigurable Computing). Originally there has been a feeling that FPGAs are too slow, power-hungry and expensive for many embedded applications. This has

Table 2.2 List of acronyms

Acronym	Meaning	Acronym	Meaning	Acronym	Meaning
ALU	Arithmetic/logic unit	ESL	Electronic system-level design	MIPS	Mio instructionsp.second
ASM	Auto-sequencing memory	EU	European Union	MoPL	Map-oriented PL
CE	Computer engineering	FIR	Finite impulse response	PC	Personal computer
CS	Computer science	FPGA	Field-programmable GA	PL	Programming language
CHREC	NSF Center for High-Performance Reconfigurable Computing	FPL	Field-programmable logic	PLA	Programmable LA
CLB	Configurable logic block	FPLA	Field-programmable PLA	PLD	Programmable logic device
CPU	vN central processing unit	GA	(Routable) gate array	PROM	Programmable ROM
DNA	Deoxyribonucleic acid	GPP	General purpose processor	PU	Processing unit
DPA	Detapath array	GPGPU	General-purpose computation on graphics processing units	RAM	Random access memory
DPU	Detapath unit	GPU	Graphic processing unit	RC	Reconfigurable computing
DSP	Digital signal processing	HPC	High performance computing	rDPA	Reconfigurable DPA
EDA	Electronics design automation	HPRC	High performance RC	rDPU	Reconfigurable element
EPLA	E-Programmable LA	HWD	Hard-wired device	rE	Reconfigurable element
EPP	Extensible programmable platform	ICT	Information and communication technology	ROM	Read-only memory
ePROM	E-Programmable PROM	IEA	Intern'l energy agency	R&D	Research & development
ERC	Electrical rules checker	IT	Information technology	SMP	Symmetric multiprocessor
ESA	Embedded software automation	LA	(Compact) logic array	TM	Transactional memory
		LUT	Look-up table	VLSI	Very large scale integrated (circuits)
		MAC	Multiply/accumulate unit	vN	von Neumann

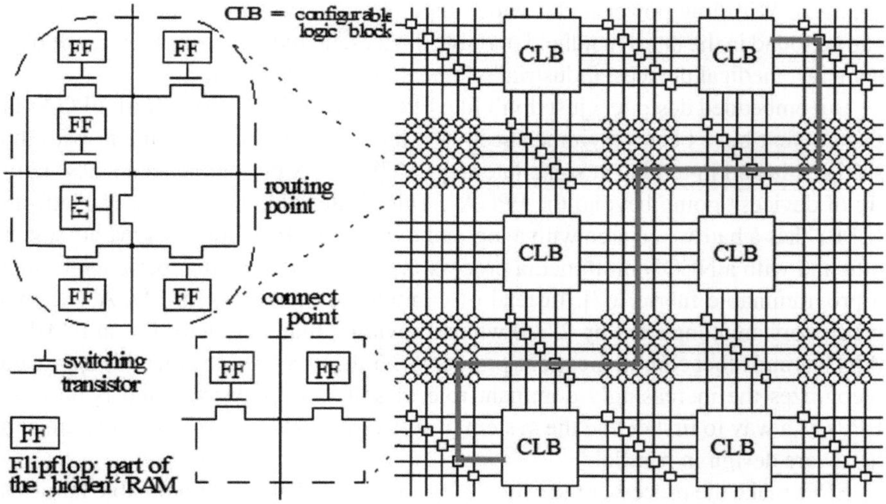

Fig. 2.5 Interconnect fabrics example of a routable GA; grey line: example of one routed wire connecting 2 CLBs with each other

#	g_{ab}				f(A, B)
	g_{00}	g_{01}	g_{10}	g_{11}	
0	0	0	0	0	0
1	0	0	0	1	A and B
2	0	0	1	0	B disables A
3	0	0	1	1	A
4	0	1	0	0	A disables B
5	0	1	0	1	B
6	0	1	1	0	A exor B
7	0	1	1	1	A or B
8	1	0	0	0	not(A or B)
9	1	0	0	1	A coin B
10	1	0	1	0	no (B)
11	1	0	1	1	B implies A
12	1	1	0	0	not (A)
13	1	1	0	1	A implies B
14	1	1	1	0	not(A and B)
15	1	1	1	1	1

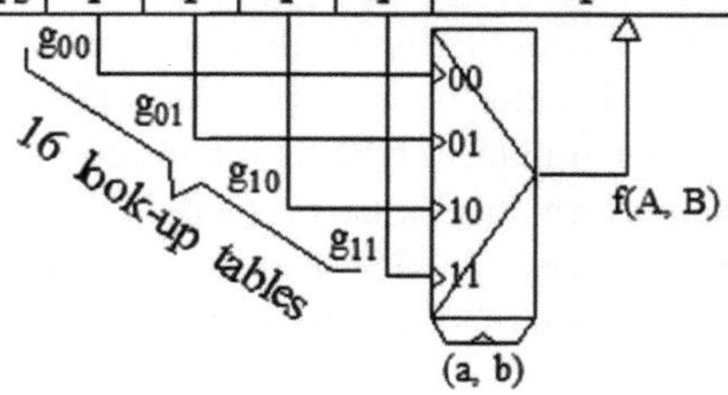

Fig. 2.6 LUT example

changed. With low power and a wide range of small packages, particular FPGAs can be found in the latest handheld portable devices, including smartphones, eBooks, cameras, medical devices, industrial scanners, military radios, etc.

But embedded designers just don't like FPGAs with CPUs inside [66]. FPGAs in this context have been very much seen as a hardware engineer's domain, with the softies allowed in to play at some late stage [67]. Xilinx pre-announced a new family of devices "going beyond the FPGA". This "Extensible Programming Platform (EPP)" has a hardwired area with a top-end twin-core ARM Cortex-A9M processor unit and with a NEON multimedia processor, memory interfacing, peripherals, and a programmable fabric [67]. Instead of communicating across an FPGA, the two processors are connected by 2,500 wires, providing much capacity for an AMBA-AXI bus and other communications protocols. Xilinx was stressing that this approach recognizes the increasingly dominant role of software in systems and is pushing EPPs as a way to first define the system in software and then carry out software and hardware design in parallel.

EPPs make the processor the centre of the device with the programmable fabric as an extra. And this, argues Xilinx, now puts the software engineer first with the hardies following behind. In EPPs the FPGA logic and the CPU will be programmable separately. FPGA configuration will be handled by the processor(s) directly, not by a serial ROM. In other words, you have to tell the FPGA you want it configured. That's very un-FPGA-like [66]. That's EPP-like. The approach of using both a processor and programmable fabric allows design to start at high level and the system to be implemented as software [67].

Xilinx's first attempt at this was an FPGA with a processor inside. This time around, it's a processor with an FPGA grafted on. That's not just semantic hair-splitting: it's the big difference between these chips and the old ones. The new chips will boot up and run just like normal microprocessors, meaning there's no FPGA configuration required at all [68].

EPPs are a result of the new research topic Network-on-Chip (NoC) [69], which is a new paradigm for designing the on-chip hardware communication architecture based on a communication fabric, also including on-chip routers. NoC CAD tool flows also support mapping applications into NoC.

Apart from ERC (Embedded Reconfigurable Computing) we have another reconfigurable computing scene: HPRC (High Performance Reconfigurable Computing), This last one is a relatively new area, but has attracted a lot of interest in recent years, so much so that this entire new phrase has been coined to describe it [71]. HPRC uses FPGAs as accelerators for supercomputers [72]. Large HPC vendors are already supplying machines with FPGAs ready-fitted, or have FPGAs in their product roadmaps. What are the benefits of using FPGAs in HPC? Also here the first and most obvious answer is performance. HPC is renowned as that area of computing where current machine performance is never enough. A problem yet to be solved here is programmer productivity [73, 75]. It is an educational challenge, that programmers with the needed mix of skills are hardly available. Will FPGAs have a tough road ahead in HPC?

2.2.2 The Reconfigurable Computing Paradox

Technologically FPGAs are much less efficient than microprocessors [68, 70]. The clock speed is substantially lower. The routable reconfigurable wiring fabrics cause a massive wiring area overhead. There is also massive other overhead: reconfigurability overhead, since of 200 transistors e. g. maybe about five or even less of them (Fig. 2.9 in [70]) serve the application, whereas the other 195 are needed for reconfigurability (Fig. 2.6). Often there is also routing congestion, so that not all CLBs can be used, what causes further degradation of efficiency. Software to configware migration yield massive improvements in speed and power consumption, although FPGAs are a dramatically worse technology. We call this the "Reconfigurable Computing Paradox". by orders of magnitude better performance with a drastically worse technology? What is the reason? It's the von Neumann paradigm's fault. The next subsection goes into details.

2.2.3 Why von Neumann Is So Inefficient

The von Neumann paradigm has been criticized often [78–81]. Peter Newman had for 15 years each month the highly critical „computers at risk" back pages of Communications of the ACM [80]. Nathan's law (by Nathan Myhrvold, a former CTO of Microsoft) said that software is a gas, which fills any available storage space: on-chip memory, extra semiconductor memory located outside the processor chip, as well as hard disks. A lady (I forgot her name) said that it even fills the internet. Nicklaus Wirth's pre-manycore interpretation of Moore's law is, that "software is slowing faster than hardware is accelerating" [78].

Why is von Neumann so inefficient? It is the von Neumann syndrome [82] caused by the fact, that instruction streams are very memory-cycle-hungry. We can distinguish two different reasons: algorithmic complexity required by the von Neumann paradigm, and, architectural issues. There are also other attempts to explain at least particular symptoms of this syndrome (Fig. 2.8 [83]) [84]. A well known architectural problem is the memory wall [85, 86] (Fig. 2.7): the access time to RAM outside the processor chip is slower by a factor of about 1,000, than to on-chip memory [86]. This difference is growing by 50% every year. It is a dramatic software engineering issue, that multiple levels of instruction stream overhead leads to massive code sizes which hit the memory wall [86]. However, rDPUs and rDPAs do not suffer from Nathan's law, since at run time no instruction streams are running through.

How data are moved is a key issue. CPUs usually move data between memories requiring instruction streams (first line, Table 2.3). This means the movement of data is evoked by execution of instructions due to the von Neumann paradigm. Also the execution of operations inside a CPU requires reading and decoding of instructions (Fig. 2.8 gives an idea of the overhead of the main components for contemporary CPUs). However, after a full migration to static reconfigurable computing an

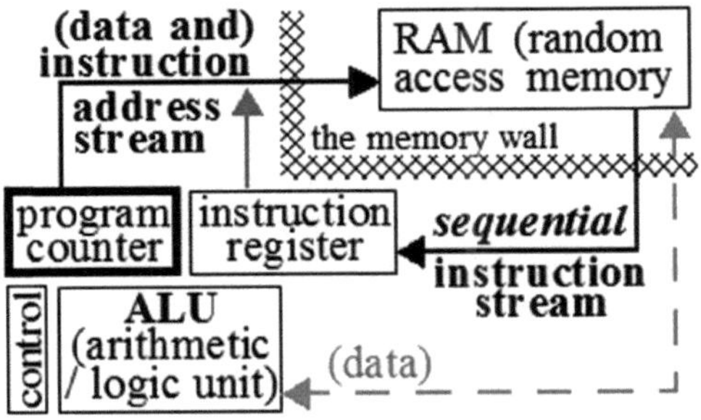

Fig. 2.7 von Neumann principles

Table 2.3 Twin paradigm fundamental terminology

#	Term	Controlledby	Machine paradigm	State register Type	Location
1	Software	Instruction streams	von Neumann	Program counter	in CPU
2	Configware	(Configuration memory)	None	None	(Hidden)
3	Flowware	Reconfigurable address generator	Data stream machine	Data counter	In ASM memory block

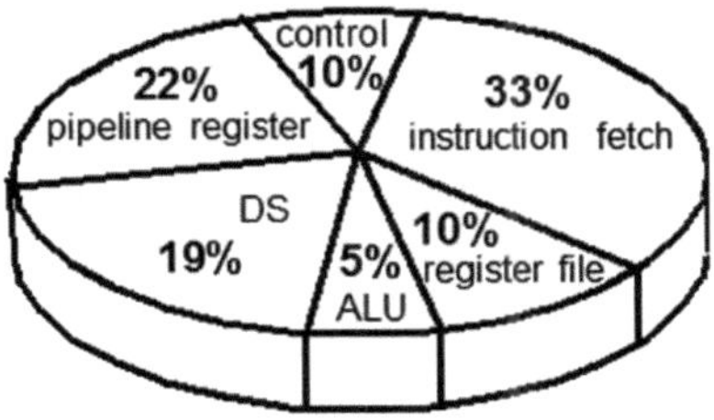

Fig. 2.8 All but ALU is overhead: ×20 efficiency [83]

algorithm is run by data streams only. Instead of a hardwired program counter reconfigurable data counters are used which do not require instruction sequences for address computation.

Also, how data are moved inside the data paths is a key issue, and pipe network structures to interconnect rDPUs avoid moving data through memory blocks because data are moved directly from DPU to DPU [87]. This means, that operation execution inside a DPU (not having a program counter) is "transport-triggered" (second line, Table 2.3). It is triggered via handshake by the arrival of the data item, not needing an instruction to call it. Not looking at dynamically reconfigurable systems ([88] only for advanced courses) we see, that reconfigurable fabrics don't perform any instruction sequencing at run time.

Beyond such fine grained reconfigurability the progress of Moore's law leads to higher abstraction levels with "coarse-grained reconfigurability" featuring also

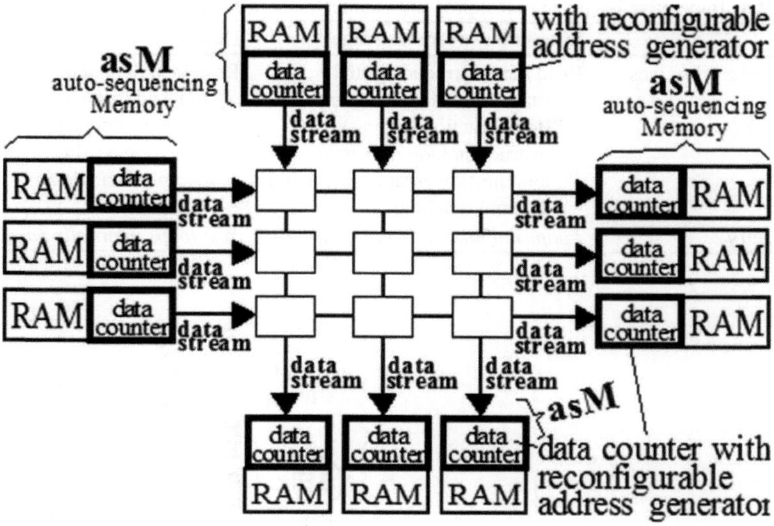

Fig. 2.9 Example for data stream processor principles

CFBs (configurable function blocks), which may be adders, multipliers and/or many other functions. The next step is coarse-grained "platform FPGAs", which also include one or several microprocessors, like the PowerPC in earlier platform FPGAs from Xilinx.

But by a migration sometimes also the amount of data streams may be minimized by changing the algorithm. Here an illustration example for reducing the algorithmic complexity is given by the migration of the well-known O(n2) complexity bubble sort algorithm away from von Neumann. The algorithmic complexity turns from O(n2) into O(n) [8]. In a similar manner, other well-known algorithmic methods can be transformed to explore parallelism and locality, like in dynamic programming as presented in [88]. The combination of these effects leads to massive speed-up and massive saving of energy.

Of course, the data entering or leaving such an array (Fig. 2.9) have to be stored. The datastream machine paradigm uses auto-sequencing Memory blocks (asM). Each asM has a reconfigurable address generator and data counter inside. so that no instruction streams are needed for address computation. All this data streams can be programmed via data-imperative languages [90], being a kind of sisters of classical instruction-imperative programming languages. Data-imperative languages are easy to teach since both classes of imperative languages use the same primitives (Table 2.4). But there is one exception: data-imperative languages also support parallelism inside loops (Table 2.4). This also contributes to the benefit by reconfigurable computing. The simultaneous use of both classes of languages we call "twin-paradigm approach" (Table 2.4 and Fig. 2.10).

Table 2.4 Imperative language twins: program counter vs. data counters

	Instruction stream languages	Data stream languages
sequencing primitives	Read next instruction goto (instruction address) jump (to instruction address) instruction loop instruction loop nesting escapes instruction stream branching	Read next data item goto (data address) jump (to data address) data loop data loop nesting escapesdata stream branching
asymmetry	*Loops not internally parallel*	*Yes: loops internally parallel*

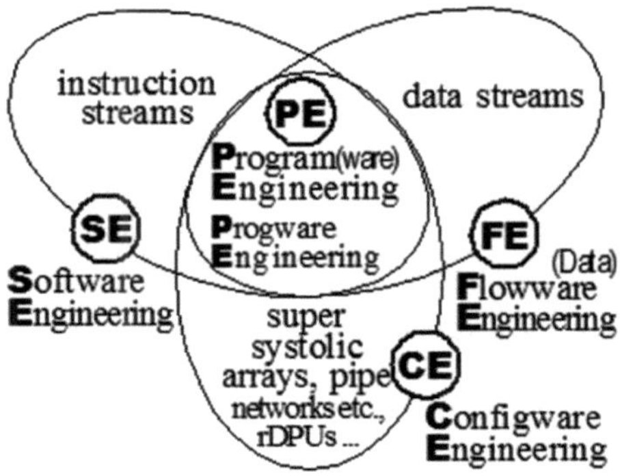

Fig. 2.10 New CS world model image

2.3 Why We Need to Reinvent Computing

We cannot afford to relinquish RC. We will urgently need this technology to cope with threatening unaffordable operation cost by excessive power consumption of the entirety of all von Neumann computers world-wide. We need to migrate many application packages from software over to configware. This is a challenge to reinvent computing to provide the qualifications needed since disruptive developments in industry have caused the many-core programming crisis. Intel's cancellation the Tejas and Jayhawk processors indicated in May 2004 the end of frequency scaling's dominance to improve performance. "Multicore computers shift the burden of software performance from VLSI designers over to software developers" [89]. For Gary Smith from GS-EDA the three biggest disruptions are not only (1) many-core silicon, but also (2) non-vN architectures, and (3) parallel software, and the center of gravity shifts from EDA to programming (not "software" how Gary named it [93], compare Table 2.3 and Fig. 2.10).

To use manycore we need to rewrite our software: our biggest problem. RTL verification moves up to ESL. EDA Industry and ESA need to merge. Especially for Embedded Software Automation (ESA) we need tools to develop parallel software. He calls for an approach of using both, processor and programmable fabric, allows design to start at high level and the system to be implemented as progware (programware, see Fig. 2.10), e. g. with tools like LabView, MatLab, or others.

From Xilinx and ARM hardware and software IP and tool chains are available useful to speed up time-to-market and reducing risk. It's easier to take advantage of accelerators than to try to integrate and program more processors.

To rewrite the software the qualified programmer population is not existing: a huge challenge to provide new educational approaches to qualify for heterogeneous systems including both, parallel software and configware. This requires much more than just bridging the traditional hardware/software chasm in education [91]. We need robust and fast implementations of adequate compilers and design tools, e. g. automated by formal techniques based on rewriting [92]. The biggest payoff will come from Putting Old ideas into Practice and teaching people how to apply them properly [94]. Dimensionality-rooted scaling laws favor reconfigurable spatial computing over temporal computing. Time to space mapping even dates back to the early 1970s and even the late 1960s, years before the first hardware description languages came up [91, 95, 96]. "The decision box (in the flow chart) turns into a demultiplexer. This is so simple! Why did it take 30 years to find out?" [97].

The impact is a fascinating challenge to reach new horizons of computer science. We need new generations of talented innovative scientists and engineers to start the second history of computing.

2.3.1 The "Parallel Programming Problem"

Parallel computer programs are difficult to write: performance is affected by data dependencies, race conditions, synchronization, and parallel slowdown. The problem is: how to optimize parallel Computing despite Amdahl's Law? The "parallel programming problem" has been addressed high performance computing for more than 25 years with very disappointing results [100–102]. Programming languages research has stalled for several decades [103]. Informal approaches are not working. For the multi-core era, we must adopt a systematic approach informed by insight into how programmers think [104].

I do not agree. We have to teach them how to think. We have to teach programmers how to "think parallel", how to find concurrent tasks, how to synchronize free from deadlock and race conditions, how to schedule tasks at the right granularity, and, how to solve the data locality problem. Perhaps a new visual programming paradigm is required [68]. What are the right models (or abstractions) to avoid typical problems [105]: multi-core version applications running slower, problems with race conditions, and strategies for migrating code to multi-core. We see a promising new horizon: a model-based twin-paradigm methodology to master the hetero of all 3: Single-core, Multicore, and Reconfigurable Computing.

We need a different way of thinking. "The shift to multicore processor architectures really is stressing existing programming models" said Richard C. Murphy at Sandia aiming at redesigning memory systems to move computation as close to memory as possible to eliminate the traditional load-store approach where big systems use more resources moving data around than for actually computing [106]. Since Linpack doesn't measure performance for actual problems in many application areas, Sandia has proposed Graph500 as a new rating system for testing skills in analyzing large, graph-based structures that link huge numbers of data points [107]. Studies show that moving data around (not computations) will be the dominant energy problem.

2.3.2 *Why FPGAs Should Win*

Most ASIC design world-wide has stopped [108]. Only 3% of all design starts are ASIC designs (Fig. 2.3) from mega-funded companies with gigantic-volume products that can afford latest generation custom SoC development, and niche players that continue doing ASIC design with older-generation processes. The enormous 97% gap can be filled best by hybrid FPGA/hard-core devices – by FPGA companies with the required technology and infrastructure to sell and support them.

"FPGAs have become incredibly capable with respect to handling large amounts of logic, memory, digital-signal-processor (DSP), fast I/O, and a plethora of other intellectual property (IP)" [109]. At 28-nm, FPGAs deliver the equivalent of a 20–30-million gate application-specific integrated circuit (ASIC). At this size, traditionally used FPGA design tools begin to break down and can no longer design and verify these devices in a reasonable amount of time.

This positions FPGA companies in the best place they've ever been – a place from where they could capture huge segments of the standard parts and ASSP business with semi-standard parts that include FPGA fabric for application-specific customization. Instead of today's still-very-general-purpose FPGAs, we'll see more devices with a narrower application focus without too much general-purpose overhead on the die. With each passing process generation, the cost of that overhead shrinks, and the cost of developing the traditional alternative solutions goes up.

Industry abandoned the "pure" FPGA [110]. Countless failed FPGA start-ups have proven that the magic is not in the fabric. The real keys are tools, IP, and support for enabling the customer/designer to get the fabric to do what they want as easily as possible, and with minimal risk.

It turns out that the solution is a mixture of FPGA fabric and hard logic coming from the FPGA companies. Instead of putting FPGA fabric in our custom SoC designs, we are getting custom SoC in our FPGAs. Today's FPGA are hybrid parts with optimized hardwired cells (like multipliers, memory, IO and even processors) and FPGA fabric living on the same die.

It is an important factor that there is an order of magnitude more software engineers than hardware guys. Usually it is the software community that selects the processor, not the hardware team. To gain the approval of software engineers the FPGA vendors realized that promoting "a processor with FPGA accelerators" is more attractive than an "FPGA with a processor inside".

The FPGA business and the processor business looks like "chip" business, but actually are more "tools, software, IP, and services" businesses. With the coming together of the embedded processing world [111] and the FPGA world, we will see if FPGA companies like Xilinx can be convincing enough in their ability to support the embedded software developer, or if companies like Intel can be convincing enough in their ability to support the FPGA designer.

"Customers are increasingly turning to FPGAs and expert 3rd party providers to design progressively complex products within shrinking time to market budgets." As the industry is developing more complex designs on programmable solutions, competent and trusted providers are required to deliver key IP, software, and services to meet the tight delivery schedules of today's system companies and to allow customers to find the right qualified 3rd party provider easier and faster than before – avoiding, that for SoC designers it's a nightmare using IPs delivered by 3rd parties or internal IP teams? An Electronics IP core, a semiconductor intellectual property core, or IP block is a reusable unit of logic, cell, or chip layout design that is the (legal) intellectual property of one party.

2.3.3 Problems We Must Solve

Furthermore this chapter outlines the educational barriers we have to surmount and the urgent need for major funding on a global scale to run a world-wide mass movement, of a dimension as far reaching as the Mead-&-Conway-style microelectronics revolution in the early 1980s. Problems We Must Solve:

1. A mass migration from software to configware for the benefit of massively saving energy, of much higher performance, and, of gaining high flexibility.
2. developing a most promising migration priority list.
3. to reeducating the programmer population for such a mass movement campaign [68], and upgrading our highly obsolete curricula for three reasons:
   (a) to realize that parallel programming qualifications are a must,
   (b) to resolve the extreme shortage of programmers qualified for RC, and
   (c) twin paradigm programming skills are a must to program hetero systems (like modern FPGAs featuring all 3: reconfigurable fabrics, hardwired function blocks, and CPUs).

As a consequence we need innovative undergraduate programming courses [98] which also teach a sense for locality, not only needed for classical parallel programming, is already coming along in RC with time to space mapping required to structurally map an application to the datastream side of the twin paradigm approach. This means, that teaching the structural programming of RC also exercises the sense of locality needed for traditional parallel programming. The extension of the non-sequential part of education should be optimized not to scare away undergraduate students. Twin-paradigm lab courses should be model-based, may be MathWorks-supported, mainly at the abstraction level of pipe networks [99].

2.3.4 *How to Introduce Reconfigurable Computing*

Since software has to be rewritten anyway, this is the occasion for the twin-paradigm approach to massively reduce the energy consumption of our computing infrastructures.

Meanwhile FPGAs are also used everywhere for high performance in scientific computing, where this is really a new computing culture – not at all a variety of hardware design. Instead of H/S codesign we have here software/configware co-design (SC co-design), which is really a computing issue. This major new direction of developments in science will determine how academic computing will look in 2015 or even earlier. The instruction-stream-based mind set will loose its monopoly-like dominance and the CPU will quit its central role – to be more an auxiliary clerk, also for software compatibility issues.

An introduction to Reconfigurable Computing (RC) [90, 112, 113] should regard the background to be expected from the reader. This chapter of the book mainly addresses a bit IT-savvy people in the public and its mass media, as well as "software engineers". Here an introduction is difficult, since in both communities people typically know nothing or almost nothing about RC. To move RC from its niche market into mainstream massive funding is needed for R&D and to reinvent programming education. To yield the attention of media and the politicians we need a highly effective campaign by mass media.

RC should urgently become mainstream. Several reasons have prevented RC from truly becoming mainstream [114]. The execution model is inherently different from the traditional sequential paradigm were we can reason about state transition sequences much better than in a hardware or a concurrent execution model. As a consequence, the development and validation of tools is substantially a traditional hardware mind set.

Tools are limited and above all fairly bridle. This means programmers must master the details of not only software development but also of hardware design. Such a set of skills is also not taught as part of major electrical engineering courses severely constraining the pool of engineering with the "right" mindset for programming RC to a selected few. Moreover the recent evolution of FPGAs and to some extent coarse-grain RC architecture make programmer and performance portability difficult at best.

One of the objectives of the REFLECT project (Chap. 11) is lowering the barrier of access of RC to the average programmers, by retaining the "traditional" imperative programming mindset in a high-level language such as MATLAB and rely on the concepts of Aspects to provide a clean mechanism (at the source code level) for the advanced user to provide key information for a compilation and synthesis tool to do a good job in mapping the computation to hardware. The approach should be by no means fully automatic [114]. Instead, we have the programmer involved but controlling the high-level aspects of the mapping while the tools takes care of the low-level, error-prone steps.

We extend the "traditional" imperative programming mindset (for software) by a twin-paradigm imperative mind (subject of Sect. 2.4.4.) also including an imperative datastream programming methodology (for "flowware" – for terminology see Table 2.3) [90]. We obtain an almost fully symmetric methodology: the only asymmetry is intra-loop parallelism, possible for data streams, however not for instruction streams (Table 2.4). The semantic difference of these machine paradigms is the state register: the program counter (located with the ALU) for running the instruction streams in executing software, and data counter(s) (located in memory block(s) [60, 61]) for running data streams in executing flowware.

2.3.5 Toward a New World Model of Computing

The traditional CPU-centric world model of the CS world is obsolete. It resembles the old Aristotelian geo-centric world model. Its instruction-stream-based software-only tunnel view perspective hides structural and data stream aspects – massively threatening the progression of system performance, where we have to confront a dramatic capability gap. We need a generalized view, comparable to the Copernican world model not being geo-centric. We need a hetero model which also includes structures and data streams and supports time to space mapping, since scaling laws favor reconfigurable spatial computing over temporal computing. Exercising time to space mapping, also by programming data streams and by software to configware migration, provides important skills: e. g. locality awareness, understanding and designing efficient manycore architectures and their memory organization being essential to cope with bottlenecks caused by bandwidth problems.

This new direction has not yet drawn the attention of the curriculum planner within the embedded systems scene. For computer science this is the opportunity of the century, of decampment for heading toward new horizons, and, to preserve the affordability of its electricity consumption. This should be a wake-up call to CS curriculum development. Each of the many different application domains has only a limited view of computing and takes it more as a mere technique than as a science on its own. This fragmentation makes it very difficult to bridge the cultural and practical gaps, since there are so many different actors and departments involved. We need the new CS world model to avoid the capability gap caused by that fragmentation. Computer Science should take the full responsibility to merge Reconfigurable Computing into CS curricula for providing Reconfigurable Computing Education from its roots. CS has the right perspective for a trans-disciplinary unification in dealing with problems, which are shared across many different application domains. This new direction would also be helpful to reverse the current down trend of CS enrolment.

Not only for the definition of the term "Reconfigurable Computing" (RC), it makes sense to use a clear terminology – not only to improve education about how to reinvent computing. It is a sluttish use of terms if "soft" or "software" is used for everything, which is not hardware. The term "software" should be used only for instruction streams and their codes. However, we generalize the term "programming" (Fig. 2.6) such, that procedural programming (in time domain) creates sequential code, like instruction streams (software), or data streams, which we call "flowware", and, that "structural programming" (programming in space) creates "structural code", which we call "configware", since it can be used for the configuration of FPGAs (Field-Programmable Gate Arrays) or other reconfigurable platforms. Summary: Table 2.3.

Table 2.5 Confusing terms which should not be used

Term	Once introduced for
Dataflow	Indeterministic exotic machines
Firmware	Nested von Neumann machines
Microcode	
Microprogram	
Software or "soft"	No use other than for instruction streams

This established terminology reveals (see Table 2.3 for the terms we should use and Table 2.5 for the terms that usually make some confusion), that a software to configware migration means a paradigm shift, away from the traditional programmer's CPU-centric world model of computing, resembling the geo-centric Aristotelian world model. To reinvent computing we need a multi paradigm hetero system world model of computing science (Fig. 2.9), which models the co-existence of, and the communication between: (1) the traditional imperative software programming language mind set with the CPUs running by software (instruction streams), (2) the reconfigurable modules to be structurally programmed by configware, and (3) an imperative datastream programming language mind set with [90] data stream machines programmed by flowware for generating and accepting data streams (asM in Table 2.3 stands for "auto-sequencing Memory", also containing the data counter inside a reconfigurable address generator). We obtain an almost fully symmetric methodology: the only asymmetry is intra-loop parallelism, possible for data streams, however not for instruction streams (Table 2.4). The semantic difference of these machine paradigms is the state register: the program counter (located with the ALU) for running the instruction streams in executing software, and the data counter(s) (located in memory block(s) [60–66]) for running data streams in executing flowware.

Figure 2.10 illustrates this triple-paradigm "Copernican" world model replacing the von-Neumann-only-centric obsolete "Aristotelian" narrow tunnel view perspective of classical software engineering, which hides almost everything, which is not instruction-stream-based. (The term "supersystolic" in Fig. 2.10 stands for the generalization of the systolic array: non-linear and non-uniform pipes are allowed like spiral, zigzag and any excessively irregular shapes.) This generalized model will help us to come up with a new horizon of programmer education which masters overcoming the hardware/software chasm, having been a typical misconception of the ending first history of computing. The impact is a fascinating challenge to reach new horizons of research and development in computer science. We need a new generation of talented innovative scientists and engineers to start the beginning second history of computing, not only for the survival of our important computer-based cyber infrastructures, but also for developing and integrating exciting new innovative products for the transforming post PC era global information and communication markets [90].

Not yet discussed in this paper is the accelerator role of GPUs (graphics processors [115]) which for some authors seem to be the FPGA's competitor w. r. to speed-up and power efficiency [116–118]. Meanwhile the very busy hype on the accelerator use of GPGPU seems to be over-exaggerated [117]. FPGAs from a new

Xilinx 28 nm high-performance, low-power process, developed by Xilinx and TSMC-optimized for high performance & low power are massively better off than GPUs. E.g., for the Smith-Waterman algorithm the following normalized performance is reported: 584 for FPGA, 25 for GPU, and, 1 for GPP [119]. Since a compute-capable discrete GPU can draw much more than 200 W, other authors call this massive power draw a serious roadblock to the adoption, not only in embedded systems, but even for data centers [120].

But going hetero by interweaving instruction stream parallelism and structural parallelism is a massive challenge requiring to master many difficult problems. The existence of thousands of languages did not prevent a stall of language research in the past two decades [104]. Being speaker in seven tutorials at Supercomputing 2010 [121] Tim Mattson of Intel complaints about what he calls "choice overload" and calls to arms. Another design tool problem is hitting the moving target of the complex value chain in SoC design: the rapidly growing segment of the electronics industry called "Electronics Intellectual Property" or "Electronics IP", where currently the designers have a nightmare using IPs delivered by 3rd parties or internal IP teams [122]. Masses of highly qualified new kinds of jobs must be created to meet the fascinating challenges of reinventing computing sciences, following the wide horizon of the new world model [121].

2.4 Conclusions

This chapter has emphasized that Reconfigurable Computing (RC) is a critical survival issue for computing-supported infrastructures worldwide and has stressed the urgency of moving RC from niche to mainstream. Since a qualified programmer population does not exist we need to use Reconfigurable Computing to Reinvent Computing (R2R) and to Rewrite Textbooks for R2R (RT4R2R), and many of us should become Reinvent Computing Evangelists (RCE). We urgently need a worldwide mass movement of R&D and education to be more massively funded and supported than the Mead-&-Conway VLSI design revolution in the early 1980s, which so far has been the most effective project in the history of modern computing science. This chapter urges acceptance of the massive challenge of reinventing computing, away from its currently obsolete CPU-processor-centric Aristotelian CS world model, to a twin-paradigm Copernican model.

For energy cost reasons, a massive software to configware migration campaign is needed. First this requires clever planning to optimize all its aspects. We also need to develop plans deciding, which software packets need to be migrated, and, which of them should be migrated first. All this requires many years, probably a decade of massive R&D and education efforts. We cannot afford to hesitate. Lobbying for the massive funding should be started right now. We should address politicians at all levels: community level, state level, national level, and European Union level. To explain all this to politicians is very, very difficult. Since politicians always watch the sentiment of their voter population, we efficiently have to teach the public, which is a dramatic challenge. Without the support by a strong tailwind from the media a successful lobbying does not seem to have any chance. All this has to be completed as soon as possible, as soon as we can still afford such massive activities. To succeed with such a challenging educational campaign the foundation of a powerful consortium to be funded at all levels is needed for running an at least Europe-wide project.

References

1. John von Neumann: First Draft of a Report on the EDVAC; University of Pennsylvania, June 30, 1945
2. P. Wallich: Profile: Lynn Conway—Completing the Circuit; Scientific American Magazine, December 2000.
3. L. Conway: The MPC Adventures: Experiences with the Generation of VLSI Design and Implementation Methodologies; Microprocessing and Microprogramming - The Euromicro Journal, Vol. 10, No. 4, Nov 1982.
4. National Research Council: Funding a Revolution; National Academies Press; January 8, 1999
5. N. N.: "The book that changed everything", Electronic Design News, Feb. 11, 2009
6. D. Kilbane: Lynn Conway - A trailblazer on professional, personal levels." Electronic Design, October 23, 2003
7. R. Hartenstein: The Paramountcy of Reconfigurable Computing; in: Young-Choon Lee, Albert Zomaya (editors): Energy Aware Distributed Computing Systems; Wiley, 2011
8. R. Hartenstein: The Grand Challenge To Reinvent Computing; XXX Congress of the SBC, 20–23 July 2010, Belo Horizonte, MG, Brazil,
9. http://wiki.answers.com/QWhy_are_computers_important_in_the_world
10. G. Fettweis: ICT Energy Consumption - Trends and Challenges; WPMC'08, Lapland, Finland, 8 –11 Sep 2008
11. D. Estrin et al: "Internet Predictions," IEEE Internet Computing, vol. 14, no. 1, pp. 12–42, Jan./Feb. 2010, doi:10.1109/MIC.2010.12
12. R. H. Katz: Tech Titans Building Boom; IEEE Spectrum, February 2009
13. R. Nuez: Google is Going into the Electricity Business; The Huffington Post, JUNE 6, 2010; http://www.huffingtonpost.com/ramon-nuez/google-is-going-into-the_b_417035.html
14. http://en.wikipedia.org/wiki/Pelamis_Wave_Energy_Converter
15. A. Zomaya et al.: Interweaving Heterogeneous Metaheuristics Using Harmony Search; Int'l Parallel and Distr. Processing Symp., May 23–29, Rome, Italy
16. J. Rabaey: Low Power Design Essentials; Springer Verlag, 2009
17. L. A. Barroso: The Price of Performance; ACMqueue, Oct 18, 2005 - http://queue.acm.org/detail.cfm?id=1095420
18. Wu-chun Feng: On the Second Coming of Green Destiny?; International Conference on Energy-Aware HPC; Sep 16–17, 2010, Hamburg, Germany,
19. Peter Kogge (editor): ExaScale Computing Study: Technology Challenges in Achieving Exascale Systems; Sep 28, 2008 - http://www.cse.nd.edu/Reports/2008/TR-2008-13.pdf
20. W. Blendinger: Post Peak – Abstieg vom Peak Oil; ASPO annual conference, Berlin, 18 Mai 2010
21. K. Aleklett: Post Peak – The Future of the Oil-Production; ASPO annual conference, Berlin, 18 Mai 2010
22. J. S. Gabrielli de Azevedo: Petrobras e o Novo Marco Regulatório; São Paulo, Dec 1, 2009
23. F. Birol: Leave oil before it leaves us; THE INDEPENDENT (UK) March 2, 2008 - http://www.youtube.com/watch?v=m377Is4tGF0

24. H. Ginsburg: Unterm Wasser liegt die Zukunft; Wirtschaftswoche 23, June 7, 2010
25. H. Simon: Leibniz-Rechenzentrum, TU Munich, 2009, Garching, Germany
26. J. Rabaey: Reconfigurable Processing: The Solution to Low-Power Programmable DSP, Proc. ICASSP 1997
27. T. Claasen: High Speed: Not the Only Way to Exploit the Intrinsic Computational Power of Silicon; ISSCC-1999, pp. 22–25, Feb. 1999
28. R. Hartenstein: Why we need Reconfigurable Computing Education; 1st Int'l Workshop on Reconfigurable Computing Education (RC education 2006), March 1, 2006, KIT Karlsruhe Insitute of Technology, Germany
29. Y. Gu, et al.: FPGA Acceleration of Molecular Dynamics Computations; FCCM 2004 http://www.bu.edu/caadlab/FCCM05.pdf
30. A. Alex, J. Rose et al.: Hardware Accelerated Novel Protein Identification; Proc. FPL 2004, Aug. 29 - Sep 1, 2004, Antwerp, Belgium,
31. N. N. Nallatech, press release, 2005
32. H. Singpiel, C. Jacobi: Exploring the benefits of FPGA processor technology for genome analysis at Acconovis; ISC 2003, June 2003, Heidelberg, Germany; http://www.hoise.com/vmw/03/articles/vmw/LV-PL-06-03-9.html
33. N. N. (Starbridge): Smith-Waterman pattern matching; National Cancer Inst., 2004
34. A. Darabiha: Video-Rate Stereo Vision on Reconfigurable Hardware; Master Thesis, Univ. of Toronto, 2003
35. R. McCready: Real-Time Face Detection on a Configurable hardware Platform; Master thesis, U. Toronto
36. T. Fry, S. Hauck: Hyperspectral Image Compression on Reconfigurable Platforms; IFCCM 2002
37. P. Buxa: Reconfigurable Processing Design Suits UAV Radar; COTS J Oct 2005 http://www.srccomp.com/ReconfigurableProcessing_UAVs_COTS-Journal_Oct05.pdf
38. http://helios.informatik.uni-kl.de/RCeducation/
39. R. Porter: Evolution on FPGAs for Feature Extraction; Ph.D. thesis; Queensland U. of Technology, Brisbane, Australia, 2001,
40. E. Chitalwala: Starbridge Solutions to Supercomputing Problems; RSSI Reconf. Syst. Summer Inst., July 11–13, 2005, Urbana-Champaign, IL, USA
41. S. D. Haynes, P. Y. K. Cheung, W. Luk, J. Stone: SONIC - A Plug-In Architecture for Video Processing; FPL 99
42. M. Kuulusa: DSP Processor Based Wireless System Design; Tampere Univ. of Technology, Publ. No. 296; http://edu.cs.tut.fi/kuulusa296.pdf
43. B. Schäfer et al.: Implementation Of The Discrete Element Method Using Reconfigurable Computing; 15th ASCE Engineering Mechanics Conf., June 2–5, 2002, New York
44. G. Lienhart: Beschleunigung Hydrodynamischer N-Körper-Simulationen mit Rekonfigurierbaren Rechensystemen; Joint 33 rd Speedup and 19th PARS Workshop; Basel, Switzerland, March 19–21, 2003
45. O. O. Storaasli, D. Strenski: Experiences on 64 and 150 FPGA Systems; Reconfig. Syst. Summer Inst., July 7–9, 2008, Urbana-Champaign, IL, USA
46. A. A. Gaffar and W. Luk: Accelerating Radiosity Calculations; FCCM 2002
47. M. Gokhale et al.: Acceleration of Traffic Simulation on Reconfigurable Hardware; 2004 MAPLD Int'l Conf., Sep 8–10, 2004, Washington, D.C., USA
48. J. Hammes, D. Poznanovic: Application Development on the SRC Computers, Inc. Systems; RSSI Reconfigurable Systems Summer Institute, July 11–13, 2005, Urbana-Champaign, IL, USA
49. W. Roelandts (Keynote): FPGAs and the Era of Field Programmability; Proc. FPL 2004, Aug. 29 - Sep 1, 2004, Antwerp, Belgium,
50. F. Dittrich: World's Fastest Lanman/NTLM Key Recovery Server Shipped; Picocomputing 2006
51. K. Gaj, T. El-Ghazawi: Cryptographic Applications; RSSI Reconfigurable Systems Summer Institute, July 11–13, 2005, Urbana-Champaign, IL, USA http://www.ncsa.uiuc.edu/Conferences/RSSI/presentations.html

52. R. Jacobi, M. Ayala-Rincón, L. Carvalho, C. Llanos, R. Hartenstein: Reconfigurable systems for sequence alignment and for general dynamic programming; Genetics and Molecular Research 2005
53. M. Feldman: In Fermi's Wake, a Place for FPGAs? HPCwire, Oct. 15, 2009
54. http://www.chrec.org/~george/tmp/ADG.jpg
55. M. Feldman: The FPGA crowd reacts; HPCwire Oct. 15, 2009
56. N. Conner: FPGAs for Dummies - FPGAs keep you moving in a fast-changing world; Wiley, 2008
57. R. Hartenstein (invited paper): The Microprocessor is no more General Purpose: why Future Reconfigurable Platforms will win; Proc. Int'l Conf. on Innovative Systems in Silicon, ISIS'97, Austin, Texas, USA, Oct 8–10, 1997
58. M. Herz et al.: Memory Organisation for Stream-based Reconfigurable Computing; IEEE ICECS 2002, Sep 15–18, 2002, Dubrovnik, Croatia; http://www.michael-herz.de/publications/AddressGenerators3.pdf
59. http://xputer.de/fqa.html#anchor81257
60. R. Hartenstein, A. G. Hirschbiel, M. Weber: MOM-map-oriented machine - a partly custom-designed architecture compared to standard hardware; Proc. IEEE CompEuro, Hamburg, Germany, May 1989
61. R. Hartenstein, A. Hirschbiel, M.Weber: A Novel Paradigm of Parallel Computation and its Use to Implement Simple High Performance Hardware; Proc. InfoJapan'90, Tokyo, Japan, 1990
62. R. Hartenstein et al. (invited reprint): A Novel Paradigm of Parallel Computation and its Use to Implement Simple High Performance Hardware; Future Generation Computer Systems, no. 7, pp. 181–198 (North-Holland)
63. M. Weber et al.: Automatic Synthesis of Cheap Hardware Accelerators for Signal Processing and Image Preprocessing; 12. DAGM-Symp. Mustererkennung (Pattern Recognition), Oberkochen-Aalen, Germany 1990
64. http://tams-www.informatik.uni-hamburg.de/applets/hades/webdemos/42-programmable/10-pla/pla.html
65. http://www.eecs.berkeley.edu/IPRO/Software/Description/platools.html
66. J. Turley: How Many Times Does CPU Go Into FPGA? Embedded Technology Journal - June 8, 2010
67. Selwood: EPP - A Platform to Bridge a Gap? Emb. Technology J. June 8, 2010
68. J. Turley: Whither Embedded Part Deux; Emb. Technology J. May 25, 2010 http://www.techfocusmedia.net/embeddedtechnologyjournal/feature_articles/20100525-esc2/
69. Jih-Sheng Shen, Pao-Ann Hsiung (editors): Dynamic Reconfigurable Network-On-Chip Design: Innovations for Computational Processing and Communication; Information Science Pub, April 2010
70. R. Hartenstein: Implications of Makimoto's Wave; http://hartenstein.de/Implications MakimotosWave2.pdf
71. R. Baxter et al.: High-Performance Reconfigurable Computing – the View from Edinburgh; AHS 2007, Aug 5–8, 2007, Edinburgh, UK
72. http://www.chrec.org/
73. E. El-Araby et al.: Exploiting Partial Runtime Reconfiguration for High-Performance Reconfigurable Computing. ACM Transactions on Reconfigurable Technology and Systems (TRETS) 1(4): (2009)
74. E. El-Araby et al.: Productivity Of High-level Languages On Reconfigurable Computers: An HPC Perspective; IEEE Int'l Conf. on Field-Programmable Technology (ICFPT 2007), Japan, December 2007.
75. E. El-Araby et al.: Comparative Analysis Of High Level Programming For Reconfigurable Computers: Methodology And Empirical Study; III Southern Conf. on Programmable Logic (SPL2007), Mar Del Plata, Argentina, Feb 2007
76. J. Backus: Can programming be liberated from the von Neumann style? C.ACM Aug 1978
77. E. W. Dijkstra: The Goto considered harmful; Comm ACM, March 1968

78. N. Wirth: A Plea for Lean Software, IEEE Computer, 28, 2, (Feb. 1995)
79. Arvind et al.: A critique of Multiprocessing the von Neumann Style; ISCA 1983
80. Peter G. Neumann 1985–2003: 216x "Inside Risks" (18 years in the back cover of each issue of C_ACM)
81. R. Hartenstein, G. Koch: The universal Bus considered harmful; Workshop on the Microarchitecture of Computer Syst, June 23–25, 1975, Nice, France
82. R. Hartenstein: The von Neumann Syndrome; Stamatis Vassiliadis Memorial Symp., Sep 2007, Delft, NL - http://helios.informatik.uni-kl.de/staff/hartenstein/Hartenstein-Delft-Sep2007.pdf
83. R. Hameed et al.: Understanding Sources of Inefficiency in General-Purpose Chips; 37th ISCA, June 19–23, 2010, Saint Malo, France
84. T. C. Chen et al., "Analysis and architecture design of an HDTV720p 30 frames/s H.264/ AVC encoder," Circuits and Systems for Video Technology, IEEE Transactions on, vol.16, no.6, pp. 673–688, June 2006
85. S. McKee: Reflections on the memory wall; Computing Frontiers, Ischia, Italy, 2004
86. J. L. Hennessy, D. A. Patterson, Computer Architecture: a Quantitative Approach, Morgan-Kaufman, San Mateo, CA, 1990
87. Kiran Bondalapati, Viktor K. Prasanna: Mapping Loops onto Reconfigurable Architectures; 8th Int'l Workshop on Field-Programmable Logic and Applications, Sep 1998, Tallinn, Estonia
88. M. Huebner, D. Goehringer, J. Noguera, J. Becker: Fast dynamic and partial reconfiguration Data Path with low Hardware overhead on Xilinx FPGAs; Proc. RAW 2010, Atlanta, USA, April, 2010
89. J. Larus: Spending Moore's Dividend; Comm ACM, May 2009
90. Juergen Becker, et al.: Data-procedural Languages for FPL-based Machines; FPL'94, Prague, September 7–10, 1994, Prague, Czechia.
91. C. Bell et al: The Description and Use of Register-Transfer Modules (RTM's); IEEE Trans-C21/5, May 1972
92. M. Ayala, C. Llanos, R. Jacobi, R. Hartenstein: Prototyping Time and Space Efficient Computations of Algebraic Operations over Dynamically Reconfigurable Systems Modeled by Rewriting-Logic; ACM TODAES, 2006
93. http://webadmin.dac.com/knowledgecenter/2010/documents/SMITH-VIRTUAL-ABK-FINAL2.pdf
94. D. Parnas (keynote): Teaching for Change; 10th Conf Softw Engrg Education and Training (CSEET), April 13–16, 1997, Virginia Beach, VA, USA
95. Fig. 13, R. Hartenstein: The History of KARL and ABL; in: J. Mermet: Fundamentals and Standards in Hardware Description Languages; Kluwer, 1993
96. M. Barbacci: The ISPS Computer Description Language; Carnegie-Mellon Univ., Dept. of Computer Science, 1977
97. N. N. (I forgot the celebrity's name): Computer Magazine, 1970 or 1971
98. R. Hartenstein (keynote): Reconfigurable Computing: boosting Software Education for the Multicore Era; IV Southern Programmable Logic Conference (SPL 2010), Porto Galinhas Beach, Brazil, 24–26 March 2010
99. Dick Selwood: Drag-and-Drop vs. HDL? FPGA and Programmable Logic J.; http://www.techfocusmedia.net/fpgajournal/feature_articles/20100803-ni/
100. K. Asanovic et al.: A view of the parallel computing landscape. Comm. ACM 52,10 (2009)
101. 4th Workshop on Programmability Issues for Heterogeneous Multicores (MULTIPROG 2011), January 23, 2011, Heraklion, Crete, Greece - http://multiprog.ac.upc.edu/
102. S. Moore: Multicore Is Bad News For Supercomputers; IEEE Spectrum,
103. M. Broy, R. Reussner: Architectural Concepts in Programming Languages; Computer Oct 2010
104. D. Selwood: Showing off in San Jose; Embedded Technology J., June 8, 2010
105. M. Anderson: Understanding Multi-Core in Embedded Systems; The PTR Group, Falls Church VA, USA, June 15, 2010

106. Rick Merritt: Next computing target—exascale systems; EEtimes, Nov. 15, 2010,
107. Neal Singer: New standard for supercomputing proposed; SANDIA LAB NEWS, Nov 5, 2010, http://www.sandia.gov/LabNews/101105.htm
108. Kevin Morris: Why FPGAs Will Win; FPGA and Programmable Logic Journal Update (techfocus media). 30 Nov 30, 2010
109. Garrison Jeff: What! How big did you say that FPGA is? (Team-design for FPGAs) (EETimes Design Article). 27 Sept. 2010
110. S. Z. Ahmed, G. Sassatelli, L. Torres, L. Rougé: Survey of new Trends in Industry for Programmable Hardware; 20th FPL, Milano, Italy, Aug 31 - Sep 2, 2010
111. Alberto Sangiovanni-Vincentelli: Quo Vadis, SLD? Proc. IEEE, March 2007
112. N. Voros, R. Nikolaos, A. Rosti, M. Hübner (editors): Dynamic System Reconfiguration in Heterogeneous Platforms - The MORPHEUS Approach; Springer, 2009
113. Ch. Bobda: Introduction to Reconfigurable Computing - Architectures, Algorithms, Applications; Springer Verlag, 2007
114. personal communication from authors of other chapters of this book
115. http://gpgpu.org/
116. V. W. Lee et al.: Debunking the 100X GPU vs. CPU myth; 37th ISCA, June 19–23, 2010, Saint-Malo, France,
117. R. Vaduc et al.: On the Limits of GPU Acceleration; USENIX Workshop HotPar'2010, June 14–15, 2010, Berkeley, CA, USA
118. R. Bordawekar et al.: Believe it or Not! Multicore CPUs can Match GPUs for FLOP-intensive Applications! IBM Research Report, April 23, 2010
119. P. Lysaght: The Programmable Logic Perspective; FPL-2010, Sep 2010, Milano, Italy
120. Thomas Scogland, Heshan Lin, Wu-chun Feng: A First Look at Integrated GPUs for Green High Performance Computing; In'l Conf. on Energy-Aware HPC; Sep 16–17, 2010, Hamburg, Germany, http://ena-hpc.org/index.html
121. N. N.: A Call to Arms for Parallel Programming Standards; HPCwire, Nov 16, 2010, http://sc10.supercomputing.org/?pg=tutorials.html
122. Russ Henke: The State of IP; EDACafé, August 16, 2010; http://www10.edacafe.com/nbc/articles/view_weekly.php?articleid=850357

Chapter 3
HiPEAC: Upcoming Challenges in Reconfigurable Computing

Ioannis Sourdis and Georgi N. Gaydadjiev

Abstract The new developments in semiconductor technology cause significant problems in chips' performance, power consumption and reliability, indicating that the "golden" CMOS era is long gone. Technology scaling does not deliver anymore significant performance speedup, the increasing power density poses severe limitations in chips, while, transistors become less reliable. The above introduce great challenges for reconfigurable computing; that is to provide the answer to the performance, power-efficiency and reliability quest posed by current technology trends. Reconfigurable Computing has the potential to achieve such a goal; however, several improvements are required to be performed first. In this chapter, we discuss a number of issues which need to be addressed in order to make Reconfigurable Computing a widely used solution for future systems.

Project name:	**HiPEAC NoE:** High Performance and Embedded Architectures and Compilation Network of Excellence
URL	www.hipeac.net
HiPEAC Partners:	RWTH Aachen, Chalmers, FORTH, INRIA, TU Delft, U Edinburgh, U Ghent, BSC, ARM, IBM, ST France.
No. HiPEAC members full/affiliated/Phd:	233/218/398 in 23 European countries
Number of HiPEAC Reconfig. Computing cluster members:	83
Period:	2008–2011
EU Program:	7th Framework Program, Network of Excellence (NoE)
Global EU Funding:	4.8 M€

I. Sourdis (✉)
Computer Engineering, Chalmers University of Technology, 412 96 Gothenburg, Sweden
e-mail: Sourdis@chalmers.se

J.M.P. Cardoso and M. Hübner (eds.), *Reconfigurable Computing: From FPGAs to Hardware/Software Codesign*, DOI 10.1007/978-1-4614-0061-5_3,

3.1 Introduction

Although a one-to-one comparison between fixed integrated circuits and reconfigurable hardware shows that the formers are better in terms of area, time (performance), and power, it has been repeatedly shown that reconfigurable computing can offer substantial benefits over ordinary general purpose machines, i.e., in the works of W. Luk [1] and M. Flynn et al. [2, 3]. It is a well known fact that reconfigurable hardware has roughly three orders of magnitude higher Area*Time*Power product compared to ASICs [2, 3]. About an order of magnitude more resources are required to realize a circuit, which then has roughly an order of magnitude higher delay, and about an order of magnitude higher power consumption. Still, for a wide range of applications, reconfigurable computing can offer faster execution times and better power/performance ratio than general purpose machines. A reason for this is that, as opposed to reconfigurable hardware, in current general purpose architectures only a small fraction of the chip is dedicated to carry useful computations [2], as illustrated in Fig. 3.1; the remaining chip-resources are spent in the memory hierarchy and other modules which only indirectly improve performance (e.g. branch predictor, complex pipeline control). The above design choice is probably unavoidable in fixed integrated circuits where the diverse application requirements need to be predicted and met at design time. On the other hand, a reconfigurable device has the advantage of post-fabrication on-demand customization which allows exploiting most of its resources to directly support computations.

Reconfigurable Computing is an emerging paradigm which combines the flexibility of software with the high-performance of hardware. Reconfigurable devices enable a massive number of programmable computational nodes that execute in parallel, providing substantial **performance benefits** over conventional general-purpose machines. Their **power/performance ratio** can often be considerably better than software as they adapt to the specific needs of an application supporting customization of the datapath and control-flow of hardware. On demand reconfiguration can be further exploited to isolate and correct defective blocks offering an excellent solution for **fault–tolerance**. For these reasons, reconfigurable hardware is becoming popular in embedded systems and application-specific design, while it can also provide an attractive solution for high-performance computing. In the past few years, increasingly more commercial solutions are found, such as the ones of Maxeler [4] and Convey [5], which exploit reconfigurable hardware to speedup computationally intensive tasks. However, in order to deploy this technology at a larger scale, there is a need to address several challenging issues.

In the remainder of this chapter we first discuss current trends in semiconductor technology and identify promising opportunities for Reconfigurable Computing. We then identify a number of challenging problems that need to be addressed in order for reconfigurable computing to be more widely used in future systems. Finally, we draw our conclusions.

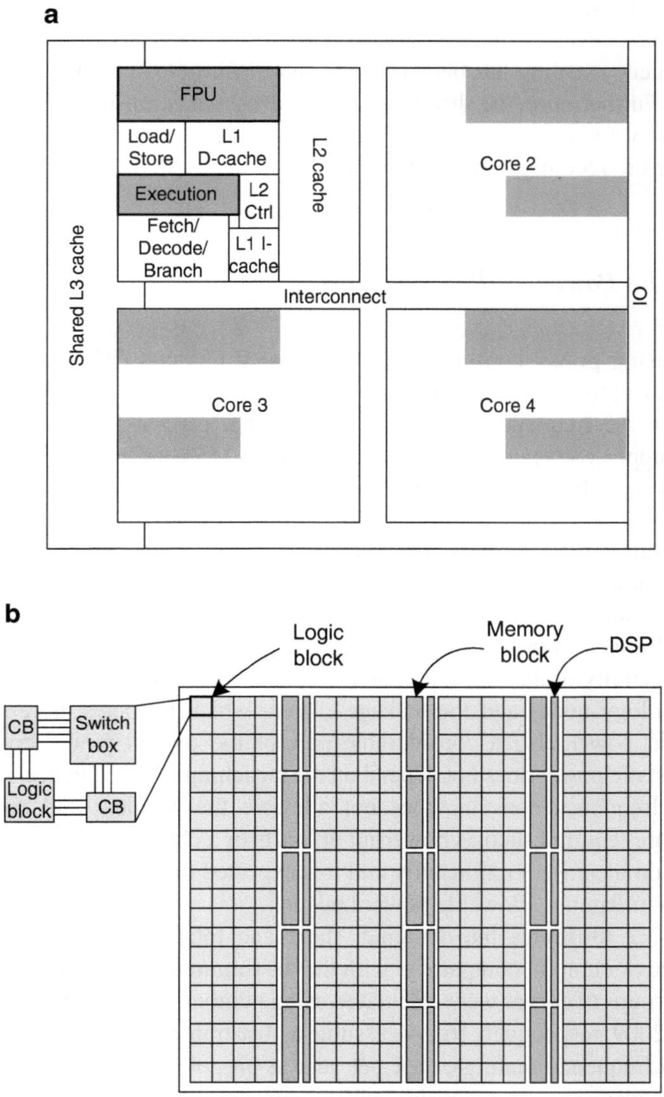

Fig. 3.1 A general purpose quad core multiprocessor chip vs. an FPGA device: the resources explicitly used for computations are significantly lower in the GPP case (FPU and execution modules) than in the FPGA device (logic blocks, DSP blocks) (**a**) A quad core multiprocessor chip. (**b**) An FPGA device

3.2 Technology Trends

Over the past few years semiconductor technology trends have changed. CMOS scaling becomes more difficult causing severe problems in building future chips. According to the ITRS [6], transistor count per chip is expected to continue doubling with each new technology generation for the next 10 years. On the other hand, power consumption does not scale well anymore. As a consequence, the clock frequency scaling has been reduced significantly to keep power at a reasonable level. Furthermore, the shrinking feature sizes make transistors less reliable. In this section, we give an overview of the current technology trends and discuss their implications to reconfigurable computing.

3.2.1 The Power Wall

The increasing power consumption of Integrated Circuits (IC) is one the major problems in current and upcoming semiconductor technologies. Dynamic power depends on the frequency of transistor switching, capacity, and squared supply voltage. Supply voltage used to reduce more than 15% per technology generation, in order to keep the power consumption low and at the same time let the operating frequency increasing with a rate of about 20–30% per annum. However, in the past few years supply voltage does not scale that well causing an increase of the power consumed on a chip. Consequently, power consumption is rapidly approaching the maximum amount of power that can be dissipated form a given chip area (roughly about 90 W/cm^2). In order to keep within this power budget one needs to limit the switching activity of the transistors on a chip. The bad scaling of supply voltage in new technology nodes and the leakage current which becomes more significant, make static power also a considerable part of the overall power consumption. However, static power does not depend on the switching activity; hence, keeping the operating frequency constant does not alleviate the problem and more drastic measures are required such as powering off parts of a chip. Overall, the total power consumption increases significantly and it is expected that in a few years chips will reach their maximum allowable power budget.

Figure 3.2 illustrates the dynamic, static and total power consumption of a Microprocessor chip in comparison with the maximum allowed power budget. Two different groups of curves were calculated using numbers retrieved from ITRS 2008, and ITRS 2009, respectively [6]; we assume a 40% transistor switching at any given time and a frequency scaling of 8% per year as proposed by ITRS [6]. ITRS 2008 predictions were showing that the leakage current will be increasing in the next 15 years, causing a substantial increase in static power. This combined with the slow supply voltage reduction and the increase in transistor count makes the dynamic and overall power consumption exceeding the maximum allowed power consumption on a chip (130–140 W) by 2016. ITRS 2009 predictions show a constant leakage current for the upcoming technology generations and a slightly better scaling of the supply voltage. This has as a consequence to predict a significantly lower static power and a slightly better dynamic power. Still, after 2019 it is expected to reach and exceed the maximum available power budget.

As mentioned above, in order to stay within the power budget the switching activity would need to be restricted or even parts of the chip may need to be powered

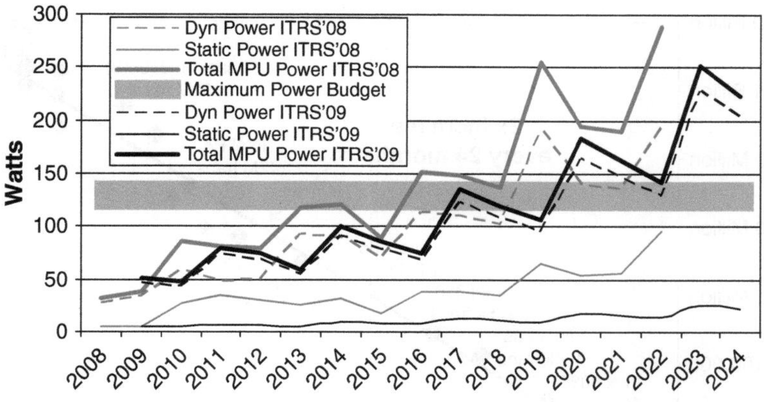

Fig. 3.2 Dynamic, Static and total power consumption of a Microprocessor chip vs. the maximum available power budget, based on the ITRS 2008 and 2009 projections

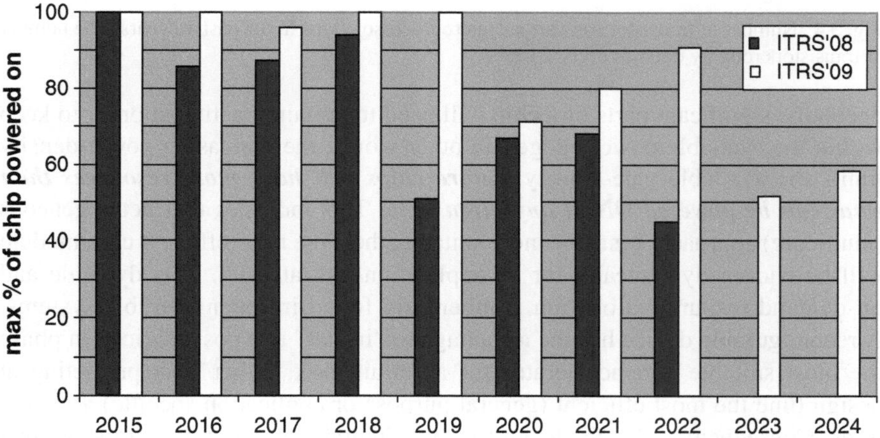

Fig. 3.3 Percentage of a (microprocessor) chip that can be ON in order to keep within the available power budget

off. Based on the above ITRS numbers, Fig. 3.3 shows the percentage of a chip that can be ON at any given time assuming 40% activity for the transistors which are ON, e.g. when 50% of the chip is ON, then, dynamic power is calculated assuming 20% overall transistor activity. As expected, after 2016, significant parts of the chip would be required to be OFF assuming an 8% annual increase in frequency. By 2023, it is expected that half of the chip would need to be powered-off.

We can draw two conclusions based on the above analysis. Firstly, ***power consumption is becoming a severe problem for future integrated circuits***, and therefore power efficient solutions will be very important. Reconfigurable computing may have an opportunity to provide such solutions for future systems.

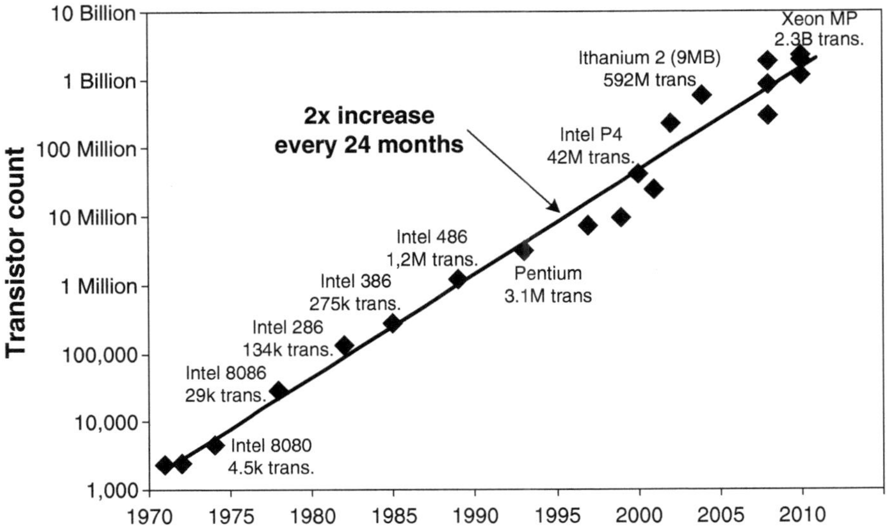

Fig. 3.4 Number of transistors on various microprocessor chips in the past 40 years. The number of transistors roughly doubles every 2 years

Secondly, significant parts of a chip will need to remain inactive in order to keep within the available power budget. In other words, the increasing power-density limits the available gate-density; ***future chips will have more resources than what can be powered ON at any given time***. This indicates that heterogeneous (multicore) approaches may be more suitable, then, the most efficient core/module will be chosen dynamically for an application/task at hand. This dynamic and on-demand resource allocation is inherently found in reconfigurable systems. A reconfigurable device has the advantage to "install" at a post-fabrication phase the most suitable core/accelerator for a certain task, rather than predicting at design time the most efficient (general purpose or application specific) architectures. Consequently, the challenge for reconfigurable computing is to show that customization and massive parallelism of reconfigurable hardware can overcome its power consumption overhead over ASICs providing power-efficient solutions. An advantage towards this for reconfigurable technology is that significant parts of a reconfigurable substrate often remain unused, while even the used parts may have lower switching activity compared to an ASIC as many of the transistors support the configuration of the hardware.

3.2.2 Transistor Count

As shown in Fig. 3.4, for the past 40 years transistor count keeps doubling every 2 years. In the 1970s a chip had circuits of a few thousands transistors; after 35–40 years we are currently able to integrate billions of transistors on a single chip.

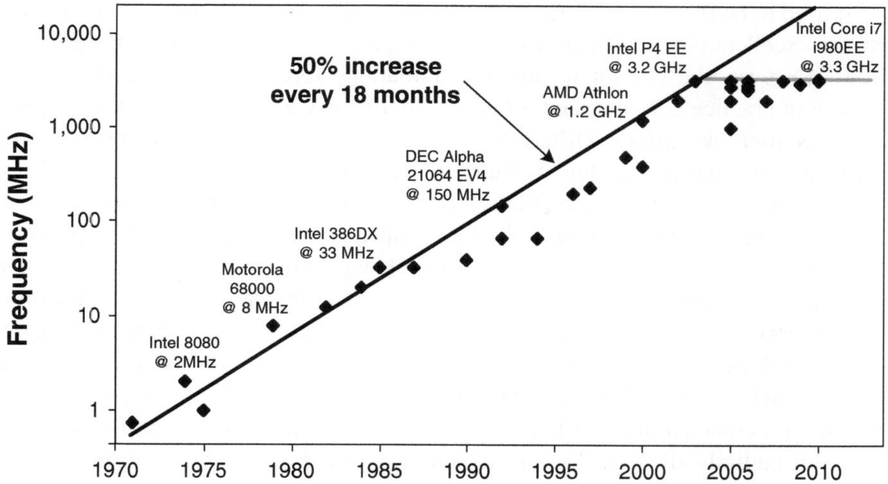

Fig. 3.5 Operating frequency of various microprocessor chips in the past 40 years. Until 2003 the increase in frequency was about 50% every technology node, however after that the increase is negligible

ITRS expectations indicate that this trend will continue for the next 15 years, offering increasingly more resources on future chips.

The rapid increase in transistor count minimizes the area cost of circuits. This is a significant turning point for Reconfigurable Computing. So far, reconfigurable hardware had a significant area cost; as explained in the introduction section, about an order of magnitude more resources are required to realize a circuit in reconfigurable hardware compared to ASICs. However, the rapid increase of on-chip devices, which currently account in billions of transistors as well as the large number of metal layers, shows *for the first time* that *resources get "cheaper"* and thus the area cost of reconfigurable hardware is not anymore a limiting factor.

Furthermore, due to power limitations, explained above, not all resources can be active at the same time; such resources can be then used to offer reconfigurability and flexibility on a chip targeting fault-tolerance, better performance or even lower power consumption of certain computations.

3.2.3 Clock Frequency

Although the technology scaling allows gates to be faster with an estimated rate of 10–13% per year according to ITRS, it is expected that due to the increasing power consumption clock frequencies will scale at best with a rate of 5–8% annually. The current ITRS proposed scaling rate for clock frequencies is 8% per annum. As depicted in Fig. 3.5, in practice the operating frequency increased rapidly for decades and then flattened off a few years ago.

It used to be the case that significant performance improvements were delivered by the excellent clock frequency scaling brought by each new technology generation. However, the current technology limitations force the clock frequency to remain in practice constant, and consequently ***there is no more performance gain from technology scaling***. Different design solutions are required to provide the missing performance-speedup. The current trend in computer systems design is to seek performance increase in parallelism and multicore architectures. Reconfigurable computing can fill, at least partially, the above gap in the missing performance speedup. For many years now, reconfigurable devices have shown their performance gains over general purpose and even application specific solutions (e.g. Network Processors). It is the challenge of Reconfigurable computing to find its role in future (high performance, but also embedded) systems and provide fast execution times, either as stand-alone solutions or as accelerator blocks attached to cores in a multiprocessor parallel system. The regularity of the reconfigurable substrate may substantially alleviate the design complexity of chips that integrate billions of transistors. Its flexibility and on demand customization can be a powerful advantage for supporting a wide range of applications and achieving fast execution times. Efficient tools and methodologies to support such reconfigurable solutions is another great challenge required for the success of the proposed approach.

3.2.4 Reliability

In the coming nanoscale era chips are becoming less reliable, while manufacturing reliable chips is becoming increasingly difficult and costly [7, 8]. Prominent reasons for that are the shrinking device features, the sheer number of components on a given area of silicon, as well as the increasing complexity of current and future chips. It is expected that a significant number of devices will be defective at manufacture time and many more will degrade and fail over the expected lifetime of a chip [7]. Furthermore, process variations as well as the increasing number of soft errors introduce additional sources of errors for future chips [7].

Although ITRS targets for a constant defect rate (1,395 defects/m^2) in order to keep the chip yield constant [6], such target is expected to substantially increase the manufacturing cost of the chips in future semiconductor technologies; alternatively, designing a new chip would need to tolerate an increasing number of defects. Apart from defects at manufacture time, aging effects are becoming more severe; transistors degrade faster, while degrading is further accelerated by the heavy testing processes (e.g. burn-in). Process variations cause devices to operate differently than expected; such variations are random dopant fluctuations, heat flux, as well as lithography problems due to the shrinking geometries. Finally, as transistor count increases, the number of soft-errors (e.g. bit-flips) per chip increases exponentially [7].

As feature size continues to shrink, transistors become less reliable. The cost for delivering reliable chips is expected to grow for future technology nodes; that is overhead in *performance*, *power* consumption and/or *energy*, as well as extra *resources*, required to perform redundant computations in time or space. The reconfigurable substrate can provide an excellent solution for fault tolerance; it can be used to adapt to faults on demand, isolate and correct defects, as well as to provide spare resources to substitute defective blocks.

3.2.5 How the Technology Trends Affect Reconfigurable Computing?

As expected, technology trends show that power consumption and reliability become a significant concern, while performance does not scale anymore with the new technology nodes. This poses an increasing demand for new more efficient solutions for improving performance as well as for providing better power efficiency and fault tolerance. *Reconfigurable computing has a unique opportunity to deliver such solutions.* The use of reconfigurable substrate is favored by the fact that resources are becoming cheaper. Transistor count keeps increasing, while power density limits the resources used at any given time to only a fraction of the chip. *The suggestion is that the "surplus" of these available unused resources can be used to provide reconfigurability.* In so doing, the use of reconfigurable substrate in future chips can offer the flexibility of on demand customization and adaptation to the system and application requirements. Such flexibility can then be used to speedup or reduce the power requirements of certain functions, as well as to tolerate faults. However, in order to accomplish the above several challenging issues need to be addressed as described in the next section.

3.3 Challenges in Reconfigurable Computing

Reconfigurable Computing (RC) has the potential to deliver solutions for the severe performance, power, and reliability problems posed by the current semiconductor technology trends. The area overhead of the reconfigurable substrate is becoming less significant due to the power density and the increasing number of devices per chip. However, in order for reconfigurable computing to exploit the above facts and become a mainstream solution several issues need to be addressed. In order to identify the RC challenges and put them in the right context, we first describe four abstraction layers and four (targeted) properties of a reconfigurable system.

A reconfigurable system can be divided in the following abstraction layers:

- Applications;
- Tools, Methods and Runtime Support;
- Systems architecture, micro-architecture, and implementation;
- Realization/Technology.

The application layer is related to the development of an application to be ported to a system. The next layer consists of design tools and methods for reconfigurable hardware and (as a lower sub-layer) adaptive runtime system support for reconfigurable systems. Subsequently, (from top-to-down) system architecture, micro-architecture, and implementation are all elements of a system built (partially or entirely) in reconfigurable hardware rather than the elements of the reconfigurable device itself. Finally, Realization/Technology is the bottom layer referring to the technology of a reconfigurable platform, built in silicon or other future fabrication technologies.

A reconfigurable system has or is desired to have the following properties:

- Power efficiency;
- Reliability;
- (expected) Performance;[1]
- Reconfiguration.

Power, Reliability and (expected) Performance are design constraints and system requirements which become more challenging with the current technology trends, while Reconfiguration is a desired property through which a reconfigurable system may deliver significant advantages over fixed IC approaches.

Table 3.1 depicts eight RC challenges putting them in the context set above. These challenges span across the aforementioned RC abstraction layers and properties and are the following:

1. Application domain extensions;
2. Improved run-time reconfiguration;
3. Power-efficient computation;
4. Fault Tolerance;
5. Tools, Methods and Runtime Support;
6. Targeted Systems (Embedded/High Performance-Multicores);
7. RC Specific Memory architectures;
8. Realization/Technology.

Most of these challenges constitute a row (abstraction layer) or a column (RC desired property) on the table where additional effort is required in order for reconfigurable computing to be used in a larger scale. Challenge 7 (RC Specific Memory architectures) is an important issue in the architecture of reconfigurable systems. Finally, Challenge 6 (Targeted Systems), although it overlaps with most of the other challenges, it is meant for identifying the different requirements, characteristics, and issues of the embedded and high performance systems which exploit reconfigurable hardware. As illustrated in Table 3.1, challenges may overlap with each other, showing that in such cases they are interdependent. Below we explain in detail each challenge.

---

[1] With the term "expected performance" we mean performance that meets the application and system requirements; that depending on the system can be: high performance, real-time performance, guaranteed latency or throughput, etc.

Table 3.1 Future challenges in reconfigurable computing: placed in a frame defined by the abstraction layers and (desired) properties/goals of a reconfigurable system

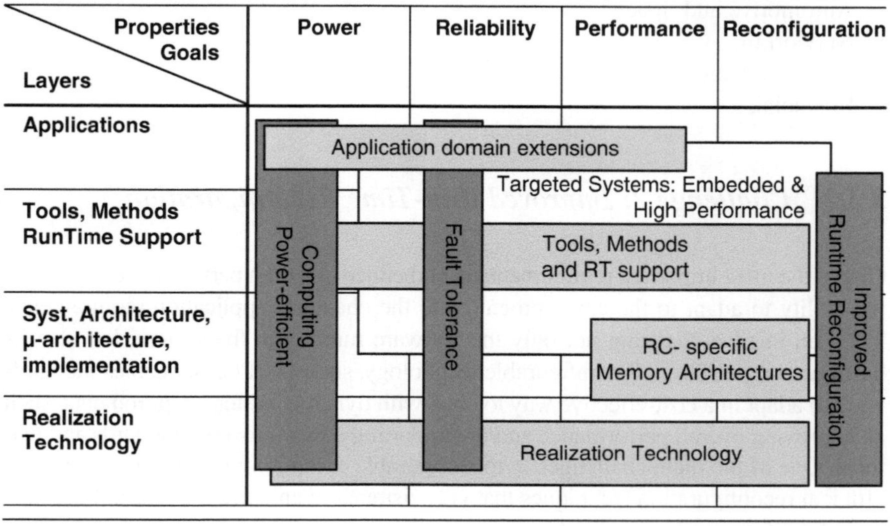

3.3.1 Challenge 1: Application Domain Extensions

An important challenge in RC is combining the high-performance and flexibility of reconfigurable devices with high reliability, durability and robustness required in many application domains by exploiting the properties inherent in FPGA devices. Applications that can benefit from such advances include: medical and transportation (reliability and safety), scientific and engineering (robustness and high-performance), and high-end consumer electronics (robustness, durability and tradeoffs in speed/size/power). Bioinformatics is another application domain where RC can be suitable since it requires the architecture to follow the algorithm evolution and increasing datasets size. This challenge involves not only improving the hardware fabric to support the desired traits, but also the methods and tools to automatically exploit these capabilities.

A. DeHon and S. Hauck analyzed in [9] the application characteristics that suit the use of reconfigurable hardware. In summary, in most cases these are applications that have one or more of the following main characteristics:

- heterogeneous computations/ require adaptation;
- Streaming;
- Parallel.

Some of the application-domains in Embedded and High-Performance computing that can benefit from reconfigurable computing are the following:

- Physics,
- Biology,
- Astronomy,

- Earth sciences,
- Medical,
- Automotive and Space,
- Networking
- High-end Consumer Electronics,
- Financial.

3.3.2 *Challenge 2: Improved Run-Time Reconfiguration*

One of the most important requirements in embedded and high performance systems is the ability to adapt to the environment or to the changing application requirements. This can involve changing not only the software running in the system, but also its hardware organization. Reconfigurable technology, such as FPGAs, enables the hardware to adapt in a cost-effective way to cope with dynamic system requirements, such as improving overall performance and/or supporting new protocols, standards and features. One of the main challenges in reconfigurable computing is the improvement of efficient reconfiguration techniques that is envisioned to enhance the areas of applications for RC (already listed in Challenge 1). For instance, it is essential the improvement of reconfiguration speed in order to keep real-time requirements for high performance and embedded applications. It is also essential to investigate in configuration interfaces which can be identified as bottlenecks in actual devices, e.g. for improving future multi-context reconfigurable architectures. In addition, another challenge is to support reconfiguration issues in the system design methodology. Finally, we need to seek efficient mechanisms (i.e., runtime support) and techniques for dynamic, reliable (partial) reconfiguration under very tight time constraints. We summarize next the challenging points regarding reconfiguration in RC:

- Speedup configuration/hide latency.
- Optimal reconfiguration granularity.
- Merging the functional interconnects with the reconfiguration infrastructure
- Low overhead modules/context switching (Multi-context capabilities/micro-architectural techniques)
- Runtime systems support, (i.e. expose RC specific properties to the programmer, OS extensions for RC management.)
- Dynamic Reconfiguration and resource scheduling at Runtime/Compile-time (compiler – runtime system interaction) for changing application and system requirements (power, performance, QoS, reliability, resource management, temperature, etc.).

3.3.3 *Challenge 3: Power-Efficient Computation*

Since RC power consumption until now is often a barrier for some applications, the associated challenging topics are: (a) the reduction of power consumption,

(b) the power management strategies and (c) the combination of both of them. Traditional microprocessor based architectures may in some cases have lower power consumption and therefore cannot be substituted in some real industrial application scenarios. Reducing power can make RC applicable to a wider set of Embedded Systems (i.e. mobile devices). Challenging here is to find novel solutions for the reduction of power consumption on physical as well as on system level, e.g. finding the suitable granularity mix and datapath structures of the underlying heterogeneous blocks. This includes the analysis of future multi-/many-core architecture solutions as well as corresponding design and programming methods/tools.

- Reduce power consumption at all levels, from System Level down to Realization/ Technology;
- Methods and Tools for power-efficiency;
- Optimization of Power/performance;
- RC aware Power management for runtime-systems /dedicated embedded OS for RC platforms;
- Power aware (dynamic) customization: e.g. granularity mix, datapath structure
- Coarser grain reconfigurable components;
- FPGA chips variability.

3.3.4 Challenge 4: Reliability

Reconfigurable platforms can play an important role to support fault tolerant computing. Their inherent flexibility can be deployed to correct, isolate, and/or replace faulty parts of a system. Efficient techniques can be developed and when combined with runtime support may result in a fault tolerant system that can recover from transient faults or cope with permanent faults preserving critical functionalities of the system. Another challenging issue is the fault tolerance of a reconfigurable platform itself. In our advantage is the fact that reconfigurable hardware is inherently redundant, still we lack however efficient methods to exploit this characteristic and provide a fault tolerant reconfigurable platform. We list below several challenging RC points regarding reliability:

- Failure redundancy by dynamic alternative implementations;
- Spatial and Temporal Fault Tolerance at the physical layer;
- Exploit reconfiguration for complete or partial fault recovery;
- Improve reliability, safety, security for NANO technology;
- Self-adaptation and self-recovery;
- Yield and reliability management;
- FPGA-specific defect analysis, defect-mapping and self-healing;
- Tools and methods for reconfigurable fault tolerant computing (i.e., runtime support).

3.3.5 Challenge 5: Tools, Methods and Runtime Support

Efficient tools, methods, and runtime (RT) support are fundamental for the success of any computing system. In case of a Reconfigurable system this is even more challenging due to the HW polymorphism.

Design methods and tools, classified as either synthesis or analysis, are the key to support designer productivity and quality in order to exploit complex and large reconfigurable systems. The increasing complexity implies huge exploration design space and therefore, there is a need for efficient tools to support system/architecture exploration at the every design phase. These tools have impact on the designer's productivity. Synthesis tools focus on automatically mapping high-level descriptions into efficient hardware implementations according to one or more performance metrics, such as speed, size, and power consumption. Analysis tools characterize the design into several abstraction layers, and can support verification of functionality and other properties such as timing and testability. The current challenge in developing tools that target RC systems is to create a coherent compilation tool-chain that covers all the main steps of synthesis and analysis, including capturing domain-specific knowledge, profiling, design space exploration, multi-core partitioning, system partitioning, data representation optimization, static and dynamic reconfiguration, optimal custom instruction set generation, and functional simulation. Partitioning an application between software and reconfigurable hardware, programming mixed software/reconfigurable hardware systems, and ensuring effective communication across the boundaries is another challenging issue. Finally, more efficient debugging mechanisms are needed as debugging reconfigurable systems can be significantly more complex compared to conventional computing systems.

Runtime support is essential in a Reconfigurable system. Being able to adjust the underlying hardware on demand -for performance, power, reliability, or other requirements- can be achieved more efficiently by a runtime system rather than statically at design/compile time. Adapting the system dynamically at runtime can be one of the primary RC advantages over conventional computing systems. Currently, we know how to develop a runtime system for a fixed hardware platform, however building a runtime system for dynamically changing hardware platforms remains an open issue. The following challenges are related to the runtime support of a reconfigurable system:

- Online monitoring;
- Load balancing;
- HW dependable SW;
- Visualization;
- Runtime resource management and scheduling;
- Extremely fast re-layout for dynamic reconfiguration;
- Managing adaptive dynamic routing.

3.3.6 Challenge 6: Targeted Systems

Reconfigurable Computing is becoming popular in the domain of embedded systems, while it can provide an attractive solution in the area of high performance computing and multiprocessor parallel systems.

Embedded Systems (ES): Reconfigurable Computing is an exciting approach for embedded systems in general and application-specific designs in particular. The main advantage of RC in such systems is that they can adapt to static and dynamic application requirements better than those with fixed hardware. Furthermore, the power/performance ratio for the reconfigurable hardware can be significantly better than that of general-purpose processors. As opposed to the traditional design workflow for ASIC embedded systems, where the application was implemented entirely in hardware, modern methodologies tend to implement only the computationally intensive parts in custom hardware (i.e. Reconfigurable) and the rest are implemented as software tasks running on a GPP. This new scenario gives foundation to the area of Hardware-Software Co-Design, which is of paramount importance for the design of reconfigurable embedded systems. Another challenging issues in ES is developing generic embedded platforms to improve productivity and reusability. Finally, applications in the ES domain may have important requirements such as being energy efficient and/or safety critical which makes their design more challenging. A reconfigurable system that provides the above characteristics is far from trivial to develop.

High Performance, Multicores, Parallel Systems: Customizing the data path and control-flow of reconfigurable hardware to the needs of the application/problem at hand can be proven beneficial in High Performance Computing (HPC). As explained in Challenge 1, several high performance applications (i.e., scientific, biology) may require massive parallelism and also adaptation in different computation phases. In the above cases, reconfigurable computing can be a very attractive solution exploiting hardware speed and parallelism combined with flexibility and customization. The RC scientific community has not yet explored extensively the potential benefits of RC for HPC, which remains an open challenge. Related to the HPC, multiprocessor parallel architectures can also benefit from RC, using reconfigurable hardware integrated in a multiprocessor architecture to accelerate processes. Mapping applications to very large heterogeneous parallel systems in order to exploit their massive computational power is significantly challenging. These systems can contain diverse components such as conventional CPUs, multicores, DSPs, GPUs, embedded memory blocks, FPGAs, on-chip multi-core reconfigurable devices, and interconnects. Multi-core devices can contain one or more hardwired and softcore instruction processors, have customizable instruction sets, and distributed RAMs. Mapping applications into these systems require considerable effort in design time, and so there is a need for effective design methodologies and techniques that can automatically accelerate an application, and have a mechanism to explore different tradeoffs in execution time, resource usage and power consumption. Some challenging issues regarding Reconfigurable computing for HPC and multiprocessor parallel systems are related to programmability, as well as to power efficiency, and performance.

3.3.7 Challenge 7: RC Specific Memory Architectures

The memory architecture of a system built in a reconfigurable platform and the memory structure of the reconfigurable platform itself are both essential for the efficiency of a reconfigurable system. Providing flexibility (number of banks, width, depth) in memory architectures at the board level to allow a "general purpose" reconfigurable computer which is suitable for more than one application is a challenging issue. Furthermore, matching the memory bandwidth with the computation power of the system can improve systems performance. Developing mechanisms to adjust the memory hierarchy of a system on demand based on the application needs (e.g. FLUX caches [10]) can be also proven beneficial for system. As reconfigurable hardware can be considered as a large memory the contents of which define the functionality of the hardware, the structure of the FPGA memory directly affects the performance of a reconfigurable platform. The configuration memory and the mechanisms used to reconfigure, download configuration bitstream, etc. affect the reconfiguration overhead of a system and its overall performance.

3.3.8 Challenge 8: Realization/Technology

There are two challenging issues related to the technology in reconfigurable computing. The first one pertains to the development of such platform, while the second concerns future technologies which can improve reconfigurable platforms. It is important to design faster, and more power-efficient reconfigurable devices, improve in fundamental functions of RC such as reconfiguration, and cope with the processes variability and technology faults of the coming nano-scale era. Furthermore, reconfigurable computing using actual reconfigurable CMOS devices like e.g. FPGAs are a pathfinder for novel computing architectures for nano technology devices. Related to the future architectures of nano technology the methods for computing, failure redundancy, fault tolerance and failure recovery must be prepared with the actual reconfigurable architectures in order to arrange a smooth transfer as soon as this technology more and more substitutes CMOS processes for electronic devices. Finally, the potential benefits for RC of emerging technologies, such as 3D technologies, nanowires, and optical, need to be explored.

3.4 Conclusions

Reconfigurable Computing has the potential to provide efficient solutions for better performance, power-efficiency and fault tolerance in future systems. The massive number of parallel computing resources and the hardware speed as well as the flexibility of on-demand customization can deliver better results compared to general purpose computers. In addition, the plethora of available on-chip resources reduce the area overhead of the reconfigurable substrate. There are several challenges which need to be addressed in order for reconfigurable computing to exploit this opportunity and be deployed at a larger scale. More efficient design tools and methods are required to enhance the use of reconfigurable hardware as well as adaptive runtime support for systems built on reconfigurable technologies. Application domain extensions for certain classes of applications will make their porting to reconfigurable systems simpler. Runtime reconfiguration needs to be improved and supported by System-level techniques. As power consumption is becoming a severe problem, novel solutions for power reduction at physical as well as at System level are very important. The characteristics of reconfigurable technology need to be better exploited to provide fault tolerance. Finally, the underlying reconfigurable substrate needs to be improved in order to find a suitable granularity-mix to reduce the delay, power and area RC overheads.

Acknowledgments We would like to acknowledge all the members of the HiPEAC Reconfigurable Computing cluster for the fruitful discussions regarding the upcoming challenges in reconfigurable computing, and especially thank the rest of the cluster board J. Becker, W. Luk, N. Navarro, and D. Sciuto for their feedback as well as R. Giorgi, D. Pnevmatikatos, D. Soudris, D. Stroobandt, and P. Trancoso for their valuable comments. Furthermore, we would like to thank S. Tzilis for his help on extracting the technology trends from the ITRS documents. Finally, we would like to acknowledge that this chapter is a continuation of the Reconfigurable Computing theme in the HiPEAC1 Roadmap [11] and has been influenced by the EU Workshop on Reconfigurable Computing [12].

References

1. Wayne Luk, "Reconfigurable Computing: Coordinated Customisation of Architecture and Compilation", 2nd HiPEAC Workshop on Reconfigurable Computing, Keynote Speech, Gothenborg, Sweden, Jan. 2008
2. M. Flynn, "Accelerating computations using data parallelism", CE TU Delft colloquium, Sept. 2009.
3. Flynn, M., Dimond, R., Mencer, O., and Pell, O. 2008. "Finding Speedup in Parallel Processors", In Proceedings of the 2008 international Symposium on Parallel and Distributed Computing (ISPDC), July, 2008.
4. http://www.maxeler.com/
5. http://www.conveycomputer.com/
6. International Technology Roadmap for Semiconductors: http://www.itrs.net/

7. S. Borkar "Designing reliable systems from unreliable components: The challenges of transistor variability and degradation", IEEE Micro, 25(6):10–16, 2005.
8. T. Austin, V. Bertacco, S. Mahlke, and Y. Cao, "Reliable Systems on Unreliable Fabrics," IEEE Des. Test, vol. 25, iss. 4, pp. 322–332, 2008.
9. Hauck, S. and DeHon, A. "Reconfigurable Computing: the Theory and Practice of FPGA-Based Computation", Morgan Kaufmann Publishers Inc., 2008.
10. Georgi N. Gaydadjiev and Stamatis Vassiliadis, "Flux Caches: What Are They and Are They Useful?", Embedded Computer Systems: Architectures, Modeling, and Simulation (SAMOS), pp. 93–102, 2005.
11. K. De Bosschere, W. Luk, X. Martorell, N. Navarro, M. O'Boyle, D.N. Pnevmatikatos, A. Ramírez, P. Sainrat, A. Seznec, P. Stenström, O. Temam, "High-Performance Embedded Architecture and Compilation Roadmap", T. HiPEAC 1: 5–29, 2007.
12. EU Workshop on reconfigurable computing, Brussels - December 13, 2007.

Chapter 4
MORPHEUS: Exploitation of Reconfiguration for Increased Run-Time Flexibility and Self-Adaptive Capabilities in Future SoCs

Arnaud Grasset, Paul Brelet, Philippe Millet, Philippe Bonnot, Fabio Campi, Nikolaos S. Voros, Michael Hübner, Matthias Kühnle, Florian Thoma, Wolfram Putzke-Roeming, and Axel Schneider

Abstract The exponential increase of CMOS circuit complexity has opened the way to the introduction of new capabilities and functionalities into electronic systems that have been sources of innovations in major growth markets. To pursue this trend all along the last decades, major evolutions of design methodologies and computing architectures have been necessaries to master this complexity. Now, increasing Non-Recurrent Engineering (NRE) costs have made the design of ASICs or System-on-Chips unaffordable for a broad class of applications, whose the low-volume markets are insufficient to make them economically viable. But, use of Commercial Off-The-Shelf (COTS) boards is not always satisfactory due to the low power-efficiency of general-purpose processors and the complexity of programming FPGAs. We thus advocate for a new class of System-on-Chips, composed of a mix of processors as well as very flexible and easily programmable accelerators in order to cope with increasing NRE costs and tight time-to-markets. In this perspective, reconfigurable architectures are very appealing for their trade-off between the performance of ASICs and the flexibility of general-purpose processors. This chapter presents an innovative approach of a dynamically reconfigurable heterogeneous platform, called MORPHEUS and which consists of a System-on-Chip integrating different kinds of reconfigurable accelerators controlled by a general-purpose processor.

4.1 Project Partners

1. Thales Research & Technology, France (coordinator)
2. Deutsche Thomson OHG, Germany,
3. Intracom SA Telecom Solutions, Greece

A. Grasset (✉)
Thales Research & Technology Campus Polytechnique1, avenue Augustin Fresnel, 91767 Palaiseau Cedex France
e-mail: arnaud.grasset@thalesgroup.com

J.M.P. Cardoso and M. Hübner (eds.), *Reconfigurable Computing: From FPGAs to Hardware/Software Codesign*, DOI 10.1007/978-1-4614-0061-5_4,

4. Alcatel-Lucent Deutschland AG, Germany
5. Thales Optronics SA, France
6. STMicroelectronics SRL, Italy
7. PACT XPP Technologies AG, Germany
8. Abound Logic, France
9. Associated Compiler Experts bv, The Netherlands
10. CriticalBlue, United Kingdom
11. Universitaet Karlsruhe, Germany
12. Technische Universiteit Delft, The Netherlands
13. Commissariat à l'Energie Atomique, France
14. Université de Bretagne Occidentale, France
15. Universita di Bologna, Italy
16. ARTTIC SAS, France
17. Technische Universitaet Braunschweig, Germany
18. Technische Universitaet Chemnitz, Germany

- Project Coordinator: Philippe Bonnot, Thales Research & Technology, France
- Start Date: January 2006
- End Date: September 2009
- EU Program: 6th Framework Program
- Global Budget: 16.57 M€
- Global Funding by EU: 8.24 M€
- Contact Author: Arnaud Grasset, Email: arnaud.grasset@thalesgroup.com

4.2 Introduction

The data intensive and real-time processing in embedded systems has often stringent requirements in term of performance and power consumption. Computing architectures for this class of embedded systems have gained a growing attention due to recent advances in multimedia and telecommunication applications. Autonomous and ubiquitous applications impose very high requirements on the overall system performance and flexibility. The system flexibility is a key feature as it gives the ability to make the application evolve during the design process, and all along the life-cycle of the product. The system can then be adapted to evolutions of the environment or to revisions of standards and specifications, be extended with new features or just be optimized. Furthermore, flexible computing platforms can support a broad range of applications in order to amortize the growing non-recurring engineering costs.

But architecture efficiency (i.e. performance density, power efficiency) is generally obtained by some kind of specialization that however impacts their flexibility and/or makes them hard to program. In the case of Application-Specific Integrated Circuits (ASIC), the whole architecture is tailored for a specific application. Despite their great flexibility, single-core general-purpose processor (GPP) cannot reach the power-efficiency of these dedicated hardware devices (see Fig. 4.1). During the last decade, the increase in clock frequency of single-core processor has been accompanied by a dramatic increase of power consumption [1]. In Application-Specific Instruction-set

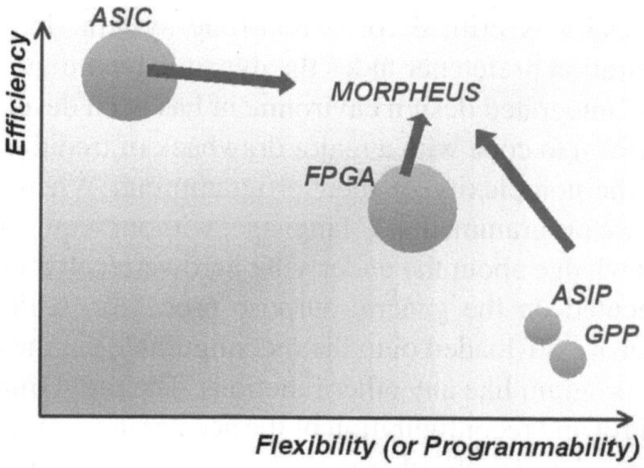

Fig. 4.1 Efficiency and flexibility of common computing architectures

Processors (ASIP), the Instruction Set of the processor is tailored for a specific application (e.g. [2]). Adding dedicated instructions improves the performance and the power efficiency of the processor at the expense of only a small loss of flexibility.

Field-Programmable Gate Architectures (FPGA) offers a trade-off between the flexibility of a software solution and the performances of dedicated hardware devices. They are very efficient to implement some kinds of dataflow kernels, without reaching, however, the efficiency of ASIC solutions. In FPGAs, interconnect resources occupies a large percentage of the chip area and lead to performance, area and power overhead. But their ability to be reconfigured is a great advantage.

Reconfigurable computing has its origin in a paper of Gerald Estrin, who proposed a new computer organization, such that the system could be temporarily tailored for a specific problem [3]. Reconfigurable processors join two different paradigms inside: computing in space as well as in time. This makes computing much more powerful and efficient at the price of more sophisticated programming. Easily programmable solutions are required to meet the tight time-to-market demands and the cost-effectiveness requirements of embedded systems.

The MORPHEUS project [4] aims to establish the foundations for a new concept of flexible embedded computing platforms, providing to the broadest range of embedded applications a cost effective compromise between the benefits of ASIC density and the flexibility of generic computing platforms. Its name is a reference to the god of dreams in Greek mythology, who has the ability to take any human's form and appear in dreams. It is based on a modular and scalable heterogeneous System-on-Chip, which integrates different but complementary dynamically reconfigurable engines and thus combines their benefits. A general-purpose processor controls the system. A prototype chip has been implemented in a 90 nm CMOS technology.

The overall platform is built as a set of heterogeneous and independent processing units connected through a Network-on-Chip (NoC). The network is a critical component as it affects the performance of the whole architecture. It has been designed to support the high bandwidth requirements of the reconfigurable engines. Dynamic reconfiguration is an essential aspect of the platform. Since the reconfiguration time can be a performance killer, minimizing the time to fetch and configure a function dynamically is critical for a real-time system. In the MORPHEUS platform, a configuration prefetcher hides the dynamic reconfiguration latency.

A dedicated and integrated design environment has been developed in concomitance with the platform to cope with a major drawback of traditional reconfigurable architectures, i.e. the complexity of their programming. When using this design flow, applications are programmed in C language, without requiring from the developer a detailed knowledge about the underlying hardware/software architecture. An application is executed on the general-purpose processor, with the exception of some functions that are off-loaded onto the reconfigurable engines. These functions are called in the C program like any other functions. The programmer does not have to manage the control and reconfiguration of the accelerators, or to program the data transfers of these accelerators. Furthermore, high-level synthesis of the accelerated functions on reconfigurable units is done from a common programming model, called Array-OL [5], for all the reconfigurable engines.

The chapter is organized as follows. In the next section, we analyze state-of-the-art reconfigurable architectures and their associated design flow. In Sect. 4.3, we explain more precisely the MORPHEUS architecture and its various elements. A prototype chip of this SoC architecture has been realized. In Sect. 4.4, we present the obtained results. They are followed, in Sect. 4.5, by the toolset presentation where each module contributes to the simplification of the programmer work. Wireless telecommunication, video film processing, networking and smart cameras application examples are then described in Sect. 4.6. This description includes the concept verification made through their implementation on the chip, using the toolset and the board for demonstration. We conclude in Sect. 4.7.

4.3 Related Work

Most competing state-of-the-art SoC are based on a single CPU, or on a combined CPU and DSP enhanced with some dedicated hardware accelerators.

By distributing demanding multimedia processing onto the accelerators, the architectures off-load the host CPU (central-processing unit), which mainly acts as an application controlling engine. (Nomadik by STMicroelectronics [6], OMAP by Texas Instruments [7], and PXA by Intel [8].)

In the case of MORPHEUS, application demands can be distributed on different units after fabrication. Its heterogeneous reconfigurable engines (HREs) and I/O facilities remain runtime configurable.

The same applications are mapped onto homogeneous multiprocessor SoCs (MPSoCs). However, heterogeneous portions, as existing in multimedia and other data intensive applications cannot be mapped as efficient as on the MORPHEUS HREs, each characterized with dedicated features. So MPSoC Implementations differ from MORPHEUS because in contrary to the heterogeneity of MORPHEUS, processing nodes of MPSoCs provide a uniform means of computation of a single granularity, similar I/O bandwidths and access patterns.

Reference [9] describes the Pleiades approach, a reconfigurable architecture template. It uses its NoC to interconnect different elements of one given reconfigurable-architecture array rather than intrinsically different, coarser reconfigurable architectures. In [10] another multi-core runtime reconfigurable SoC is introduced, where two cores are connected through a lightweight communication infrastructure. In their approach applied reconfigurable architectures are homogeneous, whereas heterogeneity is one of the major challenges of MORPHEUS. The Pleiades multi-core SoC [11] was targeted at low power consumption rather than processing efficiency.

Other reconfigurable architectures are optimized for specific applications [12], whereas MORPHEUS is capable to efficiently handle full application domains through a good balance of different heterogeneous reconfigurable engines. Others are only configurable at design time [2].

There are several languages and tools available for use with reconfigurable platforms. SystemC and SystemVerilog are mainly used for high-level descriptions of complete systems. Application specific instruction processors and accompanying tools can be developed with the help of languages and tools like Mescal, ArchC, LisaTek, Chess/Checkers and XPRES Compiler. Design space exploration and development for coarse-grained reconfigurable hardware can be done with the FELIX tool [13].

C-based languages like CatapultC, GARP, Mitrion-C, SA-C and Impulse C are available for parallelized applications but are limited to specific target architectures. A more stream-oriented approach is available with tools like Mat-lab/Simulink, Scilab, AccelDSP and languages like Array-OL, Stream-C, ASC and SNET. APIs available for different traditional languages are DRI, VSIPL++, MPI, OpenMP, OpenUH and Stream Virtual Machine API. UML, sysML and SpecEdit can be used for formal specification of applications. For a more system-level approach to application mapping tools like PTOLEMY II, R-Stream and Gaspard beside the already mentioned Matlab/Simulink can be used. The projects EPICURE and RECONF offer tools targeted especially for reconfigurable plat-forms.

4.4 Architecture

4.4.1 Introduction

In the past, reconfigurable processing architectures often suffered from a not well balanced relation of generality, flexibility and efficiency. Generality and flexibility mostly comes at the cost of efficiency and vice versa. For example, fine grain FPGAs are flexible enough to address very different application classes while coarse grained reconfigurable architectures usually provide a higher efficiency for specific application classes.

To defuse this trade-off the MORPHEUS processing platform [4] integrates three Heterogeneous, dynamically Reconfigurable processing Engines (HREs), which support different but complementary styles of reconfigurable computing, in one highly scalable platform. So, algorithms from different application domains or quite complex algorithms with heterogeneous processing requirements can be mapped onto the MORPHEUS architecture very efficiently. This explicitly includes real-time processing.

Two fundamental design decisions regarding the MORPHEUS platform architecture were made at an early stage of the project. The first decision was to use a central controller to manage the whole platform. The second decision was the usage of the Molen paradigm [14] to control the HREs from the central processor, since Molen allows parallel execution of the functions mapped to the HREs and provides a mechanism to pass parameters to the functions. From an architectural perspective the Molen paradigm uses registers, which are called exchange registers (XRs), to control the processing of the reconfigurable engines and to pass parameters.

The data flow related organization of the HREs is not predefined by the architecture. Depending on the target application different styles of HRE organization can be chosen. If the application allows that consecutive processing steps can be mapped onto different HREs, which is typical for data stream processing, an execution pipeline across multiple HREs can be created. Thereby, the difficult task is to find a well balanced split of the application to the pipeline stages, since the execution speed of the whole pipeline is limited by the slowest stage. Alternatively the same HRE also can be used after reconfiguration for consecutive processing steps. This HRE usage requires that sufficiently large memory is available to store intermediate results between the consecutive runs of the HRE. Moreover, a certain reconfiguration overhead has to be accepted, as the HRE must be reconfigured for every processing step. Since this approach uses only one processing engine the resource utilization of the other available processing engines may not be optimal. Of course, also a mixture of both processing principles is possible and can be used to find a well adjusted load balance for all available HREs. Since the data flow organization of the HREs is not fixed by the hardware, very flexible and efficient software controlled usage of the HREs is possible (e.g. dynamic load balancing controlled by an operating system or the application).

Furthermore, the analysis of the targeted applications clearly showed that special emphasis has to be put to the dynamic reconfiguration mechanism. Especially the usage of the same HRE for consecutive processing steps requires very frequent reconfigurations of the HRE. Further, if real-time processing is required, strict time limits for reconfiguration have to be met.

4.4.2 The MORPHEUS Platform Architecture

The MORPHEUS hardware architecture, which is depicted in Fig. 4.2, comprises three heterogeneous, reconfigurable processing engines (HREs) which target different types of computation:

- PACT XPP from PACT XPP Technologies is a coarse-grain reconfigurable array primarily targeting algorithms with huge computational requirements but mostly deterministic control- and dataflow. Recent enhancements also allow efficient sequential bitstream and general purpose-processing. These enhancements are

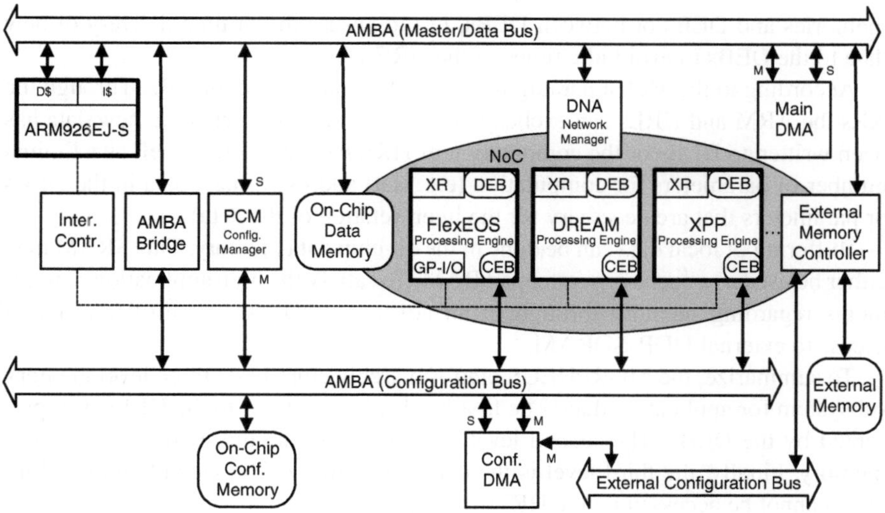

Fig. 4.2 Simplified MORPHEUS Platform Architecture

based on multiple, instruction set programmable, VLIW controlled cores which are equipped with multiple asynchronously clustered ALUs.
- DREAM is based on the PiCoGA core from STMicroelectronics. The PiCoGA is a medium-grained reconfigurable array consisting of 4-bit ALUs and 4-bit LUTs. The architecture is mostly targeting instruction level parallelism, which can be automatically extracted from a C-subset language called Griffy-C. DREAM mainly targets computation intensive algorithms that can run iteratively using only limited local memory resources.
- FlexEOS from Abound Logic is an embedded Field Programmable Gate Array (eFPGA). Thus, it is a fine-grain reconfigurable device based on LUTs. It is capable to be configured with arbitrary logic up to a certain level of complexity.

All control, synchronization, and housekeeping is handled by an ARM926EJ-S embedded RISC processor. These tasks have to be processed by the ARM with the highest priority. Other processing tasks can be mapped onto the ARM processor only if they do not compete for processing power or bandwidth with the primary tasks.

As the HREs in general will operate on differing clock domains, they are decoupled from the system and interconnect clock domain by data exchange buffers (DEB) consisting of dual ported (dual clock) memories either configured as FIFOs or double buffers with page flipping (ping-pong buffers[1]). From a conceptual point of view the HREs can access their input data only from their respective local DEBs. The ARM processor, which is in charge of controlling all data transfers between

[1] Ping-pong buffering is a mechanism to avoid idle times of the HREs while they are waiting for new data. Ping-pong buffering requires an even number of input/output buffers. If only one buffer is available it is necessary that this buffer allows parallel read and write accesses. While the HRE processes the data of the "ping" buffer, new data is pre-loaded into the "pong" buffer.

memories and DEBs or between DEBs, has to ensure the in-time delivery of new data to the DEBs to avoid idle times of the HREs.

According to the Molen paradigm each HRE contains a set of XRs. Through the XRs the ARM and HREs can exchange synchronization triggers (e.g. new data has been written to DEBs or the computation of HRE has finished) as well as a limited number of parameters for computation (e.g. start address of new data in the DEBs or parameters that are necessary for the interpretation of the data).

Buffering of local data can be done in the on-chip data memory. This SRAM may either be used as a cache or scratchpad RAM. To satisfy the high application requirements regarding memory throughput an advanced DDRAM controller provides access to external DDR-SDRAM.

To summarize, the MORPHEUS platform architecture has a three level memory subsystem for application data. The first level, which is local to the HREs, is represented by the DEBs. The second level, which is still on-chip, is the on-chip data memory. Finally, the third level is the external memory. The second and the third level cannot be accessed by the HREs directly.

As dynamic reconfiguration of the HREs imposes a significant performance load on the ARM processor, a dedicated reconfiguration control unit (PCM) has been designed to serve as a respective offload-engine. The PCM analyzes which configuration is needed for the next processing steps on the HREs. Depending on this analysis the next configurations are pre-loaded. It should be mentioned that the memory subsystem used for handling configurations also uses a three level approach as previously introduced for the application data. The configuration exchange buffers (CEB) inside of the HREs are the first layer. The second layer is the on-chip configuration memory, and the third layer is the external memory.

All system modules are interconnected via multilayer AMBA busses [15]. Multilayer AMBA busses provide a higher bus bandwidth than the standard AMBA advanced high performance bus (AHB) bus while maintaining the original AMBA AHB protocol. The higher bandwidth is achieved by using a more complex crossbar interconnect matrix. Separate busses are provided for reconfiguration and/or control and data access. As the required bandwidth for high-performance and data intensive processing might become quite high, an additional network on chip (NoC) based on ST's spidergon technology [16], with the characteristic spidergon topology has been integrated. To reduce the burden on the ARM system controller, DMAs are available for loading data and configurations. However, data transfers on the NoC also have to be programmed and initiated by the ARM processor. Similar to programming a DMA for the AMBA bus, the ARM can program the DNA (direct network access) module for NoC transfers.

Figure 4.3 shows the topology of the integrated NoC. The dashed lines in Fig. 4.3 denote the interconnections in the NoC, whereas the boxes denote the NoC nodes. The NoC provides a routing mechanism that allows exchange of data between NoC nodes that are not adjacent (e.g. DREAM to XPP In transfer via ARM node).

To avoid a possible overload of certain NoC interconnections, assumptions were made about the expected communication behavior of the modules, which are connected to the NoC. The main idea for optimizing the topology is to place NoC

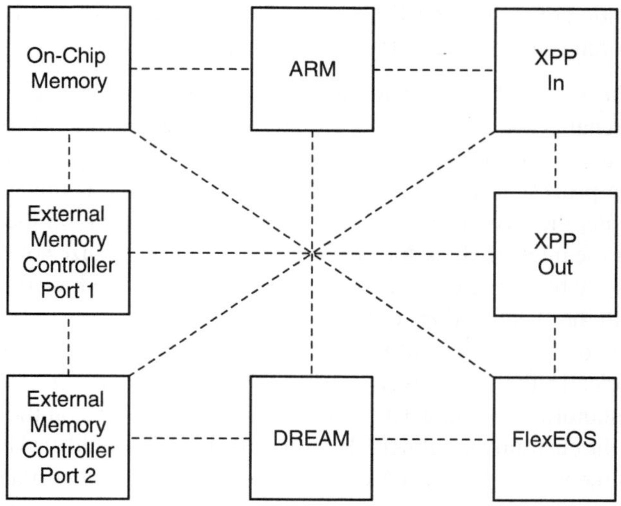

Fig. 4.3 MORPHEUS NoC topology

nodes with high inter-communication demand directly adjacent to one another, since a direct interconnection link exists between such nodes.

4.4.3 Expandability of MORPHEUS Platform Architecture

It is important to emphasize that the MORPHEUS platform architecture as it is presented in Fig. 4.2 has to be understood mainly as an architectural framework. The platform defines which modules can be part of the architectural approach and how they are integrated. But the platform for example neither defines the size of the HREs or the on-chip memories, nor the clock speed or the internal architecture of the HRE itself. Thus, in every case, before MORPHEUS technology can be used for a real chip, a customized architecture has to be derived from the general platform architecture. This process of customization (e.g. define size of on-chip memories) allows tailoring of the architecture to specific application requirements.

Modularity of the platform architecture is achieved by a unique encapsulation of the HREs. All HREs are completely encapsulated by their DEBs, CEBs, and XRs. This encapsulation leverages the exchange of certain HREs with another one from the set of available HREs. If one application would benefit from the availability of a second DREAM, a second DREAM instead of a FlexEOS or XPP could be instantiated in the final architecture. If for example an application contains two data parallel algorithmic blocks which can be mapped very efficiently on a certain HRE, the application will benefit if more than one instantiations of this HRE are available in the processing architecture.

Customization of the MORPHEUS platform architecture also can be understood as a dimensioning step for the following architectural parts:

- HREs: The size and processing parameters of HREs, which characterize the computational power of the HREs are not predefined by the platform architecture. Due to the encapsulation of the HREs both can be adapted during the customization process to the processing requirements of the application in focus. For example, the size of the coarse grain computation array in the XPP can be altered, or the internal clock speed of the DREAM and the sizes of configuration memories can be increased if necessary. Of course, such modifications have to be in line with the specifications of the respective HREs.
- Memory subsystem: Depending on the application requirements the size of all memories in the three levels can be adjusted. In particular the size of the on-chip memories and the DEBs can have strong influence on the final performance of the customized MORPHEUS architecture. It should also be mentioned that the dimensioning of the on-chip memories must consider potential limitations of the external memory data rate. Small on-chip memories may lead to an increased external memory bandwidth since intermediate results have to be stored in external memory.
- NoC: The MORPHEUS platform architecture integrates a NoC which is based on ST's spidergon technology with eight nodes. From a conceptual point of view the number of NoC nodes is not limited and thus can be increased if necessary. However, it shall not be concealed that such an adaptation will have a huge impact on other architectural components such as the DNA controller or the Configuration Manager (PCM).

However, the customization step or the MORPHEUS platform architecture requires in depth knowledge of a specific application. If this knowledge is available a customized dynamically reconfigurable architecture can be derived from the MORPHEUS platform architecture with comparable low effort.

4.5 Chip Prototype

4.5.1 Introduction: Needs (and Limitations) of a Prototype Chip

The ambition of the MORPHEUS project is to enable a viable industrial exploitation of the dynamic hardware reconfiguration concept. For this reason, the design, manufacturing and testing of a prototype of the chosen architecture template is an essential step for an assessment of the MORPHEUS concept. On the other hand, the context of a research project must necessarily be limited in terms of development time and costs to the constraints of the project itself. Other than being a significant opportunity and the only real way to prove the introduced concept, the definition of a prototype chip represented a very difficult tradeoff between the many different evaluation metrics to verify, and the available time and effort for their deployment. For this reason, the essential challenge in the MORPHEUS prototype deployment has been *the selection of the necessary metrics for the evaluation of the MORPHEUS concept, and the design of a VLSI prototype chip that could allow significant measurement of the same metrics while remaining in the allocated cost and effort budget.*

If we consider a given application context, we face a situation where the final user needs to balance the best possible tradeoff investing a given "budget" that includes constraints in terms of required computation performance and/or data throughput, hardware costs (Chip Area, Power), economic considerations related to manufacturing (Chip costs parameterized on the required market volumes, choice of technology node, yield considerations). We should define and verify with quantitative analyses on the prototype chip the introduced issues: from the start of the design process the following metrics, in order of priority, were considered as drivers for the design, manufacturing and testing of the prototype.

1. **Performance, or better said computation efficiency against the most significant cost factors**: as briefly introduced above, the MORPHEUS prototype should prove how the MORPHEUS platform can enable better cost/performance tradeoffs as opposed to alternative solutions such as Field Programmable Gate Arrays (FPGA), DSPs, Coarse Grain Reconfigurable Architectures (CGRA), multi-core signal processors, vector processors or Application Specific Standard Processors (ASSP). For this reason, we will try in the following to evaluate the prototype performance against possible competitors in comparable technology nodes in terms of sheer performance, performance per area unit, performance per power unit.
2. **Ease of design/Portability to a given application context**: In the development of such an innovative and industrially promising computation concept as Reconfigurable computing it can be acceptable to invest a significant design cost/effort for the deployment of an initial prototype chip, and the first assessment of issues/costs related to physical design and manufacturing. On the other hand, the ambition of MORPHEUS is to deliver a "General Purpose", and "Application Independent solution". For this reason, it is obvious that different application environments with very different constraints in terms of computing performance, data throughput, on-chip storage, IO bandwidth will require different tradeoffs in the manufacturing stage, related to the sizing of the main chip components (Reconfigurable areas, L1 and L2 data/configuration storage, high-speed IO interfaces). This delicate tradeoff will be evident in Sect. 4.6, where different applications of very different granularity will be compared and matched on the MORPHEUS architecture: the mapping exercises will prove how each application environment can strongly benefit in principle from the MORPHEUS concept, while similarly to any other existing embedded computing platform, this benefit can be economically viable and commercially applicable only if a variegated family of "MORPHEUS-based" devices is made available, to match the specific market constraints of each given application segment. For this very reason, while it is

acceptable to invest a significant design effort in the "development" phase of the physical design, it is mandatory that different customizations of the same architectural concept can be "composed" at low cost and effort based on the result of the first design.

3. **Design Robustness: Design for Manufacturing, Design for yield**: it is well known that technology scaling induced by the Moore's law is continuously changing the cost/performance tradeoffs related to VLSI design. On the one hand, essentially any field of signal processing applications somehow expects performance figures to scale according to Moore's law, enabling more demanding algorithms at each hardware generation. On the other hand, each scaling raises a new set of physical issues that were safely considered negligible in older technologies, such as manufacturing/lithography misalignments, power integrity and IR-drop issues, signal integrity, clock jitter, etc. In many cases, a design style that is aware of such issues may greatly help in mitigating their impact, hence easing "robustness" towards scaling with respect to alternative solutions. In the following, we will underline how reconfigurable computing in general, and the MORPHEUS concept in the specific offer specific advantages, especially as compared to RTL-to-ASIC design flows, on this respect.

4.5.2 Prototype Chip Design

According to considerations outlined in Sect. 4.3 and in the introduction to the current section, the main scope of the prototype chip design has been to:

1. Provide a scalable design style that would allow easy re-design of different configurations of the same architecture template.
2. Provide a design style that is aware and somehow mitigates physical issues related to technology scaling.
3. Merge in a single design different, strongly heterogeneous computing engines, hiding any hardware-related differences to the toolset and the user.

We chose to implement a strictly hierarchical design flow. MORPHEUS supports two "reconfigurable hardware orientations":

- LUT-oriented computation implemented by small static RAM (SRAM) and bit-level routing typical of Field-Programmable Gate Arrays (FPGA) is implemented in the Abound Logic eFPGA and in the PiCoGA reconfigurable engine that is the core of the DREAM signal processor.
- Word-level granularity typical of most Coarse Grain Reconfigurable Architectures (CGRA) is deployed in the XPP processing platform.

Typically, SRAMs and bit level routing require custom layout efforts to be area-effective, so eFPGA and PiCoGA were designed and imported as custom layout macros. Of course, the drawback of custom layout is that it is requires significant time, cost and expertise so that it makes technology scaling very economically demanding, and the adoption of regular structures makes this step easier that it would be in the custom design of arithmetic / logic structures.

Conversely, the coarse granularity of the XPP signal processor is based on 16-bit routing architectures and 16-bit ALU computation. Hence, it is more suitable to automated design, where RTL description of the device is automatically translated by Place & Route electronic design automation (EDA) tools into physical layout. This can be deployed with two options:

1. perform a flat design collapsing all the net-list hierarchy into a homogeneous "sea" of standard cells
2. provide a hierarchical flow where selected sub-entities are created independently as separate designs, and then re-used as atomic library blocks when working on a higher level.

Most ASICs exploit the flat option, often mitigated by hybrid approaches such as "regioning". In our case, due to XPP's regularity, we extensively applied hierarchical design. Manpower is focused on implementing local tricks and workarounds to solve DRC issues and close critical paths as the P&R domain is simpler and smaller. Constraining can be more aggressive, as also manual user intervention can be more invasive dealing within a smaller design space: this investment can be amortized as its gains will be repeated many times across the design. As an example, in the design of timing-critical blocks we could invest a lot of time fixing DRC errors that were generated by P&R only because the timing specs were too aggressive. In a flat design, using similar over-constraining the number of DRC errors would have increased proportionally with the area, and the only way to minimize them would have been to relax timing specs and area constraints. This "locality" is also a useful strategy to limit PVT variation: the average wire length is smaller, so most communication is local and the design is less prone to process variability, thermal variability and IR drop issues.

Whenever possible, this hierarchical approach was deployed also in the top-level design of the chip. HREs, control microprocessor, configuration controller and NoC routers were designed separately and packaged as re-usable macros. The only logic blocks that need to undergo P&R at top level are those whose structure is strictly dependent on the floorplan, such as NoC network interfaces, IO interfaces, and control/synchronization structures.

All architectural issues related to the heterogeneity of the system components were challenged and resolved exploiting the Network-on-Chip based communication strategy (Fig. 4.3). The MORPHEUS platform is divided in a set of NoC nodes that can be computation nodes (microprocessor, HREs) or storage nodes (On-Chip SRAM, external IO interfaces). Each node is plugged to a so-called Network Interface (NI) that converts data requests/responses in NoC transfers, while network Routers connect all the NIs dispatching the NoC packet appropriately across the network. The adoption of the NoC had a strong impact and significant benefits on the integration. The introduction of the Dual-Port, Dual-Clock DEBS: effectively decoupled HRE data access from the system, removing any heterogeneity in the design of global transfers that are programmed in a consistent way on all NoC nodes by setting the DMA transfer parameters. HRE-specific data access patterns are encapsulated in the HRE code libraries and only refer to the local DEB, so that they can be described only depending on the HRE functionality but do not depend on the global access patterns.

From a network topology standpoint, router instances were manually placed in the final floorplan in order to ensure even distribution of the wire length between all nodes, and avoid both wire congestion and unnaturally long paths between nodes. In case such events proved inevitable, additional routers could be added to break long paths and ease congestion, affecting only transfer latencies but not throughputs, and with no impact on the transfer programming by the toolset or the end user.

4.5.3 Prototype Chip Features

4.5.3.1 Technology Options

The MORPHEUS chip (see Fig. 4.4) was designed in CMOS 090 technology of ST Microelectronics. The chosen technology process flavour supports seven Metal layers to ensure good routability of the global lines, critical to enable NoC wires performance. Computation efficiency was considered a prioritary metric with respect to power consumption, so a High-Performance (HP) flavor of the process was selected in order to enable gate commutation speed versus leakage consumption. The reference voltage is 1 V, although the chip was designed in order to be functional in standard temperature conditions in the range [0.9 V:1.2 V].

The XPP, the controlling ARM processor, and all structures related to data and configuration transfers are implemented by Standard Cells. In order to mitigate leakage effects typical of the HP process, the design utilizes standard cell libraries supporting two different threshold voltages: specific low threshold, high speed cells are utilized only on critical paths, accounting to 11.4% of the standard cells area, while high threshold, low leakage cells were utilized in the remaining 88.6%.

The MORPHEUS chip features a total area of 110 mm, 97 M transistors including L1 and L2 memory and IOs (see Table 4.1). It features static power consumption of 230 mW. Separated, gateable clock domains were implemented on the separate reconfigurable engines (eFPGA, DREAM, XPP). Since all domains are dynamically configurable, dynamic consumption of the devices strongly depends on the computation: average dynamic consumption ranges around 700 mW with local peaks above the 3 W mark. Local and global IR-Drop phenomena, very common in similar high-performance architectures, are significantly limited by the multiple, asynchronous clock domains. The chip contains 1,961 kbits of L1 Dual Port memory plus 5,970 kbits of L2 memory shared by all engines and accessed by the NoC. In all, memory occupies 16% of chip area, IO 6%, Std cells 55%.

Timing closure in worst case conditions (wc, 0.9 V, 125 C) was at 250 MHz for the top level interconnect, 200 MHz for DREAM, 150 MHz for XPP and 100 MHz for the

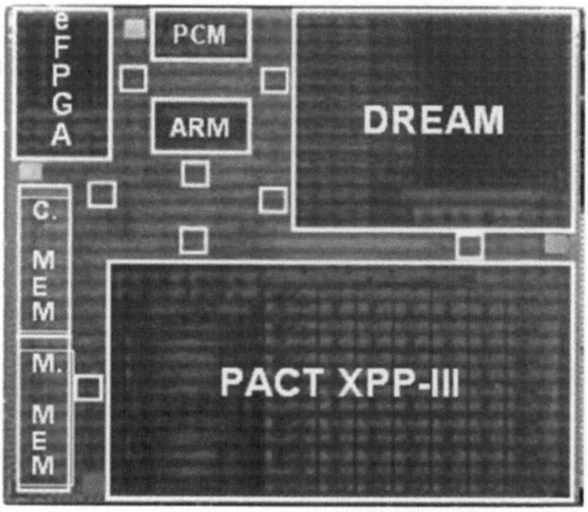

Fig. 4.4 Photograph of the MORPHEUS prototype

Table 4.1 Area distribution in the MORPHEUS prototype chip

Total Chip Area (mm^2)	10.48×10.48=110
Custom layout macros area (mm^2)	16.1 (15%)
Embedded memory (SP/DP) (kbit)	5,970/1,961
Embedded memory area (mm^2)	17.5 (16%)
IO ring area (mm^2)	6 (5.4%)
Standard cell count (Kgates)	8,507 (33%)
Overall transistor count (Millions)	97
Standard cell density	52%
High speed versus low leakage standard cells (area ratio)	11.4%

eFPGA module. In typical conditions the respective performance figures raise to 320, 260, 180 MHz respectively. Referring to worst case conditions, these figure lead to a 12.5Gbit/s throughput on the NoC. Choosing 16-bit operations as reference, the test chip is capable to deliver 60 GOPS, 0.6 GOPS/mm^2, 20 GOPS/W.

4.5.3.2 Performance Assessment

In order to evaluate the theoretical performance of the design, arithmetical operation of different bit widths were mechanically mapped over the different HREs. This mapping is not related to a specific algorithm deployment, since it is only aimed at evaluating the theoretical processing capability of the platform. As such, it does consider routing overheads internally to HREs, but does not consider data communication consistency and relative hazards related to a specific algorithm. The choice of repeating the experiment over variable bit-widths is based on the assumption that most application feature bit-width dynamics that are variable between various stages of the computation. Figure 4.5 shows how each of the computation engines varies offer performance that significantly

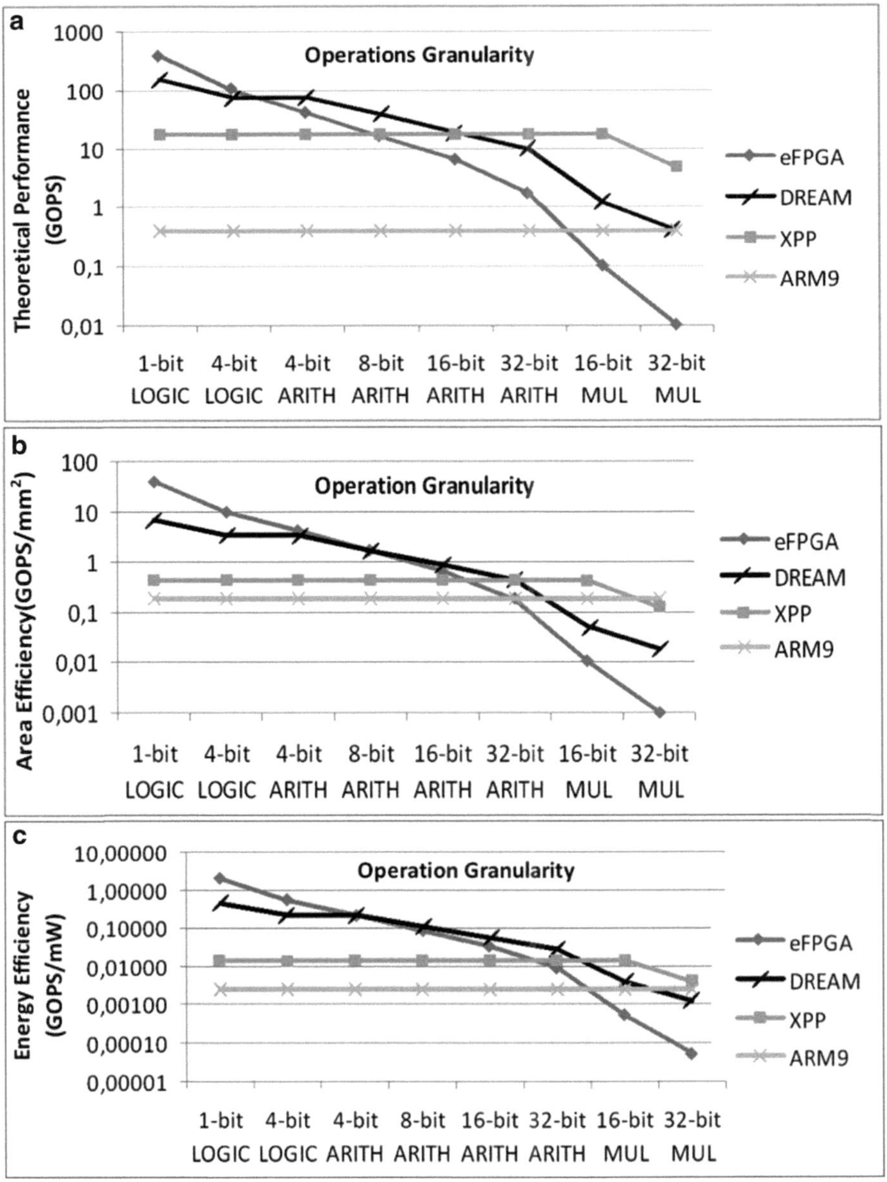

Fig. 4.5 Theoretical performance of the MORPHEUS HREs

depends on the operations bit-width: at low bit widths, predictably, the eFPGA yields larger efficiency, as it maps small operations on appropriate structures without the area overhead of 32-bit circuitry as in the case of standard processors/DSPs. In the 8- to 16-bit range, the efficiency of the DREAM engine is evident, as it exploits smart resource allocation (data granularity of 4/8-bits) with a pipelined structure that allow significant throughput. After the 16-bit mark, XPP can exploit is significant computation parallelism, offering a significant margin on sequential architectures.

Table 4.2 Comparison of MORPHEUS performance metrics against alternative computation platforms

Device	Techno	GOPS	GOPS/mm^2	GOPS/W
MORPHEUS	CMOS090	60	0.6	20
Microprocessor	CMOS090	0.35	0.15	1
ASIC	CMOS090	n.a.	~10	~800
TU iVisual (JSSC 2008)	CMOS180	77	1.16	205
KAIST Vect Proc (JSSC 2009)	CMOS130	125	3.4	214
Philips Xetal II (JSSC 2009)	CMOS090	107	1.44	170
Cell		200	n.a.	1:2
UCB Bee2	FPGA	n.a.	n.a.	2.2
Xilinx Virtex-II 8,000	FPGA	450	n.a.	3.5

The graphs clearly demonstrate the advantages enabled by the MORPHEUS approach: the user may exploit for every computational kernel the computation fabric that is most suited to its needs; wherever is possible to partition the application on a balanced pipeline between the three engines, it is possible to exploit the ideal tradeoff between hardware flexibility and computational efficiency for all stages of the application. The reported performance figures classify MORPHEUS roughly in the range of vector processors in the same technology node (see Table 4.2). On the other hand, MORPHEUS allows significant advantages against FPGAs and general purpose embedded processors, again referred to the same technology node.

4.6 The Toolset

MORPHEUS application development makes use of a number of successful tools from industry. An innovative aspect of the MORPHEUS approach is the seamless design flow from a high level description toward target executable code.

The MORPHEUS architecture is powerful and very flexible thanks to its heterogeneous embedded hardware accelerators. The drawback of such architectures is the complexity that one must challenge to optimally use it and reach its expected high performance. In addition to managing the different components, performing data exchanges through registers and DMAs and setting proper synchronization mechanisms, each accelerator comes with its own language making.

An ambition of the MORPHEUS toolset [4] is to abstract the heterogeneity and the complexity of the architecture in such a way that software designers may program it without a deep knowledge and experience on HREs (hardware architecture, programmability and so on). Reconfiguration mechanism at Run Time (RTR) makes possible multitasking and very high silicon reusability. The Predictive Configuration Manager (PCM) of MORPHEUS hides the context switching overhead and abstracts the interface of the reconfigurable engines from the designer's point of view by an intermediate graphical representation of the applications extracted at design time.

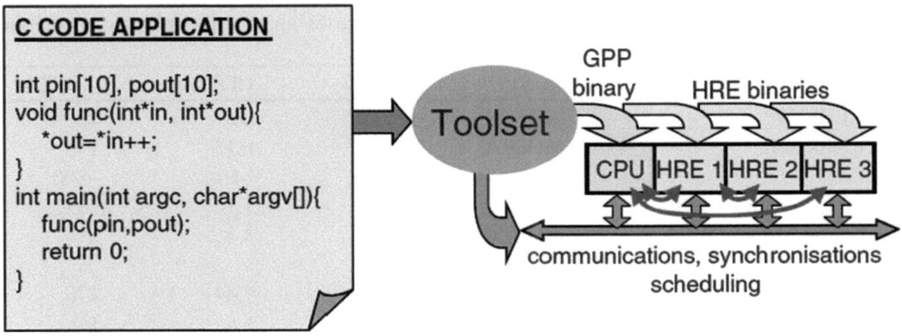

Fig. 4.6 Toolset objectives

Figure 4.6 shows briefly the MORPHEUS "Toolset objectives". The software engineer uses a C code description language to write his application. From this code, the toolset configures then the three HREs.

The MORPHEUS toolset enables the productivity improvement of application development on reconfigurable computing platform. Furthermore, the toolset objectives are not limited to the MORPHEUS architecture; they can be generally used on any reconfigurable computing platform.

The proposed toolset uses a two-level approach to optimize and map an application on the architecture:

At higher level, the user must have a global view and understanding of the whole application and focuses on mapping each function on each HRE as well as the global scheduling of the HREs, the ARM, the memory transfers and the synchronizations. At this level the user can choose which function should run on which HRE and how are stored and moved the data. This is done through configuration files and pragmas in C code. The programming model is based on the MOLEN paradigm from [14]. The Molen organization and the corresponding programming paradigm provide a solution to most of the issues that arise such as: an opcode explosion, a limitation of number of parameters, a lack of parallel execution and a modular approach. At runtime, this level is managed by the operation system by scheduling the loading of proper HREs' binary code against the whole execution and the data communication.

At lower level and for each accelerated function, the user focuses on local optimization of the code of this function on the targeted HREs. At this level, the user designs its accelerated function by graphically assembling sub-functions. Each sub-function is described in C code and then mapped on a set of HREs. Figure 4.7 shows such an assembly of sub-functions, and the description of the MORPHEUS architecture used to map each sub-function.

The accelerated functions are supposed to contain some intrinsic parallelism with data-streaming model. The platform allows also, at higher level, the programming

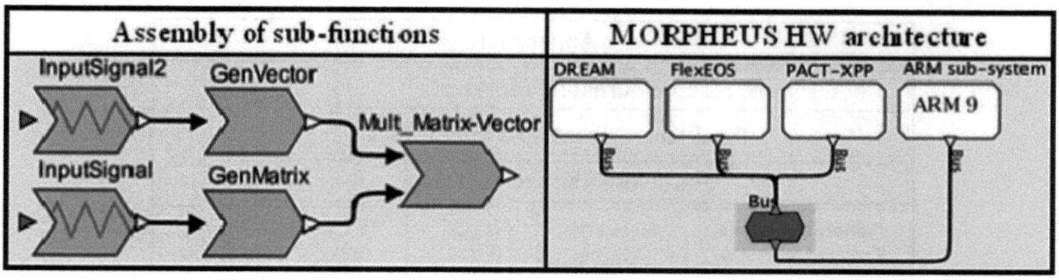

Fig. 4.7 Accelerated function low-level design

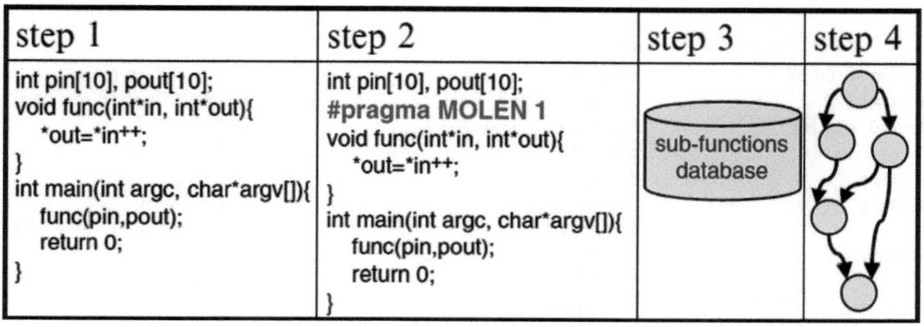

Fig. 4.8 Programming steps

of parallel threads that can run independently on different accelerators using a Open-MP compliant set of directives.

Thanks to the toolset, the development of an application on MORPHEUS is nearly as easy as writing a sequential C code for a General Purpose Processor. When implementing an application on the architecture, the designer splits it into control parts (executed on the ARM processor) and computation intensive parts (mapped on the HREs) by following these steps (Fig. 4.8):

Step 1: The application is written in standard C language. During this step, one can validate the application against use cases and test benches.

Step 2: The programmer identifies the functions that must be accelerated by setting a pragma in the application C code on the top of each of them. The programmer indicates a number corresponding to the function identifier used in step 3 which will be used in the assembler instructions (SET, MOVX, etc.).

Step 3: These accelerated functions are captured inside a graphical environment, called SPEAR [5], by connecting and assembling some building blocks, called elementary sub-functions and written in C language.

Step 4: The toolset generates the bitstream for every HREs (if used).

The four steps are described by the following figure:

The SW compilation flow and the RTOS manage the creation of a software executable running on the main processor, directly from the application code of the step 2. This executable controls the reconfiguration and the execution of HREs but also programs

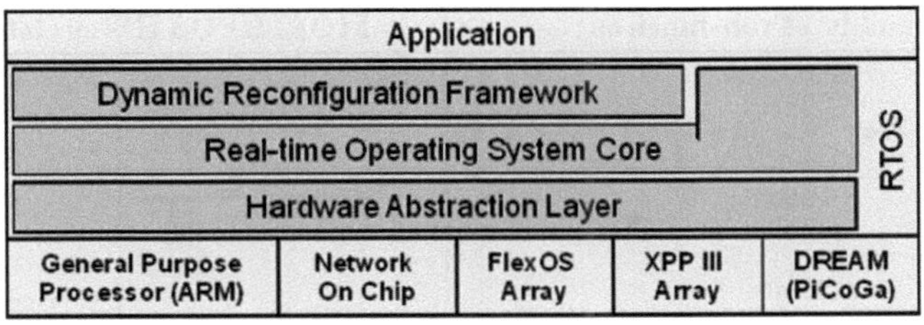

Fig. 4.9 Structure of the RTOS

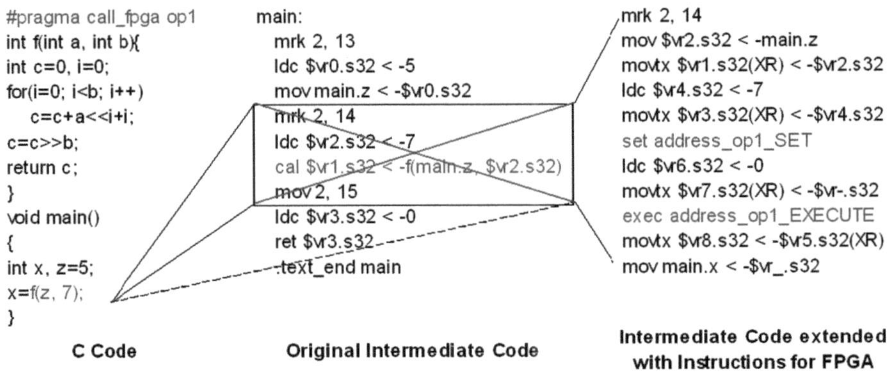

Fig. 4.10 Pragma annotated code and subsequent compiler transformations

DMA transfers. The COSY compiler [17, 18] replaces the calls to accelerated functions in the application code by the MOLEN directives, then these directives are called by the RTOS services. The Operating System handles the scheduling of the whole application and delivers services to the application for accessing the architecture.

The RTOS has a layered structure which is shown in Fig. 4.9. The bottom layer is the Hardware Abstraction Layer which provides a more uniformed access to the reconfigurable hardware and the system infrastructure. It provides virtual exchange register for the compiler which are mapped to the parameter registers in the heterogeneous reconfigurable engines. It provides also the basics for a pipeline services between the heterogeneous reconfigurable engines.

As depicted in Fig. 4.10, the accelerated functions' identification "pragmas" are set immediately before the definition of the function to be off-loaded to the HRE. The pragma's arguments are the identifier of the hardware operation that corresponds to the function and the preferred implementation's configuration of the operation.

The communications are however managed in the Spatial Design framework. This part of the toolchain handles the development of specific code for each HREs. The high level synthesis produces a configuration bitstream from an operation specification. In the tool flow, SPEAR is an interface of the high-level synthesis of operations and of the SW compilation flow, enabling their cooperation in an integrated and coherent toolset.

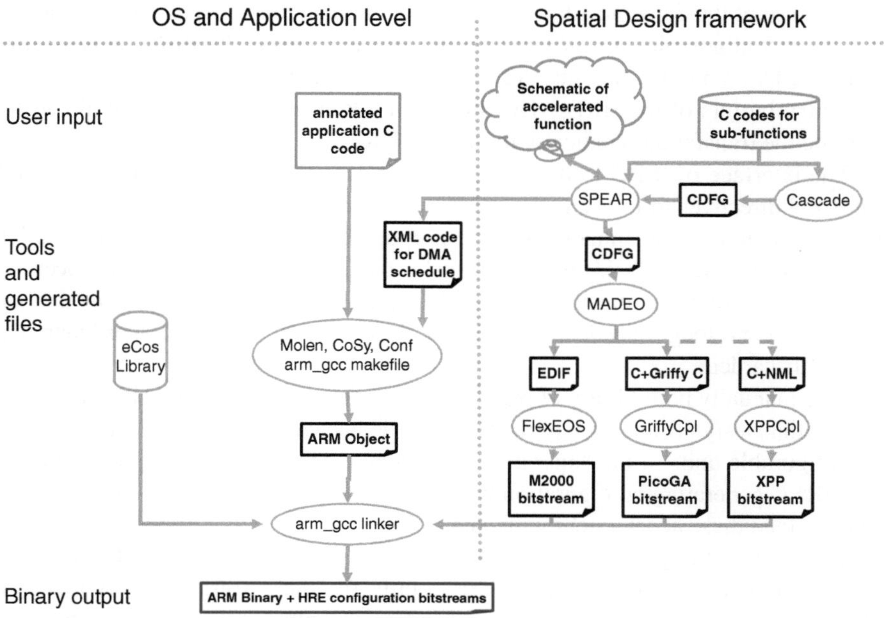

Fig. 4.11 Simplified internal view of the tool chain

SPEAR helps to capture the specification of an accelerated function, which is independent to the targeted HRE. It enables to manage, in a coherent framework, both the HW interface of the HRE (i.e. local buffers and addressing mechanism) and the data transfers. Connections of SPEAR with high-level synthesis and SW compilation tools avoid risks of incoherencies between the layout of data in local buffers and main memories, the DMA transfers, the addressing mechanisms of the HREs to access local buffers and the operation specification.

The toolset is composed of several tools (see Fig. 4.11) whose tasks are to analyze the user sub-functions, to generate data movement (and to synchronize when needed), to generate HRE configuration bitstreams, to schedule the whole application by loading the HRE in time for execution and to launch the DMA for data movements, to compile the sets of binaries and to make a MORPHEUS package. Despite the large number of tools, the toolset remains user's friendly as it is closely integrated and automated. Indeed, the programmer only works with the SPEAR graphical tool and Makefiles.

To execute multi-threaded applications, their dynamic nature requires a central management. In MORPHEUS, this is provided by the Intelligent Services for Reconfigurable Computing (ISRC) layer on top of an embedded real-time operating system (eRTOS).

ISRC handles the interwoven topics of scheduling and allocation. The schedule is dependent on the availability of the different reconfigurable units. If there are functionally equivalent implementations of the requested operation available it makes a choice at run-time depending on the platform/application status possible.

Because of the dynamic allocations, the communication between the application and the reconfigurable units is also only indirect and handled by ISRC by programming the DMA and DNA controllers to transfer data between memory and the data exchange buffers of the HREs. The linking of a transfer to an operation allows to migrate the transfer automatically to the new HRE.

The interface of ISRC to the compiler extends the SET/EXECUTE concept of MOLEN directives to load and run a function on a HRE, with BREAK to wait for the completion of the function for the purpose of synchronization with parallel operations and MOVTX/MOVFX to transfer parameters from the ARM processor to a specific exchange register of the HRE and reverse. RELEASE informs that an operation is no longer needed, e.g. after loop exit and the corresponding bitstream can be discarded.

As a manually built library of pre-defined sub-functions could not be exhaustive, the Spatial Design framework of the toolset is dedicated to the design on the various reconfigurable units of the "operation" mentioned above. It includes programming the HREs, generating configuration bitstreams but also managing the communications to feed these accelerators with data.

To this end, the goal of Spatial Design is twofold: To hide the HREs heterogeneity and to abstract the hardware details for the programmer. A Control Data Flow Graph (CDFG) format is used as an intermediate and technology independent format inside the framework. High-level synthesis techniques are used in a tool, called MADEO, that acts as a back-end for the framework [19–21]. The flow then relies on HREs proprietary tools to generate the configuration bitstreams.

A second objective is to provide domain-specific models and languages to application programmers. In this way, operations can be specified at a high-level of abstraction (improving design time and flexibility), without sacrificing performances. Operations are modeled as a directed acyclic graph (see Fig. 4.12 part "a"), in a formalism called Array-OL which is well suited to represent deterministic, data intensive and data-flow applications such as the kind of operations accelerated on HREs. SPEAR automatically generates a CDFG model of the accelerated function and some communication parameters to feed the HRE. They are forwarded to the SW compilation flow, contributing in this way to seamless hardware/software integration. The Cascade tool [22] is used to generate the CDFGs for the elementary functions.

Each tool has been developed and integrated in the flow. The whole toolset is fully functional. Several applications from signal processing and image processing have notably been developed with the platform. The flow has been validated on a SystemC simulator of the architecture, which allowed us to validate it in parallel with the development and the prototyping of the chip. In the current development status, the main limitations are located in the Spatial Design part that supports the C language structures and data types with restrictions, and does not yet target the PACT XPP HRE. The configuration bitstreams generated with the PACT proprietary tools [23] can however be easily integrated into the compilation flow of the toolset.

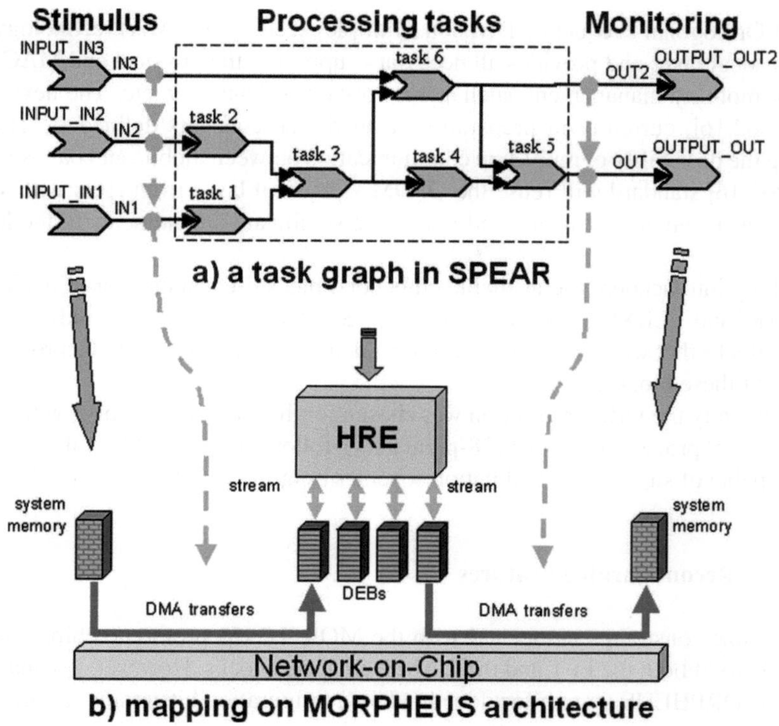

Fig. 4.12 Overview of an operation mapping on a HRE

4.7 Applications and Demonstrators

Four different case studies borrowed from complementary domains with different reconfigurations needs have been used to evaluate and validate the MORPHEUS platform and its associated toolset. Applying the proposed technology in domains like wireless telecommunications, intelligent cameras, high definition video/digital film technology and network routing systems aim at proving that MORPHEUS platform concepts can be successfully applied in real world applications of high complexity.

4.7.1 Wireless Telecommunications

4.7.1.1 Case Study Overview

The application targeted by Intracom Telecom Solutions is the emerging IEEE 802.16j standard. The latest standard currently in force from the IEEE 802.16 family is 802.16e, the basis for Mobile WiMAX technology. This standard mandates the use of Orthogonal Frequency Division Multiple Access (OFDMA) technology for the physical layer and provides all necessary support in the physical and MAC layers for mobility management, such as network entry, handover, etc. The next standard, 802.16j, currently in preparation, extends the concepts defined in .16e by adding the possibility of multi-hop communication between mobile and base station. The 802.16j standard will reuse the OFDMA physical layer from .16e, with some minor enhancements possible, and will make significant amendments to the MAC layer.

Taking into account the available embedded blocks' capacities (specifically for DREAM and FLEXOS), Intracom Telecom Solutions performed a study so as to select blocks that would a) fit in these embedded blocks and b) match the processing nature of these blocks.

Eventually the DREAM section was chosen for the target application, which was a word-level processing block (128-point FFT), followed by a QAM symbol demapper, capable of supporting modulation schemes ranging from QPSK to QAM46.

4.7.1.2 Reconfigurable Features

Due to size constraints associated with the MORPHEUS prototype chip, it is not possible to fit both the FFT and the QAM Demapper blocks. However, to capability of the MORPHEUS system for downloading configuration bitstreams for different HW accelerators at runtime offers a convenient solution. Thus, it is possible to set-up the downloading of the QAM demapper block while the already setup and running FFT block is processing the data delivered to the DEBs bound to this block. Once the QAM Demapper has run its course on the current set of data points, the FFT configuration bitstream is loaded again in DREAM and the cycle repeats.

4.7.1.3 Design Using MORPHEUS Toolset

The SPEAR tool was used for capturing the target application, in two parts: one GDFG was generated for the FFT block and one for the QAM demapper (see Fig. 4.13). The resulting CDFGs were used by the MADEO tool downstream for the generation of the configuration streams for the DREAM array, as well as the control code running on the ARM processor, responsible for downloading the configuration streams and for managing the data transfers to and from the DEBs used for communication.

The code for QAM demodulation functions is theoretically very simple. They have been considered separately (QPSK, QAM16 and QAM64). Their implementations on the HREs take less than a day effort for each of them and do not require any HREs knowledge since normal C code is required. The design on the SPEAR graphical interface can easily be built and requires a few minutes (see Fig. 4.14). Globally a few days are necessary for the implementation of the application including C code for ARM and SPEAR capture. The simplicity of the code to be written is, as in the

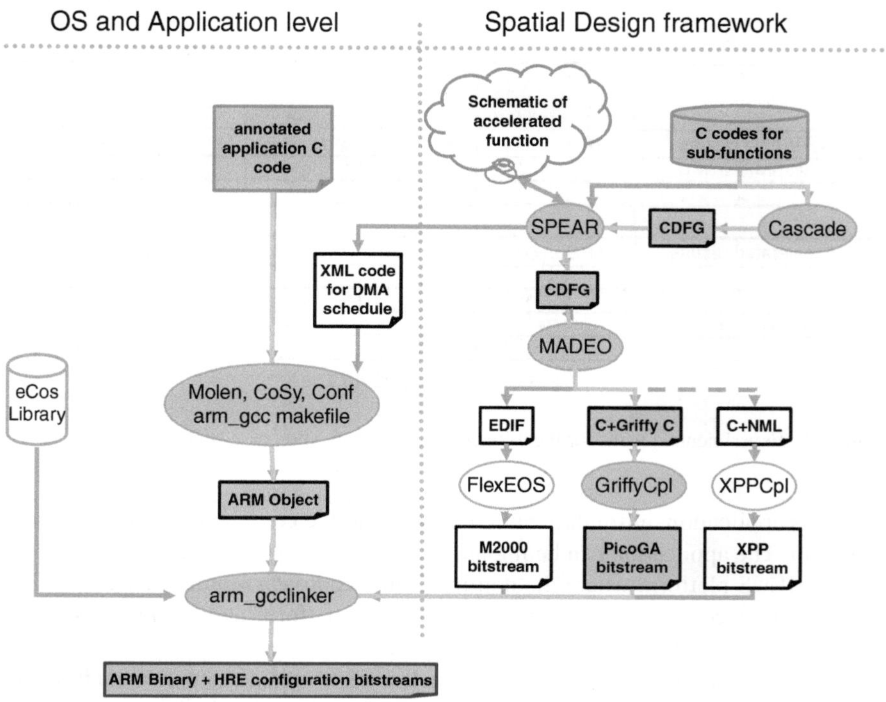

Fig. 4.13 Design flow overview for the wireless telecommunication case study

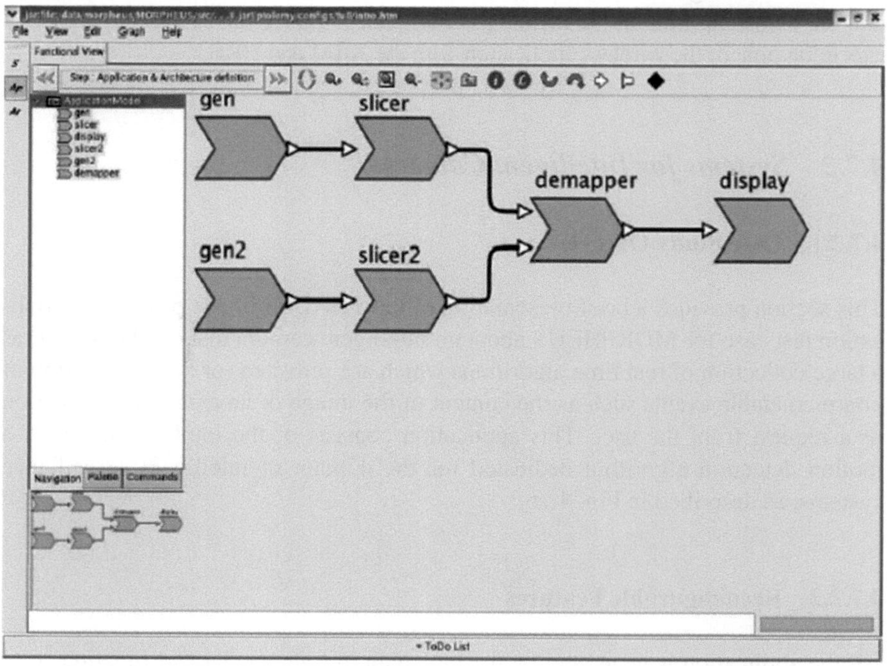

Fig. 4.14 Accelerated function graph from SPEAR

	FFT + Slicer + Demappe r QPSK	FFT + Slicer + Demappe r QAM16	FFT + Slicer + Demappe r QAM64	SLICER + Demappe r QAM64	FFT128
ARM only results	**ARM cycles**	**ARM cycles**	**ARM cycles**	**ARM cycles**	**ARM cycles**
EXECUTION (ARM ONLY)	918 617	927 578	940 197	26 982	913 215
TOTAL	925 487	931 847	947 109	28 730	918 379
Accelerated results	**ARM cycles**	**ARM cycles**	**ARM cycles**	**ARM cycles**	**ARM cycles**
TOTAL CONF_DMA	3 745	3 745	3 745	1 835	1 910
TOTAL LOAD/DELOAD	5 279	5 278	5 278	3 126	2 152
EXECUTION (WITH HRE)	6 584	8 408	10 301	6 108	4 193
TOTAL	19 751	21 574	23 467	12 754	10 713
EXECUTION TIME RESULT					

Fig. 4.15 Experimental results for the wireless telecommunication case study

previous application, absolutely not comparable to the complexity of the execution platform. The application can be implemented within approximately 1 week on the MORPHEUS platform where the non-integrated approach requires 2 weeks. If we include the necessary time required to learn the platform and required tools and languages, the productivity gain is even higher.

Figure 4.15 presents the results for the wireless telecommunication case study. In this table, we have different results corresponding to the number of cycles between the ARM only and the ARM with the HREs. TOTAL_CONF_DMA corresponds to the duration of the DMA configuration, TOTAL_LOAD/DELOAD corresponds to the loading time and the unloading time in the memory access and, to finish, the EXECUTION is the execution time of the wireless application with the ARM and HRE.

4.7.2 Systems for Intelligent Cameras

4.7.2.1 Case Study Overview

This section provides a brief presentation of the THALES image processing application test case for MORPHEUS about an intelligent camera that can be viewed as a large collection of real time algorithms which are activated (or not) in function of non predictable events such as the content of the image or an external information or a request from the user. This application consists of the implementation of a motion detection algorithm dedicated for the domain of intelligent surveillance systems, as described in Fig. 4.16.

4.7.2.2 Reconfigurable Features

All the application is generated by the toolchain in four accelerated functions (see the block diagram in Fig. 4.17): the first accelerated function corresponds to the

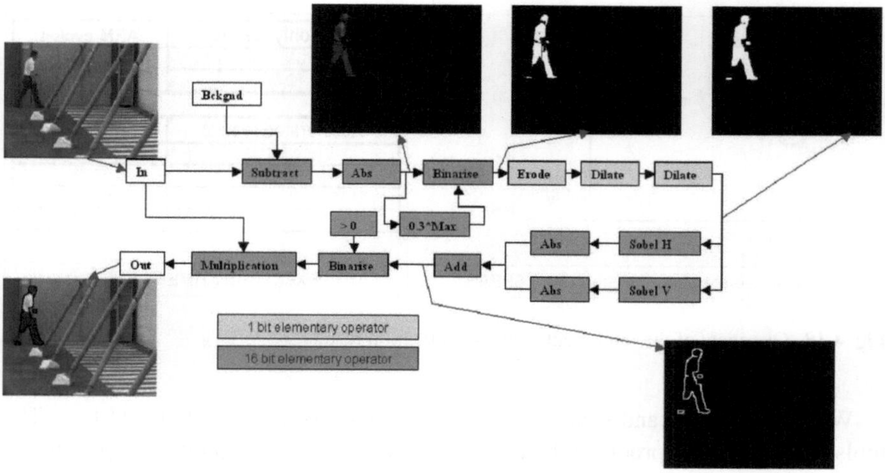

Fig. 4.16 Combination of operators for a simple motion detection algorithm

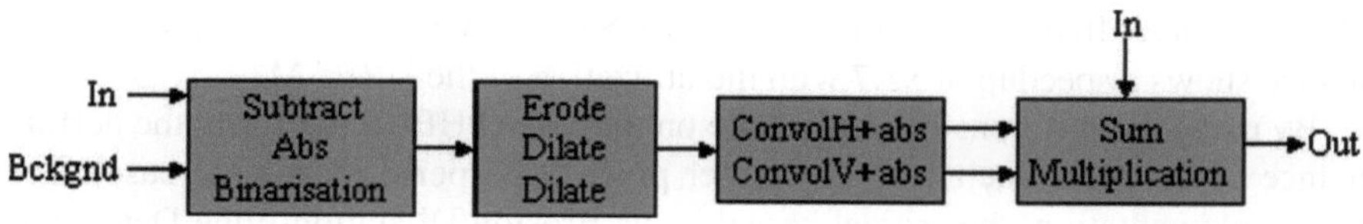

Fig. 4.17 Operator blocks organization

subtraction, the absolute value and the binarisation between two images (the background and the current image); the second accelerated function is the opening (erosion and dilatation) and a second dilatation ; the third accelerated function corresponds to the convolution H/V from a SOBEL matrix ; the last accelerated function is the sum, the absolute value and the multiplication between the current image and the result of the third accelerated function.

However, to capability of the MORPHEUS system for downloading configuration bitstreams for different HW accelerators at runtime offers a convenient solution. Thus, it is possible to set-up and load the blocks one-by-one.

4.7.2.3 Design Using MORPHEUS Toolset and Performance Analysis

The application is implemented as a sequence of elementary operators described in C code to enable reuse capability. SPEAR (Signal Processing Environment and ARchitectures) tool is then used to build automatically the interfaces of each operator described in C code (see Fig. 4.18). This step provides the modules ready to be run by the HREs. Operators could be stored into an image processing library to enable reuse capability. SPEAR is used to create the interfaces around each operator. SPEAR allows SystemC simulation of the code on PC architecture and generates the bitstream for each operator.

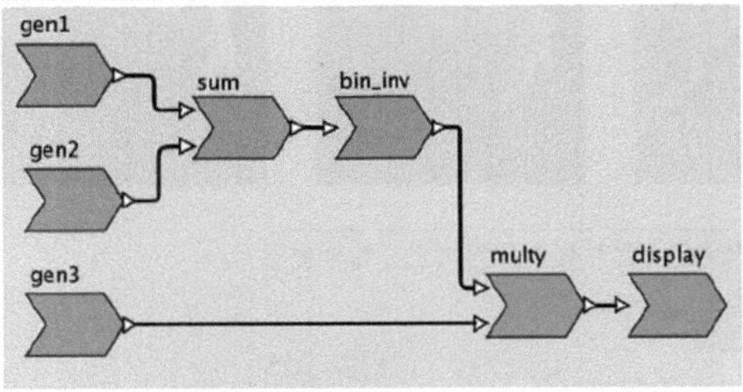

Last accelerated function Graph from SPEAR

ARM only results	ARM cycles
EXECUTION (ARM ONLY)	613 131
TOTAL	666 134

Accelerated results	ARM cycles
CONF DMA	8 038
DMA	67 721
EXECUTION (WITH HRE)	103 978
TOTAL	179 737

EXECUTION TIME RESULT

Fig. 4.18 Overview of the intelligent camera application results

Writing operators and synthesis of an application is a very automated task. The toolset provides high productivity and high reuse capabilities without in-depth knowledge of the hardware. MORPHEUS toolset is a very efficient synthesis toolset.

Among the different HREs available in MORPHEUS, the figure has been obtained by manual implementation of the application on the PiCoGA unit only (the DREAM target frequency within MORPHEUS is 250 MHz). The measured performance shows a speedup of x3.7 with the utilization of the DREAM.

By using the different HRE available on the MORPHEUS platform, the performance could even reach 1.27 cycles per pixel (20% performance increase). For this implementation, the critical kernel is the Erosion/Dilatation/Edge Detection. It has been implemented on the DREAM engine whose frequency is mentioned above.

4.7.2.4 Experimental Results

Toolset allows a very high productive work. The bar chart graph from Fig. 4.19 indicates a speed up of 3.4 between the execution on the ARM only and the same application on the ARM coupled with the DREAM accelerator. The pie chart schema shows the repartition about the time spent by the component during the execution.

The DMA communications take about the same time as the HRE execution. The DMA transfers can be improve in using the pipeline stage to decrease the time of computation by the HRE. In a case where the DMA and the HRE work would take place in parallel (e.g. when implementing the pipeline) we would gain about half of the total time (i.e. the DMA time would then no linger be added to the computation time at least in the pipeline).

Moreover, the higher level programming that is allowed by the toolset gives a programming accessibility to a larger range of programmers. Since it is not needed to get specific skill corresponding to knowledge about low level architectural considerations, even people not really aware about the platform characteristics can program the chip.

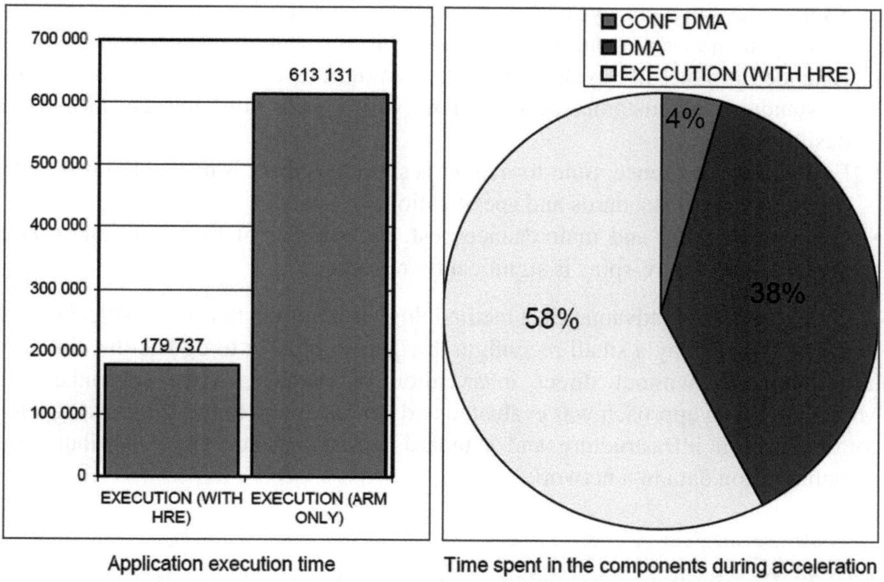

Fig. 4.19 Execution time of the intelligent camera application

4.7.3 Ethernet Based In-Service Reconfiguration of SoCs in Telecommunication Networks

4.7.3.1 Case Study Overview

The increasing complexity of communications systems, coupled with evolving standards and specifications pose tremendous pressure on the design process. Moreover, the telecommunication market is dynamic, requiring short development cycles. Manufacturers, which aim an early presence on the market, are not able to wait for standards and specifications getting stable and mature. As a consequence of standard and specification weaknesses, they have to cope with costly design respins as well as high project and business risks.

Fundamental constraints like performance, design costs, flexibility, power consumption etc. usually result in a trade-off when determining the components for the implementation platform: ASICs, FPGAs, microcontrollers, μPs, DSPs, or a mixture thereof [24, 25].

Reconfigurable Systems-on-Chip (SoC) architectures offer a trade-off between the performance of ASICs and the flexibility of programmable general purpose processors. The key advantages are as follows:

- Higher performance and lower power consumption than pure FPGA solutions.
- Increased flexibility during development: late design changes and bug fixes may be incorporated even late in the development cycle.

- Shorter development cycle: boards including hardware and software may be developed and even manufactured before chip verification is finalized.
- Extended product life-cycles: manufactured devices can be adapted to changes of standards or customer specifications which were not foreseen during the design phase.
- Early market presence: time-to-market is shorter as there is no need to wait with development till standards and specifications are stable.
- Decreased design and maintenance cost: the number of time consuming and expensive design re-spins is significantly reduced.

To leverage these advantages, a methodology is required, to extend typical signal processing ASICs by a small reconfigurable part to be able to update this SoCs in large networks without direct intervention of on-site service personnel. In MORPHEUS an approach was evaluated and demonstrated to use the existing telecommunication infrastructure and standard network protocols to distribute the reconfiguration data in a network.

4.7.3.2 Distribution of Reconfiguration Data Within a Network

Once it is required to update SoCs deployed in a telecommunication network, there are basically three ways to transfer the reconfiguration data to the systems:

- *On-site download by service personnel* – each system is updated manually. This is not a real option as it causes an enormous effort to travel to all network sites
- *Combination with a software release upgrade* – reconfiguration data are downloaded to the systems as part of a new software image. This requires internal communication channels from the main controller of a system to the various SoCs in that system. In currently deployed systems usually such channels are not in place. Further it restricts the opportunities to update SoCs to the long software release upgrade cycles. Many network providers, especially the large ones, deploy a new software release less than once a year.
- *In-band download within the communication signal* – unused bandwidth is filled with reconfiguration data and transmitted via existing data paths through the network. No additional internal interfaces are needed as the data stream goes through the SoCs anyway for signal processing. The reconfiguration data are marked as such to enable the SoCs detecting and extracting their reconfiguration data from the data stream.

We consider the in-band download as the most appropriate method. It does not require any special hardware or software modifications, except the replacement of conventional chips by reconfigurable SoCs.

In the following the update is first explained from a network level view, whereas in the next section we describe what happens inside the chip. As communication protocol we choose Ethernet, but the method may be applied to any other kind of communication protocol like SDH/SONET, OTN, IP, Fiber Channel etc. The selection of Ethernet for our demonstration was mainly driven by two factors: Ethernet is a convenient choice for proof-of-concept as it enables the usage of standard devices and is scalable down to a size reasonable for a demonstrator. Furthermore, Ethernet is one of the key protocols of future communication systems, rapidly evolving and thus providing enough weaknesses in standardization and specification of new features to justify the usage of reconfigurable SoCs.

For transmission with Ethernet, the reconfiguration data are split into several pieces, each one small enough to fit into the payload of an Ethernet packet.

The reconfiguration data is located in the payload portion of the packet. It consists of several sections:

- The *Reconfiguration Data Header* marks the payload as such.
- The *Reconfiguration Device Address* selects the device to be reconfigured. A node can contain more than one reconfigurable device. These devices may need individual reconfiguration data due to different technology (ASIC, FPGA) and/or different functions. So it is essential that every device in a single node can be addressed individually.
- The *Reconfiguration Packet Number* (Rec. Pkt #) is needed to be able to restore the correct order of the reconfiguration data portions. The reason for this requirement is the fact that the consecutive transmission of reconfiguration packets through the network cannot be guaranteed. Therefore the data must be sorted before reconfiguration can start.
- The *Reconfiguration Data Payload* contains portions of the reconfiguration data, which are collected until the entire reconfiguration image is complete. After sorting and validation of the image the reconfiguration process is initiated.

The reconfiguration packets as described above are created at a central location and transferred to the network via a distribution gateway (for which e.g. the network management gateway may be used). The gateway transmits the packets as broadcast to the whole network. Thus, in a properly configured network the reconfiguration data reach automatically all nodes with the SoCs to be updated.

4.7.3.3 SoC Architecture and Reconfiguration Mechanism

For communications systems different types of chip architectures are available with various characteristics, giving a designer a choice in selecting the most appropriate for a particular purpose. ASICs typically operate at a relatively low power and, if a large production run is involved, can be inexpensive to manufacture. In addition, they can be more densely provided in a system because of their relatively lower power consumption, reducing cooling requirements. Finally, signal processing in high-end communications systems requires high processing power to be able to handle single line data rates of up to 40Gbit/s (and more in future systems) and total capacities in the Terabit/s range.

Thus, ASICs often are the preferred choice of a designer. On the other hand, FPGAs provide much more flexibility. The implementation may be corrected or updated anytime during design and verification – and even after deployment it is still possible to reprogram FPGAs.

The approach demonstrated in MORPHEUS leverages the advantages of both approaches by combining them into one System-on-Chip. For this purpose, the Abound Logic embedded FPGA technology is applied [26]. The design parts, which are considered to be candidates for future changes ("weak parts"), are mapped to the embedded FPGA, whereas the design parts regarded as stable are implemented in the ASIC. These design decisions are essential and need to be a trade-off between keeping the embedded FPGA small and correctly identifying all weak parts.

Once an Ethernet packet is received at the input of the SoC, it is first processed by a packet filter. The filter detects the reconfiguration packets addressed to this particular device and extracts it from the regular data stream. Reconfiguration packets sent to different types of devices are ignored and forwarded like normal customer data packets. This way it is possible to have different kinds of reconfigurable devices in a network and to reconfigure them independently with separate data streams. The extracted reconfiguration packet then is duplicated. One copy is sent to the reconfiguration memory (RAM), where the reconfiguration data are collected. This RAM is additional memory dedicated for reconfiguration. Its size depends on the size of the reconfigurable core – which usually should be small compared to the ASIC part of the chip. The other copy of the packet is forwarded to the regular signal processing part of the SoC, as it is done for regular customer data. This broadcast ensures that the reconfiguration data are not absorbed by the first updated SoC but are continued to be forwarded to the rest of the network.

As soon as the reconfiguration data is complete – determined by packet number and checksum – the reconfiguration controller initiates and controls the update of the embedded FPGA.

4.7.4 Film Noise Grain Removal Application

After scanning analog films in order to digitize them, the generated digital data contains a certain kind of noise which is caused by the grain of the analog film material. The film grain removal application targets to remove –or at least to reduce– this grain. Especially if high resolution projection of the digital film material on large screens is intended, film grain removal becomes an indispensable operation to improve the reproduction quality. Further, it is obvious that removal of film grain helps to achieve a better compression efficiency of the digital film material.

The algorithm which was developed for this application combines typical picture processing algorithms with very different processing and data transport requirements. The diverseness of the different algorithmic blocks qualifies the film grain removal application for an evaluation of the MORPHEUS processing platform with its heterogeneous processing engines.

The main algorithmic blocks of the application are motion estimation, motion compensation, and a three dimensional hierarchical wavelet transformation (DWT). Analysis of these blocks showed that motion estimation and motion compensation can be processed very efficiently on the coarse grain XPP while the DWT is well suited for the mid grain DREAM. Dynamic reconfiguration of the HREs is necessary since the number and the complexity of the algorithmic blocks is too high to be mapped onto the available processing resources at the same time. Algorithmic blocks that compete for the same HRE have to share the HRE through time multiplexing. Algorithmic blocks which are too complex to be mapped as a whole onto one HRE have to be split into smaller pieces. Then, the selected HRE can be reconfigured sequentially to process the small pieces. However, since the complete film grain removal application should run in real-time, it has to be stressed that there are strict time budgets for data processing as well as for dynamic reconfiguration of the HREs.

To process one frame the XPP has to be configured two times. First the XPP has to process the motion estimation, second the motion compensation. Processing the DWT for one frame by the DREAM requires ten DREAM configurations. To relativise this number it should be mentioned that no functional sub-block of the DWT requires more than four configurations. This number can be handled very efficiently by the DREAM. More details regarding the mapping of the application onto the MORPHEUS processing platform is provided in [27].

All data transfers within the application, internal or external, utilize the available NoC. The ARM9 processor is used as main controller who schedules and triggers data transfers and the data processing by the respective HREs.

The I/O interface of the application is memory based. A real streaming interface is not implemented, but can be added later since a second interface to access the external memory is available.

The processing nodes of the MORPHEUS platform typically run with the following frequencies: The ARM runs at 200 MHz, the XPP runs at 120 MHz and the DREAM runs at 40 MHz. With these parameters the MORPHEUS demo chip is very stable. Separate tests have shown that the respective processing units can run much faster. The measured maximal clock speed for the ARM was 320 MHz, for the DREAM 240 MHz, and for the XPP 200 MHz. In all cases an increase of the core voltage is required. Using the example of the 2D-DWT, Fig. 4.20 shows the impact of clock speed increase to application performance for DREAM.

The 2D-DWT is one of three major functional sub-blocks of the DWT which have to be processed by the DREAM. The blue line (rhombus) shows the relation of execution time to clock speed for the DREAM the red line (circle) represents the processing time if the 2D-DWT is mapped onto the ARM, running with a fixed clock rate of 200 MHz. The intersection of both curves shows the clock speed of the DREAM which is required to reach the execution time of the ARM implementation which is at ~70 MHz. This is almost three times lower than the ARM clock speed.

It was already mentioned that for the mapping of the film grain removal application onto the MORPHEUS platform run-time reconfiguration of the HREs is mandatory. Measurements have shown that the configuration times for the XPP are

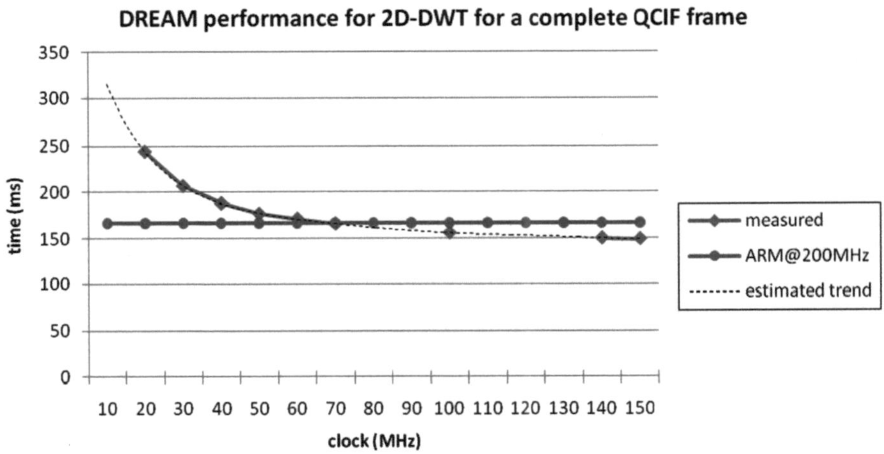

Fig. 4.20 DREAM performance at different clock speeds

approximately between 1 and 1.8 ms. For the DREAM they are between 1 and 10 ms. The variance of the measurements is caused by the size of the respective configurations (size of configurations depends on complexity of algorithmic blocks). Further, the measured times have been taken from the perspective of the central controller (ARM) of the application. This explicitly includes the transfer of the configurations from main memory to the HREs local configuration memories (CEBs) as well as the HRE internal configuration time. In fact, if the configuration process is considered in cycle accurate RTL simulation it becomes clear that the configuration times are mainly determined by the data transfer of the configurations into the CEBs of DREAM and XPP.

The measured processing time of the film grain removal application running on QCIF (176 × 144 pixel) resolution is approximately 600 ms per frame, of which about 80 ms are required for configuration. Since the application requires real-time processing with at least 25 frames per second this is about 15 times longer than the time budget for one frame. A detailed analysis of this result shows quickly that the main bottlenecks of the current implementation are the internal and external data transfers. RTL simulation discloses that DREAM and XPP have long stall periods waiting for new data or waiting until already processed data is stored. Implicitly Fig. 4.20 already shows this bottleneck. There are two possible explanations for the inverse proportional run of the blue curve (rhombus) in the figure. The first explanation, reconfiguration overhead, can be excluded since the 2D-DWT requires only four DREAM configurations. It is a feature of the DREAM to provide four configuration layers to store up to four configurations simultaneously. Switching the active configuration layer among the four configuration layers requires only very few clock cycles. The second explanation which is more appropriate is that the data transfer times for the 2D-DWT dominate the real processing times on the DREAM.

Fortunately, due to the remarkable flexibility of the MORPHEUS processing platform, there are plenty of options to improve the performance of the application in general and the performance of the data transfers is special. In the current available implementation, several features of the NoC, which is used for data transfers, are only partly used or not used at all. Examples include the parallel usage of NoC channels and the possibility to configure a NoC channel for the next transfer while the current transfer is running. Additionally, in [28] it is shown that impressive improvements in external memory bandwidth can be achieved if an advanced memory controller [29] is integrated into the MORPHEUS platform architecture. Generally, an increase of the on-chip memories will help to significantly reduce the number of necessary data transfers, as well as processing and reconfiguration overhead. Finally, with knowledge of the current performance bottlenecks, it is possible to optimize the implementation of the film grain removal application in order to avoid the current bottlenecks.

If the proposed optimizations are taken into account, the film grain removal application running on the MORPHEUS demonstration platform in real-time with the considered QCIF image resolution becomes a very realistic and attainable goal.

Regarding the development effort, the implementation time of the film grain removal algorithm on the MORPHEUS platform was approximately the same as the implementation time of the algorithm on a state of the art FPGA-based platform. When taking into account that the MORPHEUS development environment was experimental and used for the first time, this shows the impressive potential of the MORPHEUS approach to save design time. This potential will be even further improved with the finalized toolset for rapid development of heterogeneous MORPHEUS applications, which was not available at the time of development of the film grain removal application.

4.8 Conclusions

In this chapter, the different aspects and outcomes of the MORPHEUS European project have been described. A modular heterogeneous dynamically reconfigurable SoC architecture has notably been studied and developed in the frame of the project. A design flow is associated to this architecture, with the aims to improve the programming productivity of the platform and to shorten the development times of applications. This whole hardware/software approach for dynamically re-configurable platform constitutes an innovative solution for embedded computing.

The hardware architecture is composed by an ARM9 processor and three Heterogeneous Reconfigurable Engines. Naturally, the combination and the complementarity of the different HREs are key elements of the whole system. The processor is used as a central controller managing the whole platform. A prototype chip fabricated in ST 90 nm technology has allowed to evaluate the efficiency of the architecture. Taking into account the flexibility and the programmability of the architecture, the reported performance density and power efficiency figures are very interesting.

Despite the heterogeneity of the system and the necessity to handle dynamic reconfigurations, the MORPHEUS platform is also easily programmable thanks to its high-level programming model. A toolset allows a seamless flow from an high level description of an application to its implementation. Knowledge of the C language is sufficient to program the MORPHEUS chip thanks to this toolset. To address the design of the specific functions accelerated on the HREs, a description of these functions with all their inherent parallelism is easily and intuitively captured in the SPEAR tool as a graph of interconnected processing blocks. Implementations of the accelerated functions on any HRE are automatically generated from this model. Ease of programming at the system level is brought by two main elements that are the MOLEN paradigm and the ISRC library. The MOLEN directives permit to easily manage the different accelerated functions at compilation time thanks to COSY compiler, while the ISRC library provides system services for the control of the HREs at runtime.

The feasibility and the relevance of the proposed approach have been demonstrated through several case studies coming from complementary domains. Their deployment on the chip prototype has shown the good results in term of power consumption and performances of the platform.

Furthermore, many perspectives exist to make evolve and progress this kind of SoC approaches. The proposed system-level principles let opportunities to extend and enhance the platform by increasing the number of accelerators, by more powerful allocation schemes, by more efficient memory interfaces, and so on. The modularity of the platform is a key advantage to easily change the architecture reconfigurable engines and adapt the toolset accordingly.

Acknowledgments The authors would like to thank all the partners of the project consortium who were involved in studying and providing the required technologies, specifying the requirements and assessing the results. This research was partially funded by the European Community's Sixth Framework Program.

References

1. S. Borkar, Design Challenges of Technology Scaling, *IEEE Micro*, IEEE Computer Society Press, 1999, 19, pp. 23–29.
2. R.E. Gonzalez, Xtensa: A Configurable and Extensible Processor, *IEEE Micro*, 2000.
3. G. Estrin, Reconfigurable computer origins: the UCLA fixed-plus-variable (F+V) structure computer, *IEEE Annals of the History of Computing* 24, 4 (Oct. 2002), 3–9.
4. N. Voros, A. Rosti, and M. Hübner, Eds., Dynamic System Reconfiguration in Heterogeneous Platforms - The MORPHEUS Approach. Springer Netherlands, 2009, *Lecture Notes in Electrical Engineering*, vol. 40.
5. E. Lenormand and G. Edelin, An Industrial Perspective: Pragmatic high-end signal processing environment at Thales, *Proceedings of the 3rd International Workshop on Synthesis, Architectures, Modeling and Simulation (SAMOS)*, 2003.
6. M. Paganini, Nomadik: AMobile Multimedia Application Processor Platform, *2007 Asia and South Pacific Design Automation Conference*, 2007, pp. 749–750.

7. H. Mair, A. Wang, G. Gammie, D. Scott, P. Royannez, S. Gururajarao, M. Chau, R. Lagerquist, L. Ho, M. Basude, N. Culp, A. Sadate, D. Wilson, F. Dahan, J. Song, B. Carlson, and U. Ko, A 65-nm Mobile Multimedia Applications Processor with an Adaptive Power Management Scheme to Compensate for Variations, *Digital Signal Processing*, pp. 8–9.
8. L. Clark, E. Hoffman, J. Miller, M. Biyani, S. Strazdus, M. Morrow, K. Velarde, and M. Yarch, An embedded 32-b microprocessor core for low-power and high-performance applications, IEEE Journal of Solid-State Circuits, vol. 36, 2001, pp. 1599–1608.
9. T.A. Bartic, J. Mignolet, V. Nollet, T. Marescaux, D. Verkest, S. Vernalde, and R. Lauwereins, Topology adaptive network-on-chip design and implementation, *Structure*, 2005, pp. 467–472.
10. G. Smit, E. Schuler, J. Becker, J. Quevremont, and W. Brugger, Overview of the 4S Project, *2005 International Symposium on System-on-Chip*, 2005, pp. 70–73.
11. M. Wan, H.U. Zhang, V. George, M. Benes, and A. Abnous, Design Methodology of a Low-Energy Reconfigurable Single-Chip DSP System, *Signal Processing*, 2001, pp. 47–61.
12. J. Xie, K. Fan, Z. Mao, Q. Wang, C. Yang, W. Zhu, and S. Wang, A Reconfigurable Architecture Specific for the Butterfly Computing, *Context*, 2009, pp. 83–86.
13. C. Morra, J. Becker, M. Ayala-Rincon, R. Hartenstein, FELIX: Using Rewriting-Logic for Generating Functionally Equivalent Implementations, Proceedings of the 15th International Conference on Field Programmable Logic and Applications, 2006 (FPL '05), Tampere, Finnland.
14. S. Vassiliadis, K. Bertels, G. Kuzmanov, et al., The MOLEN Polymorphic Processor, *Proceedings IEEE Transactions on computers*, vol. 53, No. 11, pp. 1363–1375, 2004.
15. ARM Limited, Multi-layer AHB Technical Overview, v2.0, 2008, http://infocenter.arm.com/help/index.jsp?topic=/com.arm.doc.dvi0045b/index.html
16. M. Coppola, R. Locatelli, G. Maruccia, L. Pieralisi, A. Scandurra, Spidergon: a novel on-chip communication network, *Proceedings of the International Symposium on System-on-Chip*, pp. 16–18, 2004.
17. ACE, CoSy Compilers Overview of Construction and Operation, white paper, April 2003.
18. E. Moscu Panainte, K. Bertels, S. Vassiliadis, The Molen Compiler for Reconfigurable Processors, *ACM Transactions in Embedded Computing Systems (TECS)*, February 2007, Volume 6 , Issue 1.
19. J. Cambonie, S. Guérin, R. Keryell, L. Lagadec, B. Pottier, O. Sentieys, B. Weber, S. Yazdani, Compiler and system techniques for soc distributed reconfigurable accelerators, *Proceedings of the 4th International Workshop on Synthesis, Architectures, Modeling and Simulation (SAMOS)*, pp. 293–302, 2004.
20. J. Boukhobza, L. Lagadec, A. Plantec, J. C. Lelann, CDFG Platform in MORPHEUS, *AETHER - MORPHEUS Workshop AMWAS'07*, Paris, France, 2007.
21. D. Picard, L. Lagadec, Multi-Level Simulation of Heterogeneous Reconfigurable Platforms, *International Journal of Reconfigurable Computing 2009* (2009).
22. CRITICALBLUE, Boosting Software Processing Performance With Co-processor Synthesis, white paper, 2005.
23. PACT XPP Technologies, PACT Software Design System XPP-IIb (PSDS XPP-IIb) - Programming Tutorial. Version 3.2, November 2005.
24. T. Murgan, M. Petrov, M. Majer, P. Zipf, M. Glesner, U. Heinkel, J. Pleickhardt, B. Bleisteiner, Adaptive Architectures for an OTN Processor: Reducing Design Costs Through Reconfigurability and Multiprocessing, *ACM Computing Frontiers Conf.*, Ischia, Italy, April 2004.
25. T. Murgan, A.M. Obeid, A. Guntoro, P. Zipf, M. Glesner, U. Heinkel: Design and Implementation of a Multi-Core Architecture for Overhead Processing in Optical Transport Networks. *ReCoSoC2005*, Montpellier, France, June 2005.
26. M. Baron, M2000's Spherical FPGA Cores, *MicroProcessor Report*, December 2004.
27. S. Whitty, H. Sahlbach, W. Putzke-Röming, R. Ernst, Mapping of a Film Grain Removal Algorithm to a Heterogeneous Reconfigurable Architecture, *Proceedings of Design, Automation and Test in Europe (DATE)*, 2009.

28. S. Whitty, H. Sahlbach, B. Hurlburt, W. Putzke-Röming, R. Ernst, Application-specific Memory Performance of a Heterogeneous Reconfigurable Architecture, *Proceedings of Design, Automation and Test in Europe (DATE)*, 2010.
29. S. Whitty, and R. Ernst, A bandwidth optimized SDRAM controller for the MORPHEUS reconfigurable architecture, *Proceedings of the IEEE Parallel and Distributed Processing Symposium (IPDPS)*, 2008.

Chapter 5
hArtes: Holistic Approach to Reconfigurable Real-Time Embedded Systems

Georgi Kuzmanov, Vlad Mihai Sima, Koen Bertels, José Gabriel F. de Coutinho, Wayne Luk, Giacomo Marchiori, Raffaele Tripiccione, and Fabrizio Ferrandi

Abstract When targeting heterogeneous, multi-core platforms, system and application developers are not only confronted with the challenge of choosing the best hardware configuration for the application they need to map, but also the application has to be modified such that certain parts are executed on the most appropriate hardware component. The hArtes toolchain provides (semi) automatic support to the designer for this mapping effort. A hardware platform was specifically designed for the project, which consists of an ARM processor, a DSP and an FPGA. The toolchain, targeting this platform but potentially targeting any similar system, has been tested and validated on several computationally intensive applications and resulted in substantial speedups as well as drastically reduced development times. We report speedups of up to nine times compared against a pure ARM based execution, and mapping can be done in minutes. The toolchain thus allows for easy design space exploration to find the best mapping, given hardware availability and real time execution constraints.

5.1 Project Partners

1. Atmel Roma S.r.l., Italy
2. FAITAL S.p.A., Italy
3. Fraunhoffer-Institute für Graphische Datenverarbeitung (IGD), Germany
4. Imperial College of Science, Technology and Medicine, London, UK
5. INRIA, France
6. Leaff Engineering, Italy

G. Kuzmanov (✉)
Computer Engineering Lab, Faculty Electrical Engineering,
Mathematics and Computer Science, Technische Universiteit Delft, TUD,
Mekelweg 4, 2628CD, Delft, The Netherlands
e-mail: g.k.kuzmanov@tudelft.nl

J.M.P. Cardoso and M. Hübner (eds.), *Reconfigurable Computing: From FPGAs to Hardware/Software Codesign*, DOI 10.1007/978-1-4614-0061-5_5,

7. Politecnico di Bari, Italy
8. Politecnico di Milano, Italy
9. Scaleo Chip, France
10. Segula, France
11. Thales, France
12. Thomson, France
13. Technische Universiteit Delft, TUD, The Netherlands
14. Università di Ferrara, Italy
15. Università Politecnica delle Marche, Italy
16. Université d'Avignon, France

- Scientific Coordinator: Koen Bertels, Technische Universiteit Delft, TUD, The Netherlands
- Start Date: 2006-09-01
- End Date: 2010-03-01
- EU Program: 6th Framework Program (IST-5-035143), Integrated Project (IP), No. EU-IST-035143
- Global Budget: 17,34 M€
- Global Funding by EU: 10,15 M€
- Contact Author: Georgi Kuzmanov, Email: g.k.kuzmanov@tudelft.nl

5.2 Introduction

Europe has a strong leadership position in the Embedded Systems (ES) market. The ES market is considered to be highly segmented and delivers technology to end products in telecommunication, healthcare, avionics, automotive, and many other areas. A recent report [1], commissioned by the EU, stipulates that the Embedded Systems market is very competitive and rapidly evolving as new technologies are being introduced. It also grows at a higher rate than other ICT markets as suggested by [2]. Furthermore, [1] indicates that it is expected that embedded components, such as advanced multimedia systems in a car, will represent an increasingly larger part of the total value of the end product.

Since ES are requiring higher performance and richer functionality, the trend towards multi-core platforms leads to a number of challenges for the industry. The increased functionality of ES leads to increased complexity of the applications. One strategy chosen by the industry so far is to implement a large part of this functional complexity through software. Embedded Systems are thus becoming more software oriented. Not only are the applications becoming increasingly complex, the computing power needed to provide the necessary real time performance can only be given by multi-core platforms. The shift to multi-core computing is seen in all segments of the ICT market including desktop and servers for the same reason, namely the need for increased computing power. The increased complexity at both hardware and software level (leading to approaches such as hardware/software co-design) poses enormous productivity challenges for the industry, as there is insufficient experience or expertise available to adopt multi-core platforms in a smooth way. According to a study by Venture Development Capital (VDC) [3], the military and communications industries are the most likely to adopt multi-core technologies, as their applications require such performance. The study also suggests that OpenMP will be increasingly adopted to make legacy code ready for execution on such multi-core platforms. However, enhancement and adjustment of available commercial embedded software to enable use with processors having two or more cores are primary concerns (VDC [3], p. 6).

5.2.1 Legacy Code

One important limitation, which is imposed on any technology that the ES industry may adopt, relates to the existing code base. As pointed out by the HiPEAC roadmap [4], software has become the rigid component in modern computer and embedded systems. Hardware tends to evolve much faster than software and ES companies do not want to loose their intellectual property or go through expensive re-design processes with uncertain outcomes.

5.2.2 The ES Industry: Prepared?

When ES companies were asked whether the Embedded Systems industry was prepared to address the issues regarding multi-core computing, the overall rating was 2.06 on a scale of 1 (not prepared) to 5 (well prepared). (VDC [3], p. 7). This indicates that the ES industry does not consider itself to be well prepared to address the challenges sketched above. Recent announcements by ARM and Xilinx are further indications that the Embedded Industry is indeed going into the direction of (heterogeneous) multi-core systems. Contacts with large industrial corporations such as ABB and Siemens also confirm that, for instance, FPGAs are becoming increasingly important but only in combination with a general-purpose processor. Even for entry-level multi-core platforms, mapping applications on them is still considered problematic.

The hArtes technology relieves the system developer of this technical challenge by automating this process. If needed, the developer can always override decisions proposed by our tools. This technology also bridges the gap between hardware and software design. In case reconfigurable components are used, the hArtes tools automatically generate the hardware kernels and will make the necessary modifications to the original source code while maintaining the functional equivalence.

One of the challenges is not only to map the application(s) in the most efficient way on the available hardware components, but also to determine what hardware components are necessary to satisfy the overall design objectives. This hardware/software co-design effort assumes that a large design space can be explored to find the best mapping of the software on the hardware resources. Evidently, such design space can be huge, especially when reconfigurable components are available that can support runtime reconfigurability. Without some kind of automation to explore various design choices, such an endeavor becomes rapidly intractable.

The main objective of the hArtes project was to develop an integrated toolchain that provides (semi-) automatic support for the entire hardware/software co-design process [16]. Where in principle different design objectives can be formulated such as low power or restricted bandwidth, the hArtes project assumes that application acceleration is one of the main objectives. Starting point is an existing or new application written in C and the end point is an executable with (modified) code mapped on a multi-core platform consisting of a general purpose processor, a DSP chip and an FPGA. Moreover, the hArtes toolchain embeds configuration bitstreams for the reconfigurable components of the system, thus providing as a final result a complete and operational system supported both at the software and at the hardware levels. In short, the hArtes design flow will in a first phase restructure the code in such a way that any available task level parallelism can be exploited. The restructured application is then analyzed to see how and if these tasks can be mapped on particular hardware components. Out of the identified tasks and taking the cost of implementation, a number of those tasks are selected for mapping to hardware resources. Various back-end compilers will then insert the appropriate instructions necessary to start any of the available heterogeneous processing cores. A feedback loop ensures that certain choices can be evaluated and modified after which the same design steps can be repeated. The benefits of the hArtes toolchain are the following:

- the use of familiar programming paradigm;
- the platform complexity is hidden from the programmer;
- the tools can be used both for new and legacy code;
- the approach allows a design where, for example, already available IP-blocks can be easily integrated;
- substantial reduction of time-to-market and easy prototyping;
- complete system generation both at the software and at the hardware levels.

The remainder of this chapter is organized as follows. Section 5.3 presents an overview of the project and its main objectives. Section 5.4 describes the hArtes toolchain in more detail. Section 5.5 presents the hardware platform. Section 5.6 describes the applications and summarizes the experimental results. The chapter is finally concluded in Sect. 5.7.

5.3 Project Objectives

Embedded systems applications require increasingly more processing power for which single processor platforms are no longer sufficient. On the other hand, multi-core platforms not only find their way into the desktop and server markets but also in the embedded systems domain. Such platforms can contain any number of computing nodes, ranging from RISC processors to DSPs and FPGAs. Such platforms provide the necessary computing power that industry needs for the most demanding applications.

hArtes (Holistic Approach for Real Time Embedded Systems) was based on a clear vision: (i) future Embedded Applications require heterogeneous multi-processor platform comprising a multiplicity of RISC processors, DSPs and reconfigurable logic blocks; (ii) future applications are complex and require coordinated software and hardware development tools to follow an end-to-end (holistic) methodology capturing real time constraints and platform capabilities; (iii) the hArtes tools are to be validated on representative applications; (iv) the hArtes Consortium selected Audio and Video applications for Automotive, Consumer and Professional segments as representative Benchmarking Applications with high market potential.

The hArtes project addressed research and development issues of embedded systems. More specifically, the project addressed the processing part of an embedded system. It investigated hardware/software integration and built a general-purpose toolchain. The toolchain is general purpose as it accepts applications written in a multiplicity of commercially available languages and it produces semi automatically a "best-fit" mapping of such applications into a heterogeneous reconfigurable system. The project targeted real time systems and applications. The toolchain is intended to provide a fast development trajectory from application coding to the design of a reconfigurable embedded computing system. This embedded system is composed by readily available IPs, off-the-shelf components and reconfigurable logic. To save tool development costs, the project partners extensively adopt and integrate hardware and software components out of the shelf. Innovations (e.g. co-design for dynamic reconfigurable processing, high level abstraction design/mapping metrics, reconfigurable operation chaining) were added when necessary to develop a state of the art toolchain.

The goal of the hArtes is to allow Europe to capitalize on its dominance in embedded systems and gain market share as embedded systems and general purpose computing continue to converge. The core issues as addressed by the project can be summarized by its general and specific objectives.

5.3.1 General Objectives

hArtes was a 3 year integrated project intended to develop a methodology and a toolchain supporting the entire embedded system design flow, from high level algorithm definition to efficient modular implementation of a reconfigurable heterogeneous system (RHS). The hardware real-time reconfigurable platform considered has been composed by IPs and off-the-shelf components. Applications, written in common/readily available languages and with graphic tools, are mapped by the toolchain in order to "best fit" a given reconfigurable platform. The toolchain integrates existing components and supports development of new state-of-the-art algorithms, software and reconfigurable hardware design techniques.

A prototype reconfigurable hardware platform has been developed within the project. Its architecture is a judicious balance between hardware performance and easy interface to the toolchain. The platform allows extensive test of the complete project chain starting from algorithm and going all the way down to a working, stable and efficient application.

5.3.2 hArtes Specific Objectives

The following specific objectives have been identified:

Objective 1 Understand the requirements of the methods and algorithms that support multiple types of high-level algorithm descriptions, which can be transformed into a single description, possibly with directives about design constraints.

Objective 2 Study the automatic and semi-automatic translation of a high-level generic algorithm description into a representation that can be physically mapped onto a particular real-time reconfigurable heterogeneous system.

Objective 3 Explore innovative algorithms, software, and reconfigurable hardware that can be used to optimize the implementation with respect to user-defined metrics, to enhance performance or to reduce platform cost.

Objective 4 Investigate partitioning algorithms that can exploit dynamically reconfigurable hardware: such hardware can change its functionality at run time, to support function upgrade or to adapt to operating conditions.

Objective 5 Evaluate the proposed approach using advanced audio applications, such as immersive audio systems and in-car audio systems, which can illustrate and quantify its capabilities?

hArtes provides a framework for rapid development and optimization of reconfigurable real-time embedded systems. Some key features of the proposed toolchain within this framework are described below:

- Develop/integrate methods and algorithms for unification of multiple types of high-level algorithmic entries into a single algorithm description and a single design solution.
- Semi-Automatic translation of a high-level generic algorithm description into a representation that can be physically mapped onto a particular reconfigurable real time heterogeneous system.
- Innovative algorithms, software, and reconfigurable hardware are added to best fit the platform in terms of some metrics (e.g. performance and platform cost reductions).
- In contrast to existing hardware/software co-design tools that perform partitions for fixed function hardware, partitioning algorithms are added to exploit dynamic reconfigurable hardware changes of functionality.

- The intended project outcome is to provide a prototype tool-chain and a reconfigurable hybrid platform to allow the building of innovative designs. To provide proof of concept, the domain of audio and video applications is used to show the capabilities of the toolchain.

5.4 The hArtes Toolchain

The hArtes approach aims to simplify the use of heterogeneous multi-core platforms by providing an integrated toolchain that transforms existing and new applications to allow execution on heterogeneous computing platforms. The developed framework is, in our opinion, a lasting result of the research and development efforts. All other elements of the project collaborate and make it possible the achievement of this goal, by ensuring that the tool development effort takes into account inputs coming both from experienced state-of-the-art applications and from the constraints that state-of-the-art engineering places on the development of reconfigurable hardware. The main architectural paradigm assumed by hArtes and more specifically by the hArtes toolchain is the Molen Polymorphic Instruction Set Computer (PISC) [11–13].

5.4.1 The Molen Architectural Paradigm

The Molen reconfigurable microcoded ($\rho\mu$-coded) processor was originally proposed in [11]. More details on the Molen microarchitecture have been published in [12]. The general proposal of the Molen polymorphic processor is: *by displaying means to maintain the reconfiguration at the architectural level, to achieve high flexibility in tuning the system for the specific application.*

The operation of Molen is based on the co-processor architectural paradigm. A GPP (core processor) controls the execution and the (re)configuration of a reconfigurable co-processor, tuning the latter for specific algorithms. According the original proposal, the reconfiguration and the execution of the code on the reconfigurable hardware is done in firmware via *reconfigurable microcode* ($\rho\mu$-code). The $\rho\mu$ -code is an extension of the classical microcode, which includes reconfiguration and execution. The microcode engine is extended with mechanisms that allow permanent and pageable reconfiguration and execution microcode to coexist. *Within hArtes, the Molen concept has been extended towards systems comprising any combination of a general purpose processor, a reconfigurable co-processor, and a DSP.*

5.4.1.1 The Molen Machine Organization

The main computing components in the Molen machine organization, depicted in Fig. 5.1, are the 'Core Processor', which is a general-purpose processor (GPP), the 'Reconfigurable Processor' (RP), and the DSP. The 'Arbiter' issues instructions to

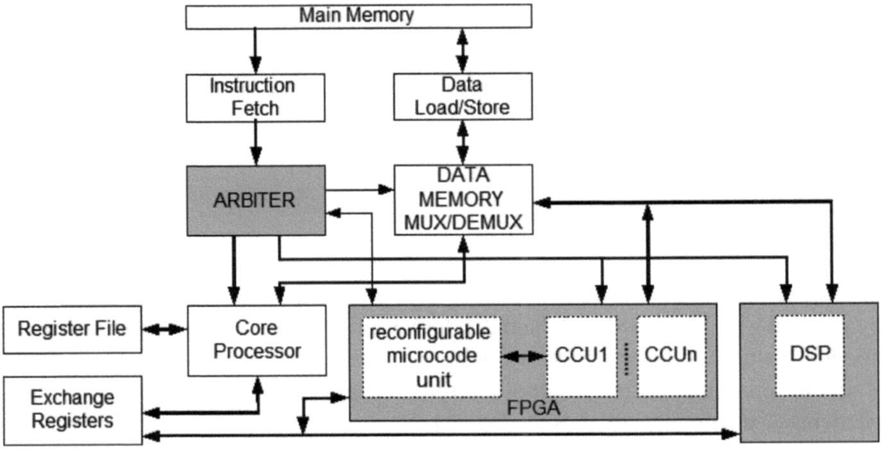

Fig. 5.1 The Molen machine organization

either of the processors by partial decoding of the instructions received from the instruction fetch unit. Data are fetched by the 'Data Load/Store' unit. The 'Memory MUX' unit allows data accesses from either the core processor or the core processors in a shared memory paradigm. The reconfigurable processor is further subdivided into the $\rho\mu$-code unit and multiple *custom computing units* (CCU).

A CCU is implemented in reconfigurable hardware, for instance FPGA, and may contain some local memory. Essentially, the CCUs are intended to support additional and future functionalities that are not implemented in the core processor. Pieces of application code can be implemented on the CCU in order to speed up the execution of the overall application code. A clear distinction exists between code that is executed on the co-processors (the RP or DSP targeted code) and code that is executed on the core processor. Data must be transferred across the code boundaries in order for the overall application code to be meaningful. Such data include predefined parameters or results, or pointers to such parameters or results. The parameter and result passing is performed through a mechanism utilizing so-called exchange registers (XREGs) depicted in Fig. 5.1. Larger pieces of data are communicated via the shared memory mechanism.

The support of operations by the reconfigurable processor can be initially divided into two distinct phases: set and execute. In the set phase, the CCU is configured to perform the supported operations. Subsequently, in the execute phase the actual execution of the operations is performed. This decoupling allows the set phase to be scheduled well ahead of the execute phase and thereby hiding the reconfiguration latency. Furthermore, no specific instructions are associated with specific operations to configure and execute on the CCU as this will greatly reduce the opcode space. Instead, pointers to $\rho\mu$-code, which emulates both the configuration and the execution of programs, are used. Consequently, two types of $\rho\mu$-code are distinguished: reconfiguration microcode that controls the configuration of the CCU; execution microcode that controls the execution of the implementation configured on the CCU. For the FPGA technology, the reconfiguration microcode can be interpreted as bitstreams that are loaded to reconfigure the chip for particular functionalities. The *set-execute* paradigm is also applicable for the DSP co-processor. During the set phase, the DSP is initialized, before an actual subroutine is executed on the DSP during the execute phase. The set-execute mechanism is supported by explicit instructions or program code primitives, included in a new programming paradigm described below.

Programming Paradigm

In order to target the Molen processor, a sequential consistency programming paradigm was proposed in [14]. The paradigm allows for parallel and concurrent hardware execution and it is intended for single program execution. It requires only a one-time architectural extension of a few instructions to provide a large user reconfigurable operation space. In [12], three variants of the instruction set extension, referred to as Polymorphic Instruction Set Architecture (πISA), were identified: *minimal*, *preferred*, and *complete*.

Minimal πISA: This is essentially the smallest set of Molen instructions needed to provide a working scenario. The four basic instructions needed are: *set*, *execute*, *movtx*, and *movfx*. By implementing the first two instructions (*set*/*execute*), any suitable CCU implementation can be loaded and executed in the reconfigurable processor. The *movtx* and *movfx* instructions are needed to provide the input/output interface between the RP targeted code and the remaining application code.

A *preferred πISA* contains a *partial* set (*p-set*) instruction and a *complete* set (*c-set*) instruction to address long configuration times and partial configuration capabilities of some modern FPGA chips. In addition, two *prefetch* instructions (*set prefetch* and *execute prefetch*) provide a way to diminish the microcode loading times by scheduling them well ahead of the moment that the microcode is needed.

Finally, the complete *πISA* is further extended with the *break* instruction. The break instruction provides a mechanism to synchronize the parallel execution of instructions by halting the execution of instructions following it. The sequence of instructions performed in parallel is initiated by a *set* or an *execute* instruction. The end of the parallel execution is marked by the *break* instruction.

In hArtes, the following πISA instructions were considered: *set*, *execute*, *movtx*, *movfx*, *break*. However, in order to address multiple processors architectures and to abstract any unnecessary details of the hardware platform from the programmer, the πISA instructions were implemented as macroinstructions (programming primitives) invoked as system calls. Furthermore, at the C-code level, a set of pragma annotations were defined, which were subsequently translated to πISA macroinstructions by the hArtes toolchain. More details on these different pragma annotations are provided in the section to follow.

5.4.2 hArtes Toolchain Description

The holistic toolchain for real time embedded systems developed in hArtes is composed of three ToolBoxes (illustrated in Fig. 5.2): Algorithm Exploration and Translation ToolBox, Design Space Exploration ToolBox, and System Synthesis ToolBox. The development of the ToolBoxes is the focal point of the project. Entry point for the toolchain is C-code that can be generated by (public-domain or proprietary) tools such as matlab and scilab, manually coded or automatically-generated code. Information regarding the target platform is available in the form of XML files. If the developer has prior knowledge with respect to the mapping of the application, simple code annotations can be used to express that knowledge. The Task Transformation Engine then analyses the entire application and decides on a mapping. The mapping could specify some parts staying on the general-purpose processor, while other parts would be executed on a DSP or on an FPGA. These decisions are guided by additional and automatically inserted code annotations. The set of compiler tools then take the annotated code and generate the appropriate binaries that are linked together. The generated executable can be immediately executed on the hardware platform. If needed, this process can be repeated until the design objectives are achieved.

The goal of the hArtes toolchain is to allow developers write their applications at a software level and exploit the latest advances of heterogeneous platforms without having expert low-level hardware knowledge. The toolchain achieves this through two complementary methods: (1) a (semi)-automatic approach where architectural decisions are introduced at each level of the toolchain by computerized means but can be overriden by developers, and (2) abstracting low-level details through source annotations. Source annotations are used primarily to describe parallelism and mapping either by developers or by the tools, and are introduced in the application through C pragmas. In the following, we describe some of the pragma annotations used:

#pragma omp: OpenMP pragmas are used to specify where parts of the application can be executed in parallel. Each section can thus be executed independently of and in parallel with the other sections in the defined OMP parallel section.

Example

```
#pragma omp parallel sections
{
#pragma omp section
  {
    fft();
  }
#pragma omp section
  {
    dct();
  }
}
```

#pragma map: this pragma can be used on top of a function declaration or a specific function call to indicate that a particular hardware component (such as a DSP or an

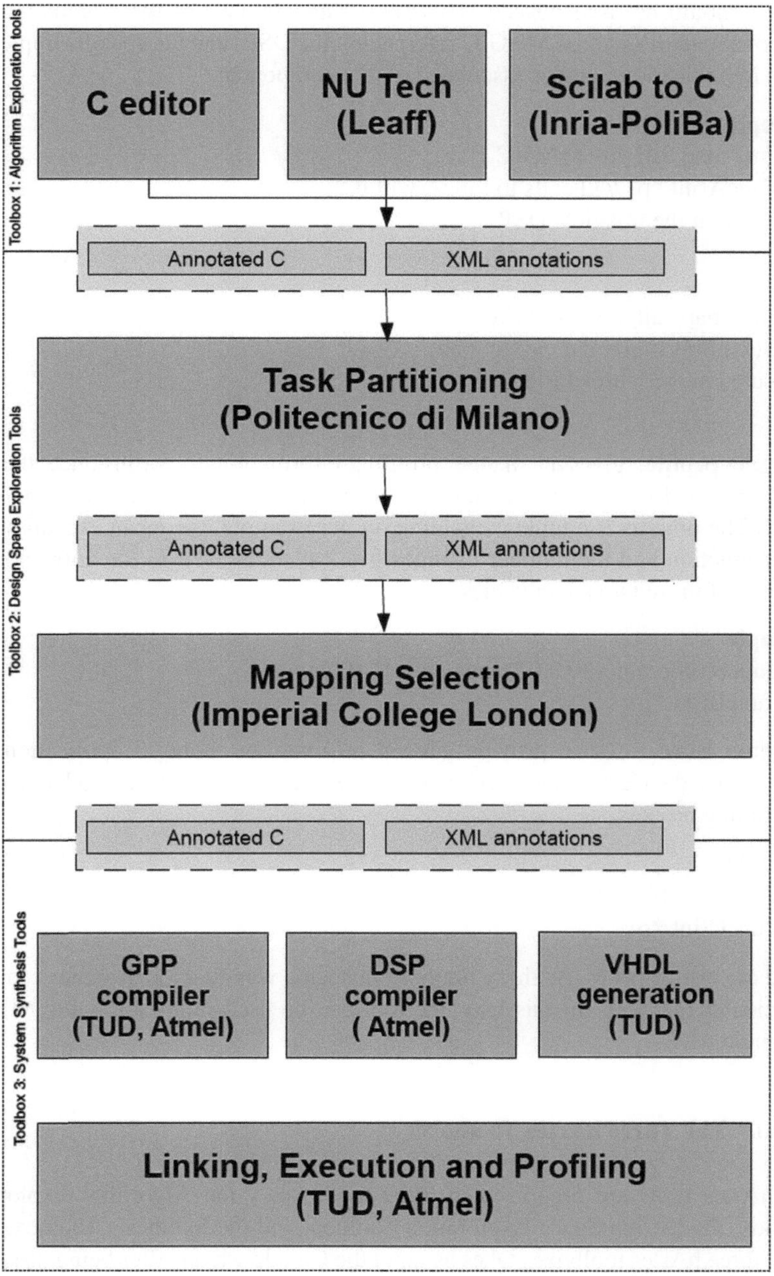

Fig. 5.2 The hArtes toolchain

FPGA) should be used instead. This source annotation has two parameters: the component name (such as MAGIC referencing the DSP) and the specific implementation identification number associated to that component.

Example

```
#pragma map call_hw MAGIC 1
void funcA(int *p);//all calls to funcA will be
//executed in the mAgicV DSP
...
{
#pragma map call_hw VIRTEX4 2
funcB(x);//this particular call will be
//executed in the Virtex4 FPGA
}
```

#pragma profile: gives the mapper profiling information on, for instance, the CPU time consumed by a particular function or on the number of times the function is called. The pragma is a tuple containing, in our example, the mean execution time of the function and the number of times it is called, as well as the corresponding variance of those two mean values.

Example

```
#pragma profile time(30,0.02) num_calls(5,0.8)
void funcB(int *p);
```

#pragma issue: used to provide general information about an application. The only type used so far is 'blackbox' which forces the routines to be excluded from the partitioning process.

Example

```
#pragma issue blackbox
void funcC(int *p);
```

There are also ways to use these pragmas in such a way that the designer can force a particular mapping. In this way, the tool can be used in an automatic or semi-automatic way.

5.4.2.1 The Three hArtes Toolboxes

The hArtes tools are organized in three toolboxes - the Algorithm Exploration Toolbox, the Design Space Exploration Toolbox, and the Synthesis Toolbox. Each of the three hArtes toolboxes takes as input the C-code which, after being processed by the tools in the box, is outputted again to the next toolbox. A high-level overview of the entire toolchain is provided in Fig. 5.2.

At the top-level of the toolchain, we find the *Algorithm Exploration Toolbox (AET)*, which contains tools that allow system designers to develop and refine their algorithms at a higher-level of abstraction. In the hArtes case, the Leaff NUTech [5] tool and Scilab [6] are used. NU-Tech is a graphical software development tool which enables developers to graphically design and simulate their application by connecting logic blocks together, after which C-code is generated. The second tool is Scilab, which is an open source equivalent of Matlab. SciLab takes Matlab code as input and then generates a C version from it. The use of the Algorithm Exploration Toolbox is optional and developers can write C code directly as an entry point to the toolchain.

The next level of the toolchain is the *Design Space Exploration* (*DSE*) Toolbox where the application undergoes various transformations. The DSE consists of two processes: partitioning and mapping. The partitioning tool is named Zebu [refs may help] and it is developed inside PandA, a hardware/software co-design framework developed at Politecnico di Milano. It aims at identifying the tasks at the proper granularity, in which the application can be decomposed to improve its performance. Zebu does so by taking into account information about the potential speed-up of each solution. Zebu starts from the C code of the application and takes into account information extracted from the XML file containing the platform specification. Through a feedback loop, information with respect to possible acceleration of certain tasks is also taken into account by Zebu. The tool behaves as a compiler in that it is composed of:

- the frontend, which creates the intermediate representation of the input code using a slightly modified version of the GNU/GCC compiler.
- the middle-end, refers to the *core* of Zebu, where the tool creates an efficient partitioning [7] exploiting internal performance estimation techniques [8]. Then, it performs transformations on the resulting task graph to determine the proper granularity of the tasks in order to take into account the additional overhead required to manage them.
- the backend, which generates the C code annotated with OpenMP and initial mapping directives.

Subsequently, the mapping tool, named hArmonic, [9], developed by Imperial College London, takes as input task partitions generated automatically by Zebu or manual partitions derived by developers, and maps them on specialized processors (DSP or FPGA) in order to accelerate the application. In short, hArmonic determines how different parts of the C application would be mapped to the available resources on the reconfigurable platform. The mapping process [10] provides two novel features. First, near-optimal solutions can be rapidly generated by combining an inference rule engine (which generates solutions by construction from any point of design space) with a heuristic search algorithm. Second, application developers can influence the task mapping process by providing directives to guide the search for a mapping solution. The mapping tool accepts as input an arbitrary number of C source-files and the XML platform specification. The mapping process then automatically verifies whether established C guidelines are satisfied and pinpoints potential problems in the source code that prevent a good or a feasible mapping. Following this stage, the mapping process selects, based on a set of filtering rules, the processing elements that can support each part of the application in order to minimize the search space for the mapping process and to comply with the hArtes platform requirements and limitations.

The final stage of the toolchain is the *Synthesis Toolbox*, developed by Delft University of Technology and Atmel, which contains all the necessary backend C

```
int main(int argc,char**argv) {
 ...
 if (DSP_SUCCEEDED (status)) {
  status = PROC_setup (NULL) ;
 }
 if (DSP_SUCCEEDED (status)) {)
  status = PROC_attach (processorId, NULL;
   if (DSP_FAILED (status)) return -1;
 } else {
   return -1;
 }
 if (DSP_SUCCEEDED (status)) {
  args [0] = strNumIterations ;
  status = PROC_load (processorId, dspExecutable_myfft,
NUM_ARGS, args) ;
  if (DSP_FAILED (status)) return -1;
 }
 if (DSP_SUCCEEDED (status)) {
  status = PROC_start (processorId) ;
  if (DSP_FAILED (status)) return -1;
 }
}
```

```
int main (int argc, char *argv[]) {
 ...
 ret = mAgicV_load_PM("myfft.bin",_m_fd_extm);
 if(ret!=0) return ret;
 ret = mAgicV_load_DM("myfft_datamem.bin");
 if(ret!=0) return ret;
 ret = mAgicV_load_XM("myfft_extmem.bin",(0x365890));
 if(ret!=0) return ret;
 ret = mAgicV_init_PMU();
 if(ret!=0) return ret;
 magicV_start();
 ...
 magicV_wait();
 ...
}
```

```
void my_fft(...) {
 ...
}

int main(int argc,char**argv) {
 ...
 #pragma call_hw MAGIC 1
 my_fft();
 ...
}
```

Fig. 5.3 Mapping code on different platforms

compilers for each type of processing element (such as GPP, DSP, FPGA) in the system. Once the mapping tool (hArmonic) generates a mapping solution [10], it generates C source code for each compiler in the Synthesis toolbox. Each source is compiled separately, and then linked together to form one single binary. The binary, which contains heterogeneous object code for different processing elements, is supported by the hArtes runtime interface, and provides the following services: platform and processing element initialization, remote procedure call (Molen programming model) and remote thread control, memory allocation, profiling and debugging support. The GPP compiler, which is based on gcc 4.3, has the role of generating the Molen instructions for the annotated functions. It also does a basic scheduling of EXECUTE and BREAK Molen calls, needed to parallelize the procedure calls found in OpenMP sections. In the context of the hArtes project, the Molen instructions are implemented as runtime library function calls to ease even more the addition of new architectures. The FPGA compiler is the Delft Workbench Automated Reconfigurable VHDL Generator (DWARV) [15] which is a C-to-VHDL generation toolset. This compiler exploits the available parallelism of algorithms and generates designs suitable for hardware/software co-execution using the Molen programming paradigm. It is composed of two modules: the data flow graph builder and the VHDL generator. The data flow graph built is based on the SUIF compiler infrastructure and has as input a standard C source from which it generates a hierarchical data flow graph. The second module, takes the data flow graph and using an 'as soon as possible' scheduling, generates synthesizable VHDL. When needed, DWARV will use already available IP blocks. One good example is the floating point library, which is based on vendor specific designs such as Xilinx IP, and used when compiling C code that contains floating point operations. The hArtes Linker produces a single executable, linking all the contributions coming from the different PE compilation chains. These contributions include executable binaries as well as configuration bitstreams. It is based on GNU Linker (LD) targeted for ARM Linux and a customized hArtes Linker script. The bitstreams and the DSP executable files are transformed accordingly so that they can be integrated into the ARM ELF file.

In order to better illustrate the power of our technology, we include some code samples that need to be written when mapping parts of an application on, in this case, a DSP accelerator. In the example, we assume that an FFT-routine will be executed on the accelerator. Fig. 5.3 Illustrates 3 different kinds of code snippets. The code examples on the left are respectively targeting either the Atmel Diopsis DSP platform or the TI Omap L138. The smaller code sample on the right provides exactly the same functionality as either examples on the left but then using the hArtes approach. Using the latter, one can abstract away from the lower level platform details as this is taken care of by the back-end tools of the hArtes toolchain.

5.5 Hardware Platform

In order to validate and evaluate the hArtes design approach, a dedicated hArtes Hardware Platform (hHP) was developed. The hHP system is composed of a certain number (two in the present design) of independent blocks, each of which includes a general-purpose RISC processor, a DSP processor and an application-specific configurable block, that also incorporates a RISC processor. We call this basic building block (RISC processors + configurable element) a Basic Configurable Element (BCE). The top-level organization of the hHP with two BCEs is shown in Fig. 5.4. The developed system has two independent heterogeneous and configurable processors that communicate among each other and with an audio I/O subsystem – the latter supports several ADAT channels. All BCEs have the same organization, they are able to run any selected thread of a large application, as partitioned by the hArtes toolchain. The BCEs share data and synchronization information via a direct data link with high bandwidth of 200 Mbyte/s. However, private (distributed among all its components) memory is also available on each BCE, for faster access to local data. This way, complex applications sharing global data segments can run on the hHP with little porting effort. However, tuning for performance is possible, as data segments that do not need to be shared are moved to local memory banks within the appropriate BCE. Dedicated hardware for massive data streaming is also available on the system, as shown in Fig. 5.4 (the "Audio I/O subsystem" block). Up to 64 input audio channels and up to 64 output audio channels are available on 8 + 8 ADAT interfaces, a standard that defines a digital link designed for high-quality audio transfers.

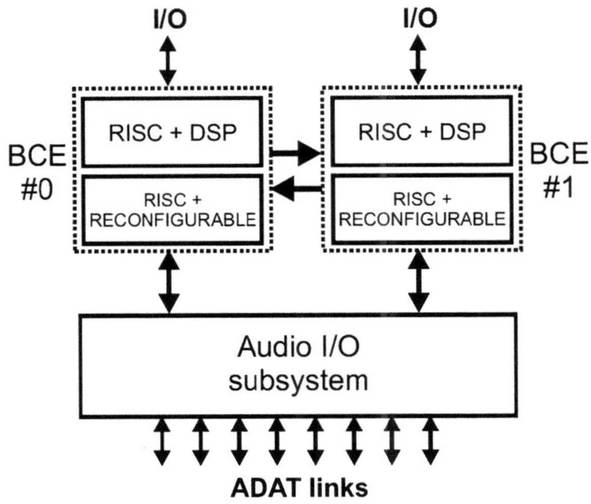

Fig. 5.4 Top level organization of the hArtes hardware platform

Figure 5.5 provides an overview of the BCE, which is the basic building block of the platform, supporting several processing architectures. One or more BCEs work in parallel to support an hArtes application. There are two main blocks, namely the RISC/DSP processor (the D940HF component produced by Atmel) and the reconfigurable processor based on a high-end FPGA (Xilinx Virtex4-FX100), that carries also two PowerPC 405 processing cores. The block to the left, the "DCM board", and the block on the right, the "FPGA module", are logically part of the same BCE; they are implemented as two separated boards, both placed on a common main board. This configuration was chosen to cut development time, since the DCM board was already available; no significant limitation comes from this solution. This choice adds flexibility to the platform itself, as modules may be replaced by a different subsystem with a different set of functions (e.g., a video processor for multimedia applications).

Each of the two elements (DCM module and FPGA board) has a private and independent memory bank ("Mem 1" and "Mem 3" in Fig. 5.5) boosting overall memory bandwidth, needed to sustain a large computational throughput. Also, two independent FLASH memory banks are present, to keep user data and programs. The FPGA module has also a shareable memory bank ("Mem2" in Fig. 5.5). This bank can be shared at several levels, i.e., among all the computing elements of the BCE itself (the D940HF component, the reconfigurable fabric of the FPGA and the PowerPC CPU inside the FPGA), and by the processors of any other BCE available on the system. This feature is supported by specific functional blocks configured inside the FPGA. Memory sharing or any other program-controlled pattern of BCE to BCE communication are handled by the large bandwidth links connecting the BCEs, making data transfers of short data packets very efficient. Finally, each BCE

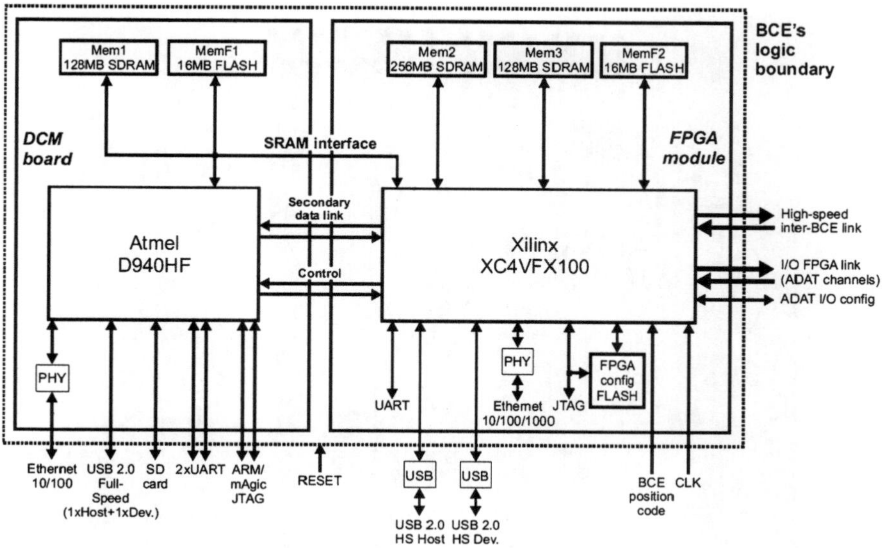

Fig. 5.5 Block diagram of the Basic Configurable Element (BCE) of the hHP

has several standard input/output interfaces (Ethernet, USB, UART), located both on the DCM board and on the FPGA module.

The D940HF component on the DCM board contains two cores: one ARM processor and one mAgic DSP core – a floating-point core, optimized for DSP. The connection of both of these computational elements to the rest of the BCE goes through a common static memory interface. A memory access arbiter within the FPGA controls memory requests coming from the DSP core, from the PowerPC cores, and from the reconfigurable fabric. A software lock mechanism has to be implemented to guarantee memory coherency, since neither the ARM/DSP cores, nor the PowerPC core implement native cache coherency facilities.

The hHP main board contains the basic infrastructure to support the two BCEs, their interconnections and the I/O connectors. The key components connected to the main board are the two BCEs shaped by two FPGA modules and two DCM boards. All the appropriate power conditioning and distribution elements are located on the main board, as well. The system clock is distributed to all daughter boards with low skew, except for the DCMs that have a private clock. The main board has SD card connectors for the DCMs, Fast-Ethernet, USB Host/Device and serial ports. For audio applications, eight ADAT optical decoders and eight optical encoders are available. A medium-sized FPGA (Virtex4-LX40) converts the audio samples and connects all ADAT decoders with all BCEs. This FPGA contains also the logic needed to control the configuration of the ADAT encoders and decoders.

The entire hHP developed in the project is depicted on Fig. 5.6. The two BCEs use two daughter boards each, one for the D940HF processor and one for the FPGA

Fig. 5.6 The hArtes hardware platform

based infrastructure. These daughter-boards are located in the center of the mother board from Fig. 5.6. The ADAT interfaces and several standard I/O connectors are clearly visible at the top and at the bottom of the picture, respectively.

5.5.1 Special Features for Audio Processing

One specific feature of the hHP is its strong support for high-level data streaming. This support includes dedicated I/O audio ports, which completely offload the CPUs on board from I/O tasks, so they can handle heavy computational workloads. At the basic hardware level, the platform has 8 input and 8 output ADAT interfaces, a widely-used standard that uses optical-fibers and packs eight audio streams on each link. As a result, the system supports up to 64 stereo audio channels. Any combination of ADAT input channels can be bidirectionally interconnected with any of the two BCEs via a dedicated FPGA, using software accessible configuration registers. Audio data, after reaching the target BCE, is processed by a dedicated engine, implemented inside the FPGA. Each channel has two independent memory buffers and data from any incoming stream is automatically copied to the corresponding buffer – a "ping-pong" approach. When a (ping or pong) buffer is ready, the engine raises an interrupt. Finally, for each ADAT channel integer to floating point and vice-versa conversions are performed on-the-fly (note that the ADAT standard uses 24-bit signed integers).

5.5.2 The Basic Infrastructure

The hHP consists of a static, non-configurable partition and a dynamically configurable portion. The latter can be augmented at run-time by partially configuring a subset of the FPGA on-the-fly to perform application-specific functions. Such functions are identified, prepared, compiled, and loaded into the FPGA by the hArtes tool-chain, ready for execution.

The static FPGA configuration is loaded onto the device at power-up. A non-exhaustive list of the statically implemented features includes the following:

- shared access to the common memory area (*M*em2) by the D940 processor and any processor, implemented in the FPGA;
- massive data transfers between the ADAT channels and the memory buffers;
- on-the-fly data re-formatting for the audio streams;
- message passing interface between the two BCEs;
- control for partial on-the-fly reconfiguration of the FPGA.

A dynamic FPGA configuration is reconfigured "on-the-fly" employing specific features of the Xilinx FPGAs utilized. The Virtex 4FX FPGA is divided in two logically separated domains. One domain includes the PowerPC cores and all static functional blocks needed for basic system support described above. The second, dynamic, domain is reserved for reconfigurable functions, identified and implemented in hardware by the hArtes toolchain. The dynamically reconfigurable FPGA domain changes its logic structure as required by the running applications under software control using the Molen programming primitives.

5.5.3 System Software Support of the hHP

At present, the hHP runs a Linux Operating System, booted on the D940 processors; if both BCEs are in operation, each of them runs its own independent instance of the OS. One of the two BCEs is configured as the master of the whole system, which starts and manages slave processes on either of the BCEs. A limited set of key software instances has been developed to support the applications running on the hardware, such as: mapping of the BCE shared memory into the application addressing space; specific drivers to support the ADAT audio functionalities; specific drivers to initialize and perform the dynamic reconfiguration of CCU modules; programming API to support inter-BCE data transfers; automatic build of a single ELF file that contains all the executable elements making up a complete application, that are selectively loaded onto the ARM processor, the DSP core and the FPGA reconfigurable partition.

Overall, the hArtes hardware platform was built for demonstration and proof-of-concept purposes. It combined the Atmel Diopsis ARM-enabled platform with a Virtex 4FX FPGA. A Molen machine organization, having the ARM as the master processor with the DSP and the FPGA as co-processors, was implemented. The hArtes applications, described in the following section, were mapped on the hHP and were successfully executed meeting all real time constraints.

5.6 Applications and Experimental Results

Regarding applications, the primary focus of the hArtes project was on advanced audio applications, such as immersive audio systems and in-car audio systems, which can illustrate and quantify its capabilities. In this section, we shall briefly introduce the immersive audio and the in-car audio applications, considered in the hArtes project, followed by some results obtained from experiments on the hArtes hardware platform.

5.6.1 Immersive Audio

The motivation for the immersive audio subtask of the hArtes application development is to demonstrate the viability to build advanced and highly parallelized real-time immersive audio reproduction systems on the hArtes reference hardware using the hArtes toolchain. The immersive audio approach envisioned in the hArtes project emphasises on unobtrusiveness of users (i.e., no headphones or other devices necessary with the user) and allows high audio simulation accuracy. Both features in combination are possible by choosing the novel approach of wave-field synthesis (WFS) in combination with a beamforming (BF) approach for 3D audio recording. In the following, we will give a short introduction on the audio processing that is used for hArtes.

Usage scenario: There are two sides in an audiovisual transmission - acquisition and rendering. On the acquisition side, a microphone array records the audio of speakers and other sound sources in the room selectively. At the rendering side- a loudspeaker array reproduces the audio by projecting the recorded acoustic sources in spatial directions that re-enact the recording directions. In order to identify the recording directions, a visual approach has been taken: by tracking faces with a video camera. Speakers are recognized and followed, sending updates to the positions directly to the beamformer that runs the microphone array. The latter sends positional information to a wave-field synthesis system that runs the loudspeaker array.

Algorithm description: Beamforming and wave-field synthesis strongly rely on the wave character of the sounds to be reproduced in space. While in the beamformer waves are superimposed in order to create directional gain patterns, i.e., to spatially filter directional waves, in WFS, the superposition is found in every point of the listening space where an acoustic wave is synthesized.

BF is the directional filtering of spatially separated signals using sensor arrays with adaptive or fixed array processing algorithms. The output signal of such a sensor array can generally be described as a sum of the sensor signals filtered by some sensor-specific transfer function.

5.6.2 Experimental Results

To validate the Molen computational paradigm and to assess the current available toolset, different applications are mapped on the hArtes hardware platform. In this section, we shall not elaborate on all the stages of the design process, nor shall we describe all applications and experiments carried out. We shall rather report the results obtained for the key immersive audio applications. All speedups reported in this section are relative to the execution of the application on the hArtes board where the ARM processor provides the base line. We then generate mappings on either the DSP or the FPGA and report the speedup results when executing the application on the ARM with the DSP or the ARM with the FPGA. It is important to stress that the numbers reported are derived from real execution – and not simulation – of the entire application. Finally, even though the entire approach allows executing multiple kernels in parallel based on OpenMP annotations, two reasons prevented us from real parallel execution. The first is the limited area available for mapping large kernels. The second reason is related to the way multiple functions are combined into one hardware kernel. When multiple functions are mapped on the FPGA, the DWARV hardware compiler combines the Data-Flow Graphs (DFGs) of all those functions in such a way as to provide fastest execution but also to reuse available hardware blocks where possible.

It has been already noted that one of the applications used to validate the toolchain is a *wave-field synthesis* (*WFS*) application, which is a computation intensive audio application that uses multiple audio inputs and outputs to compute the spatial properties of sound sources in an immersive environment, for example a room or a car. A fully automatic use of the toolchain resulted in a mapping of 15 functions to the DSP resulting in an overall application speedup of 9.24 compared to an ARM based execution.

Another application tested is the *noise filter* application which contains a downsampler/upsampler, fast fourier transforms (FFT), and several other audio processing routines. For this application, the toolchain automatically mapped 16 functions to the DSP and 10 functions to the FPGA. The cost estimator, which predicts how an arbitrary function performs in each processing element, identified the *FFT* as a candidate for acceleration, a result which is confirmed by the dynamic profiler. Table 5.1 suggests that the kernel-only execution of the *FFT* (without taking into account communication costs between processing elements) on the DSP, results in a speedup of 4 against 7 for the FPGA. However, the hArtes board has bandwidth restrictions between the FPGA and external memory resulting in a 2.7 overall speedup of the entire application as listed in Table 5.2. When manual code modifications affecting memory management are applied, the performance of the FPGA based mapping reaches a 3.1 speedup. On the other hand, when mapped on the DSP, an overall speedup of 4.5 is obtained.

Table 5.1 Kernel-only speedup of the noise reduction FFT kernel (in μs)

	ARM	DSP	FPGA (Virtex-4)
FFT-kernel (μs)	13,055	3,261	1,843
Speedup	1	4	7

Table 5.2 Overall speedup for the noise reduction application (in μs)

	Original application code			Manually tuned application code
Execution processor	ARM	DSP	FPGA	FPGA (Virtex-4)
Noise filter (μs)	27,000	6,000	10,000	8,600
Speedup	1	4.5	2.7	3.1

The real-time requirement for this application, where 256 samples/cycle must be processed in less than 16 ms, cannot be met by the ARM processor alone. By using the hArtes toolchain with a complete automated approach and without any additional code modification, real time performance can be achieved with a substantial reduction in overall development time.

An equally important objective of the hArtes project was to asses to what extent a toolchain such as the one described here, can reduce the time-to-market and simplify the entire design process. In order to find this out, the validation was evidently done by the application partners who made a qualitative and partly quantitative comparison of the design effort with and without the toolchain. An example with the two immersive audio applications considered is illustrated in Fig. 5.7. Extrapolating the best experiences in this exercise, it is clear that a dramatic reduction of the mapping effort is noted. The entire toolchain analysis takes only a couple of minutes and the most time consuming parts are usually proprietary synthesis tools from, e.g., the FPGA vendors.

In addition, an international design contest was organised in order to get feedback from non-hArtes partners. The design contest consisted of mapping an application of their choice to the hArtes board (the DSP in this case) by using the hArtes tools. In total 11 teams from all over the world enrolled to the contest and provided their solution. In the context of the competition, the teams were asked to describe their experiences and assess how much the tools facilitated the entire development effort. The chosen application domains were very diverse ranging from solar panels management, a mobile web identification and location system to a satellite communication system. From this validation exercise, extended with the experiences of the application provider partners in the project, the following benefits of the hArtes toolchain became apparent:

- Familiar programming paradigm: developers can program in the same way as before and a simple annotation can be placed manually or by the tools to indicate where and how to accelerate the program execution. To the developer, the use of an accelerator is as simple as calling a function with this difference that now the accelerator is started. Especially in the case of FPGA-based acceleration, the

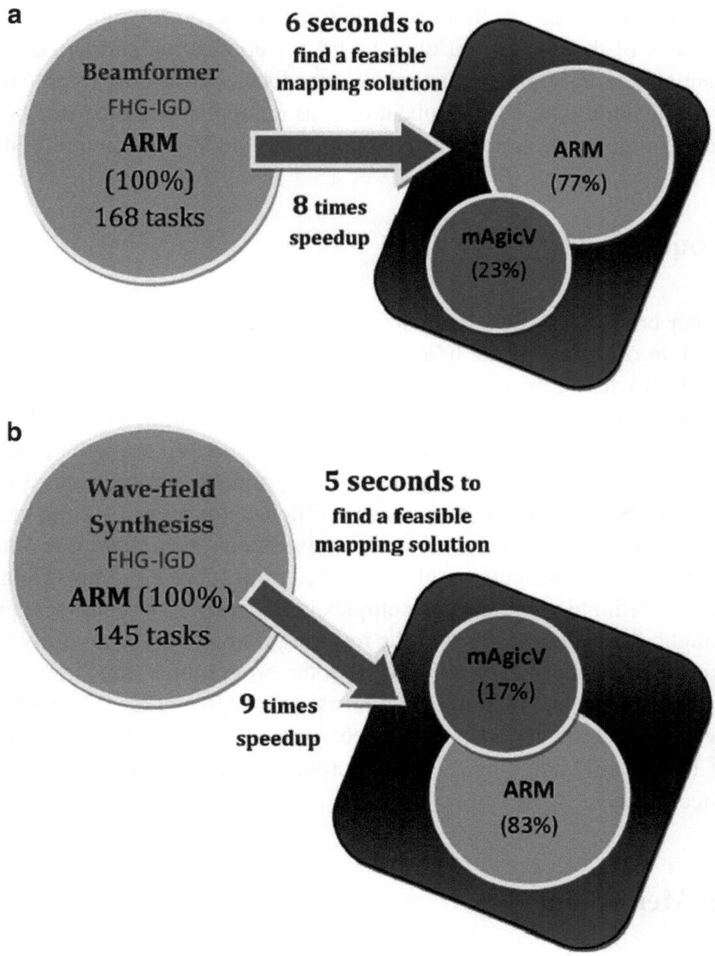

Fig. 5.7 Main results of the hArmonic tool in combination with the synthesis toolbox and the hArtes hardware platform: (**a**) Beamformer and (**b**) Wave-field synthesis. Both applications are developed by FHG-IGD

automatic generation of VHDL code proved to be of extreme importance as it is a known development bottleneck for this technology;

- Hiding platform complexity: the developer does not need to understand the complexity of the platform and does not even need to know what the exact hardware components are;
- Flexible approach for both new and legacy code: whether one wants to port existing code to such platforms or build it from scratch, the toolchain is indifferent with respect to that;
- Composable approach allowing to integrate any available IP-blocks: if the developer disposes of previously built or even purchased IP-blocks, the hArtes approach allows to easily integrate them in the application;

- Substantial reduction of time-to-market and easy prototyping: not only the participants of the contest but also the hArtes application partners experienced substantial time-to-market reduction. As anecdotal evidence, we refer to the Noise Filter application where the application was compiled to the hArtes board in a matter of minutes whereas a manual mapping would have taken multiple days.

5.7 Conclusions

This chapter briefly described the main objectives of the hArtes integrated project. It provided an overview of the hArtes design approach including the hArtes toolchain and the hArtes hardware platform. The benefits of this holistic design approach were demonstrated by experiments on immersive audio applications, which represent just a subset of all applications and experiments carried out within the project. This collective research and development effort provided a framework of tools and design approaches that allow performance efficient mapping of computationally demanding real-time applications on heterogeneous reconfigurable embedded systems. Important contributions of the hArtes project were the provisions it had developed to facilitate programming complex heterogeneous systems as if programming a single general purpose core. The resulting framework successfully hides the complexity of the underlying hardware from the programmer, thus improving programmer's efficiency and dramatically reducing times to market. To commercialize the hArtes approach and the tools developed in the project, a startup company named BlueBee Multicore Technologies was incorporated in January 2011 (http://www.bluebee-tech.com/).

5.8 In Memoriam

hArtes would not have been possible without prof. Stamatis Vassiliadis who initiated and inspired this project, but unfortunately deceased shortly after it has been launched. His ideas shaped and guided the efforts of all partners at all project stages – from the proposal preparation, which he actively worked on, till the successful completion of this challenging multinational collaboration. His vision was continuously inspiring the work of everyone from the hArtes consortium and, therefore, the final success of the project is also due to him.

References

1. Helmerich et.al., Study of Worldwide Trends and R&D Programmes in Embedded Systems in View of Maximising the Impact of a Technology Platform in the Area, 2005.
2. Future of Embedded Systems Technology, BCC Co, Inc., 2005, Report Code: IFT016B, http://www.bccresearch.com/report/IFT016B.html.

3. Venture Development Capital, Multi-core computing in Embedded Systems, white paper, 2007.
4. [Online]. Available: http://www.hipeac.net/roadmap.
5. [Online]. Available: http://www.nutech.com/.
6. [Online]. Available: http://www.scilab.org/.
7. F. Ferrandi, L. Fossati, M. Lattuada, G. Palermo, D. Sciuto, and A. Tumeo, "Automatic Parallelization of Sequential Specifications for Symmetric MPSoCs," in *Proceedings of International Embedded Systems Symposium 2007 (ISSS 2007)*, 2007, pp. 172–192.
8. F. Ferrandi, M. Lattuada, C. Pilato, and A. Tumeo, "Performance Estimation for Task Graphs Combining Sequential Path Profiling and Control Dependence Regions," in *Proceedings of ACM/IEEE International Conference on Formal Methods and Models for Codesign (MEMOCODE 2009)*, Cambridge, MA, USA, 2009, pp. 131–140.
9. W. Luk, J. Coutinho, T. Todman, Y. Lam, W. Osborne, K. Susanto, Q. Liu, and W. Wong, "A High-Level Compilation Toolchain for Heterogeneous Systems," in *Proceedings of IEEE International SoC Conference (SOCC)*, September 2009.
10. Y. Lam, J. Coutinho, W. Luk, and P. Leong, "Mapping and Scheduling with Task Clustering for Heterogeneous Computing Systems," in *Proceedings of the 18th International Conference on Field Programmable Logic and Applications (FPL08)*, September 2008.
11. S. Vassiliadis, S. Wong, S. D. Cotofana, "The Molen ρμ-coded Processor", in *Proceedings of the 11th International Conference on Field-Programmable Logic and Applications (FPL 2001)*, Springer-Verlag Lecture Notes in Computer Science (LNCS) Vol. 2147, pp. 275–285, Belfast, UK, August 2001.
12. S. Vassiliadis, S. Wong, G. Gaydadjiev, K. Bertels, G. Kuzmanov, E. Panainte, "The Molen Polymorphic Processor," in *IEEE Transactions on Computers*, Nov. 2004, Vol. 53, Issue 11, pp. 1363–1375.
13. S. Vassiliadis, G. Kuzmanov, S. Wong, E. M. Panainte, G. N. Gaydadjiev, K. Bertels, and D. Cheresiz, "PISC: Polymorphic Instruction Set Computers," in Proceedings of the International Workshop on Applied Reconfigurable Computing (ARC 2006), March 2006, pp. 274–286.
14. S. Vassiliadis, G. Gaydadjiev, K. Bertels, and E. Moscu Panainte, "The Molen Programming Paradigm," in *Proceedings of the third Int'l Workshop on Systems, Architectures, Modeling, and Simulation (SAMOS 2003)*, pp. 1–7, July 2003.
15. Y. D. Yankova, G. Kuzmanov, K. Bertels, G. N. Gaydadjiev, Y. Lu, and S. Vassiliadis, "DWARV: DelftWorkbench Automated Reconfigurable VHDL Generator," in *In Proc. of the 17th International Conference on Field Programmable Logic and Applications (FPL07)*, August 2007, pp. 697–701.
16. Koen Bertels, Vlad Mihai Sima, Yana Yankova, Georgi Kuzmanov, Wayne Luk, José Gabriel F. Coutinho, Fabrizio Ferrandi, Christian Pilato, Marco Lattuada, Donatella Sciuto, Andrea Michelotti "HArtes: Hardware-Software Codesign for Heterogeneous Multicore Platforms," *IEEE Micro* 30(5): 88–97 (2010)

Chapter 6
Smart Chips for Smart Surroundings – 4S

Eberhard Schüler, Ralf König, Jürgen Becker, Gerard Rauwerda, Marcel van de Burgwal, and Gerard Smit

Abstract The overall mission of the 4S project (Smart Chips for Smart Surroundings) was to define and develop efficient flexible, reconfigurable core building blocks, including the supporting tools, for future Ambient System Devices. Reconfigurability offers the needed flexibility and adaptability, it provides the efficiency needed for these systems, it enables systems that can adapt to rapidly changing environmental conditions, it enables communication over heterogeneous wireless networks, and it reduces risks: reconfigurable systems can adapt to standards that may vary from place to place or standards that have changed during and after product development.

In 4S we focused on heterogeneous building blocks such as analogue, hardwired functions, fine and coarse grain reconfigurable tiles and microprocessors. Such a platform can adapt to a wide application space without the need for specialized ASICs. A novel power aware design flow and runtime system was developed. The runtime system decides dynamically about the near-optimal application mapping to the given hardware platform.

The overall concept was verified on hardware platforms based on an existing SoC and in a second step with novel silicon. DRM (Digital Radio Mondiale) and MPEG4 Video applications have been implemented on the platforms demonstrating the adaptability of the 4S concept.

6.1 Project Partners

1. PACT XPP Technologies AG, Germany
2. CTIT, The University of Twente, The Netherlands
3. ITIV, Karlsruhe Institute of Technology KIT, Germany (coordinator)

E. Schüler (✉)
PACT XPP Technologies AG, Walter-Gropius-Str. 15, 80807 Munich, Los Gatos, Germany
e-mail: eberhard.schueler@t-online.de

J.M.P. Cardoso and M. Hübner (eds.), *Reconfigurable Computing: From FPGAs to Hardware/Software Codesign*, DOI 10.1007/978-1-4614-0061-5_6,

4. ATMEL Corporation, Germany
5. IMEC, Inter University Micro-Electronics Centre, Belgium
6. WMC, Twente Institute for Wireless and Mobile Communications B.V., The Netherlands
7. ASICentrum s.r.o., Czech Republic
8. Thales Communications, France
9. dicas, Dicas digital image coding GmbH, Germany
10. Harman/Becker Automotive Systems GmbH, Germany
11. Recore Systems, The Netherlands

- Project Coordinator: Eberhard Schüler, PACT XPP Technologies AG, Germany
- Start Date: January 2004
- End Date: December 2007
- EU Program: 6th Framework Program for Research and Technological Development, Information Society Technologies, Pushing the limits of CMOS and preparing for post-CMOS, Contract IST-1908. Instrument: Specifically Targeted Research Project (STREP)
- Global Budget: 9.7 M€
- Global Funding by EU: 5.4 M€
- Contact Author: Eberhard Schüler, Email: eberhard.schueler@t-online.de

6.2 Introduction

The development of energy efficient computing and software architectures for future Ambient Systems was at the core of the 4S project. Ambient Systems (also known as *ambient intelligence or ubiquitous computing)* are networked embedded systems wirelessly integrated with everyday environments and supporting people in their activities. The architecture of Systems-on-Chips (SoC) architecture suitable for Ambient Devices poses a lot of challenges: these devices have a very small energy budget, they are always operational (although quite often in a low-power mode), are small in size but require high processing performance. State-of-the-art architectures could not provide the processing power required by a fully operational Ambient Device given the tight energy limitations. To realize devices within the energy budget, flexible and highly efficient hardware *and* software architectures are needed. Moreover, without significant energy reduction techniques and energy-efficient adaptive architectures, battery-life constraints will severely limit the capabilities of these devices.

6.2.1 Heterogeneous Reconfigurable Computing

Reconfigurable systems offer the required flexibility and can adapt processing resources dynamically to the demand of applications. We distinguish several processing structures in a heterogeneous reconfigurable system, for example: bit-level reconfigurable units (e.g. FPGAs), word-level (coarse-grained) reconfigurable units, and general-purpose programmable units (DSPs and microprocessors). Unlike microprocessors and DSPs, reconfigurable units adapt the layout and functionality of hardware elements to the task to be performed. This enables exploiting the implicit parallelism of the algorithm and provides much higher performance for a given energy budget.

The adaptability of the architecture enables the system to be targeted at multiple applications. The architecture and firmware can be upgraded at any time (even when the system is already installed and running) and reconfigurability allows adaptation of the architecture for testing and maintenance purposes during and after the production phase.

Typically, some algorithms are more suitable for bit-level reconfigurable architectures (e.g. Software defined Radio PN-code generation), others for DSP-like hardware and others for word-level reconfigurable platforms (e.g. FIR filters or FFT algorithms). Most stream-based algorithms can be mapped efficiently to word-level reconfigurable architectures [10]. Application designers or, at best, high-level compilers choose the most efficient processing unit for the type of processing needed for a given application task.

This raises the question how to map various application tasks to such a heterogeneous platform and which tools will enable the application designer to benefit from a specific hardware architecture. In 4S we addressed the question by means of a runtime operating system and communication framework as described in subsequent sections.

6.2.2 Energy Efficiency

Perhaps the most significant property of ambient and unobtrusive wireless devices is their resource limitation, in terms of energy, memory and processing power. In 4S we qualified energy as the premium resource, and built time/energy adaptive architectures that adequately operate under these constraints. We applied energy-efficient design principles at all layers, including those where these considerations were not consequently applied in combination in the past. For example: analogue/digital tradeoffs, energy-efficient data processing and communication, run-time reconfiguration for efficiency, as well as high-level tools for (energy-) efficient mapping of applications to dynamic reconfigurable architectures.

Furthermore, heterogeneous reconfigurable Systems on Chip (SoC) can adapt supply voltage and clock frequency per module and so provide additional potential to save power.

6.2.3 Flexibility

Ambient Devices are able to choose from a wide range of services from various wireless access networks in its surrounding, each having their own characteristics, costs, and ownership. They operate in a heterogeneous environment where it is possible to use a combination of networks (simultaneously), each of which is optimised for some particular service. For example, Ambient Devices will have to support multiple multimedia standards, e.g. different audio and video standards (such as MP3, MPEG4, H.264). In addition, devices should be prepared to perform e.g. speech recognition/generation algorithms, signal processing algorithms for biometrics (e.g. fingerprint recognition) and security algorithms. Finally, the architecture should have the flexibility to anticipate future emerging standards.

6.2.4 Development Tools for Reconfigurable Architectures

Programming heterogeneous reconfigurable SoCs is considered to be non-trivial. Thus software designers must be supported by sophisticated tools. Furthermore, new tools must be accepted by application software designers. In 4S we developed a set of new tools on top of existing well known standards. This strategy not only provides a clearly defined migration path from state-of-the-art tools but also provides flexibility and less risk because only small steps are required to launch a new architecture for commercial products.

6.2.5 Structure of Subsequent Sections

In the following section we first introduce the 4S heterogeneous reconfigurable platform from the hardware and runtime system perspectives. The operating system section introduces the concepts of the runtime framework. Then we describe the demonstration platforms, the realisation of implemented heterogeneous hardware and the silicon implementations. The Proof of Concept section presents experimental results based on two target applications. The 4S chapter is finished with an outlook and the conclusion summarizing the achieved results.

6.3 The 4S Approach

6.3.1 The Essential Ideas

The 4S project targets the system architecture, hardware building blocks and the supporting energy efficient design and runtime tools.

6.3.1.1 Heterogeneous System Architecture

The system architecture consists of a set of *heterogeneous building blocks* like analogue blocks, hardwired ASIC functions, reconfigurable blocks, signal processors

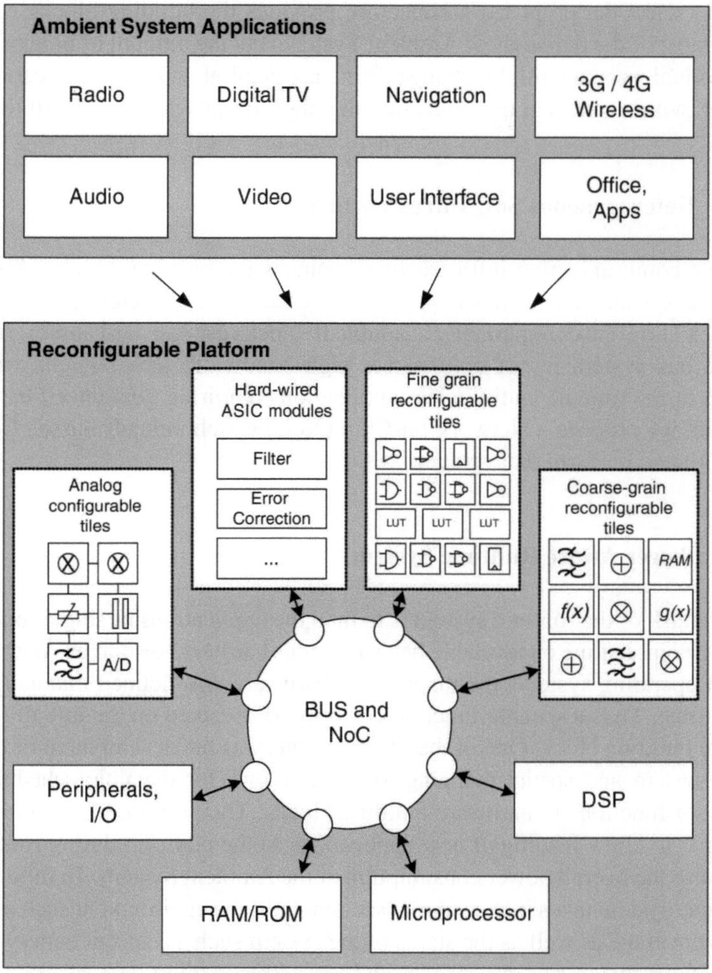

Fig. 6.1 The 4S heterogeneous hardware architecture

and microprocessors. We name the processing building blocks "tiles". The entirety of the building blocks must support a *wide spectrum* of possible applications for Ambient Devices, including wireless communication as well as audio and video processing. As those application fields have different requirements that cannot always be handled efficiently by one single tile, several tiles with different capabilities are included in the architecture.

Figure 6.1 shows the main building blocks of a universal Ambient System. For the specific realisation of an Ambient Device not all of these building blocks might be needed. If – for example – the purpose of a device is the reception of audio streams, no building blocks for video processing are needed in this special device and thus can be omitted in order to save power and silicon area. For another device that is targeted for mobile video reception other building blocks will be included or left out.

In that sense the proposed architecture provides the building blocks needed to support most of the demands of Ambient Systems but the amount of blocks used for an individual product might change from application field to application field. However, within one field of application the platform provides full flexibility.

6.3.1.2 Heterogeneous SoC Infrastructure

A flexible communication infrastructure connects the building blocks. A standard shared bus system (e.g. AMBA) provides interfaces for all heterogeneous blocks. Most of existing and commercially available IP blocks use standard busses. However, a shared bus system is not optimal for high bandwidth data streams which are common in multimedia and wireless applications requiring guaranteed bandwidth. Therefore, we propose a Network-on-Chip (NoC) which unloads busses from high bandwidth and uniform data streams.

6.3.1.3 Power Aware Runtime System

The main task of the runtime system is to map the applications or single functions at a specific point in time on available heterogeneous hardware resources. With existing standard operating systems the mapping to hardware was defined manually during compile time. Thus a specific function can only be executed on the initially planned hardware building block. One of the 4S inventions was the development of a flexible runtime system and spatial mapping tool that allows for dynamic scheduling and mapping of functions to hardware building blocks. The purpose of dynamic spatial mapping is to allow loading of new applications to the platform during runtime and to minimise the overall power consumption of the Ambient System. To fulfil this task the runtime system takes into account user demands, the environment such as current wireless coverage as well as the status of the system such as current battery status.

To support this, a runtime system and the corresponding design time tool flow have been developed in the 4S project. The development environment consists of specific tools for each tile for generation of binary code needed to configure and operate each block. The application software uses the configuration and communication mechanisms of the underlying real-time operating system and the runtime system. On top-level a XML-description specifies and characterizes the applications, the building blocks and the overall system architecture in order to gain decision criteria for the dynamic runtime mapping.

To allow for early design space exploration and concept verification, a co-simulation *framework* was developed. It allows not only to co-simulate already existing building blocks but also to verify newly developed hardware blocks, software drivers and their interaction with other parts of a system.

Figure 6.2 shows the tool flow. At top level the system properties such as the system topology, interconnect and available bandwidth are specified. The second step is describing the application in form of a task graph. The graph's nodes specify

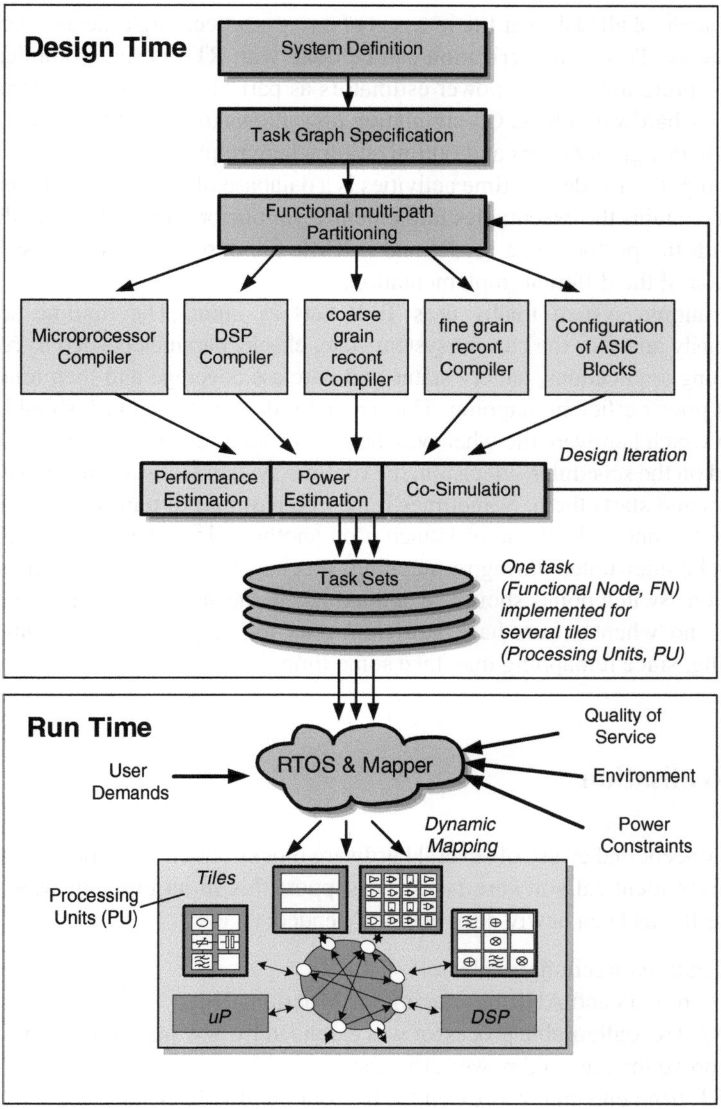

Fig. 6.2 Tools flow

the function to be performed while the edges specify the communication with other tasks. Then the designer has to decide which task should be implemented on which hardware tile. Note that in the context of the operating framework OSYRES as described later, a task is a "Functional Node" (FN) and a hardware tile a "Processing Unit" (PU). Where possible a task is implemented functional identically for several tiles. This allows the runtime system to choose among different realisations according to the actual system status and external requirements. The third step is to implement and characterize all tasks per tile in terms of energy, processing time and bandwidth requirements. This characterization can be done with RTL design compiler power tools or – preferably – with power estimators as part of the design environment of the specific hardware block. Co-simulation now allows to verify the application and to perform design iterations and optimization where required.

The output of the design-time activities is a database of so-called Task Sets. Each Task Set contains the executables implemented for one or more hardware tiles combined with the performance model and a XML-file describing the realisation and connection of the different implementations.

The runtime system finally uses Task Sets as input. The runtime scheduler continuously analyses the current system state, checks parameters such as requested and running applications, battery status and wireless coverage and then tries to find the most power efficient mapping. The scheduler defines which task should be executed on which hardware tile whenever the system status changes (spatial mapping).

Based on the scheduler's decision, the runtime system loads the tasks to the hardware tiles and starts them. Sometimes it may be required to remove a running task from one tile and to load and re-launch it on another. Since the current application must not be interrupted during re-mapping, special switch points are defined in the application. Switch points should be defined where the application state complexity is small and when buffers have sufficient data to bridge short term intermitted processing, since remapping may take some time.

6.4 Realisation

The 4S concept makes use of several hardware blocks which are capable to perform functionally identical software tasks. To support this in an efficient way existing hardware IP has been newly designed or extended:

- The *Montium* Reconfigurable core
- A NoC routers and AHB interfacing for Montium Tiles
- The *XPP* reconfigurable processor was extended by new instructions and features to improve the area and power efficiency
- The RF front end chip was extended by configurable features
- New ASIC cores for software defined radio (SDR) such as the Digital Down Converter (DDC), the Viterbi Decoder and the Signal Analyser have been implemented.

6.4.1 Montium Reconfigurable Tile

The *Montium* is a coarse-grained reconfigurable processor core and targets the 16-bit DSP algorithm domain. The Montium architecture originates from research by the University of Twente and has been further developed by Recore Systems.

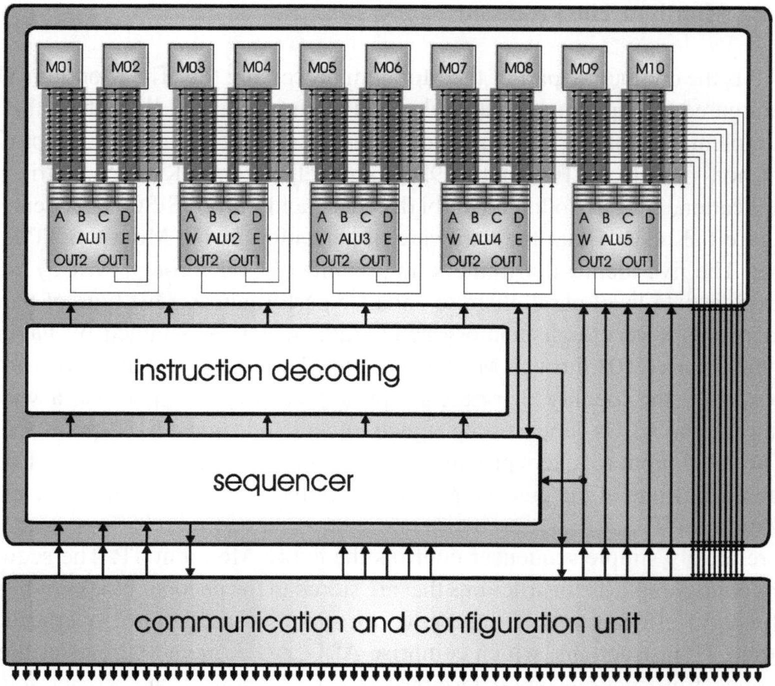

Fig. 6.3 Montium processing tile

A single Montium processing tile is shown in Fig. 6.3. At first glance the Montium architecture bears a resemblance to a VLIW processor. However, the control structure of the Montium is very different. The lower part of Fig. 6.3 shows the Communication and Configuration Unit (CCU) and the upper part show the coarse-grained reconfigurable Montium Tile Processor (TP).

6.4.1.1 Communication and Configuration Unit

The CCU implements the network interface controller between the NoC and the Montium TP. The CCU provides configuration and communication services to the Montium TP. The definition of the network interface depends on the NoC technology that is used in a SoC in which the Montium processing tile is integrated [3].

The CCU enables the Montium TP to run in "streaming" as well as in "block" mode. In "streaming" mode the CCU and the Montium TP run in parallel. Hence, communication and computation overlap in time. In "block" mode, the CCU first reads a block of data, then starts the Montium TP, and finally after completion of the Montium TP, the CCU sends the results to the next processing unit in the SoC (e.g., another Montium processing tile or external memory).

6.4.1.2 Montium Tile Processor

The TP is the computing part of the Montium processing tile. The Montium TP can be configured to implement a particular DSP algorithm. DSP algorithms that have been implemented on the Montium are, for instance, all power-of-2 FFTs upto 2,048 points, non- power-of-2 FFT up to 1,920 points, FIR filters, IIR filters, matrix vector multiplication, DCT decoding, Viterbi decoders, and Turbo (SISO) decoders [4].

Figure 6.3 reveals that the hardware organization of the Montium TP is very regular. The five identical arithmetic logic units (ALU1 through ALU5) in a tile can exploit data level parallelism to enhance performance. This type of parallelism demands a very high memory bandwidth, which is obtained by having 10 local memories (M01 through M10) in parallel. The small local memories are also motivated by the locality of reference principle. The data path has a width of 16-bits and the ALUs support both signed integer and signed fixed-point arithmetic. The ALU input registers provide an even more local level of storage. Locality of reference is one of the guiding principles applied to obtain energy efficiency in the Montium TP.

A relatively simple sequencer controls the entire Montium TP. The sequencer selects configurable TP instructions that are stored in the decoder blocks of Fig. 6.3. For (energy) efficiency it is imperative to minimize the control overhead. The Montium TP instructions, which comprise ALU, AGU, memory, register file, and interconnect instructions, are determined by a DSP application designer at design time. All Montium TP instructions are scheduled at design time and arranged into a Montium sequencer programme. By statically scheduling the instructions as much as possible at compile time, the Montium sequencer does not require any sophisticated control logic which minimizes the control overhead of the reconfigurable architecture.

The Montium TP has no fixed instruction set, but the instructions are configured at configuration time. During configuration of the Montium TP, the CCU writes the configuration data (i.e., instructions of the ALUs, memories and interconnects, sequencer and decoder instructions) in the configuration memory of the Montium TP. The size of the total configuration memory of the Montium TP is about 2.6 kByte. However, configuration sizes of DSP algorithms mapped on the Montium TP are typically in the order of 1 kByte. For example, a 64-point fast Fourier transform (FFT) has a configuration size of 946 bytes, which typically takes about 500 clock cycles required for configuring the Montium TP. Hence, the Montium TP can be configured for FFT-64 in less than 5 μs (assuming a configuration clock of 100 MHz). By sending a configuration file containing configuration RAM addresses and data values to the CCU, the Montium TP can be configured via the NoC interface. The configuration memory of the Montium TP is implemented as a 16-bit wide SRAM memory that can be written by the CCU. By only updating certain configuration locations of the configuration memory, the Montium TP can be partially reconfigured.

6.4.1.3 Montium Design Methodology

The Montium development tools start with a high-level description of an application (e.g. in C/C++ or MATLAB) and translate this description to a Montium TP configuration. The Montium design methodology to map DSP applications on the Montium TP is divided into three steps:

1. The high-level description of the DSP application is analyzed and computation intensive DSP kernels are identified.
2. The identified DSP kernels or parts of the DSP kernels are mapped on one or multiple Montium TPs that are available in a SoC. The DSP operations are programmed on the Montium TP using an embedded C language, called MontiumC.
3. Depending on the layout of the SoC in which the Montium processing tiles are applied, the Montium processing tiles are configured for a particular DSP kernel or part of the DSP kernel. Furthermore, the channels in the NoC between the processing tiles are defined.

6.4.2 Circuit Switched Network-on-Chip

The Network-on-Chip was designed as a circuit switching NoC, i.e. communication between any two tiles in the SoC goes via pre-configured routes. These routes are configured in the routers using a dedicated configuration interface of the routers [5]. Typically, in the applications considered in this project data streams are fixed for a relatively long time. Therefore, a connection between two tiles is required for a long period (e.g. seconds or longer). A large amount of the traffic between tiles will need a guaranteed throughput, which can be easily guaranteed in a circuit-switched connection. Current SoC architectures have a large amount of wiring resources that give enough flexibility for streams with different bandwidth demands. Internally, a circuit switching router has a minimal amount of control in its data pad (e.g. no arbitration). This increases the energy- efficiency per transported bit and the maximum throughput.

The network-on-chip consists of two circuit switched routers. It enables concurrent communication between the four Montium processing tiles and the rest of the SoC via an AHB bridge.

The network router interface consists of two identical unidirectional physical channels, each containing four 20-bit wide lanes. A lane consists of 16-bit data, 2-bit flit type (FT), 1-bit data valid and 1-bit acknowledge signals.

Each Montium processing tile is connected to a router via two unidirectional physical channels. In order to send data to the network a tile has to send a 20-bit network packet onto one of its output lanes. Up to four packets can be sent in parallel by a single tile, by using all outgoing lanes.

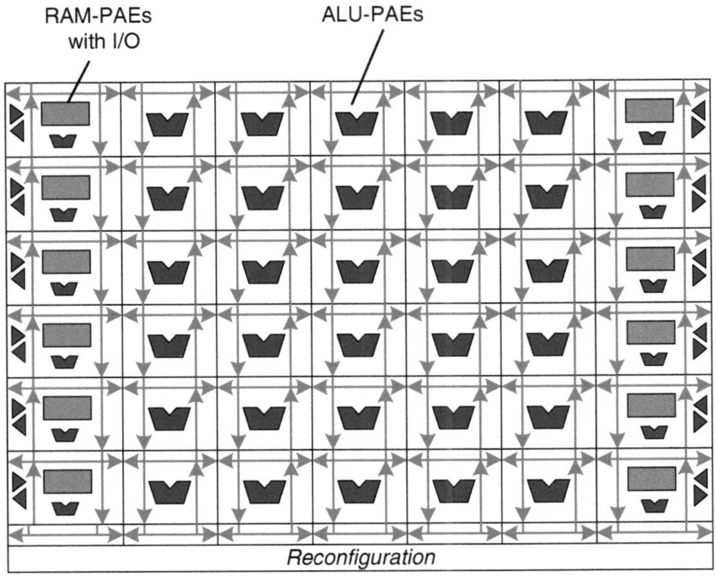

Fig. 6.4 XPP Array

6.4.3 XPP Reconfigurable Tile

The XPP Array is a coarse grained reconfigurable processing unit designed by PACT XPP Technologies [6]. The XPP Array (Fig. 6.4) is built from a rectangular array of two types of Processing Array Elements (PAEs): Those in the centre of the array are *ALU-PAEs*. At the left and right side are *RAM-PAEs* with *I/O*. An ALU-PAE contains three integer ALUs, two in top-down direction and one in bottom-up direction. A RAM-PAE contains two ALUs, a small RAM block and an I/O object. The I/O objects provide access to external streaming data sources or destinations.

The horizontal routing busses for point-to-point connections between XPP objects (ALUs, RAMs, I/O objects, etc.) are integrated in the PAEs. Separate busses for 16-bit data values and 1-bit events are available. Furthermore, vertical routing connections are provided within the ALU-PAEs and RAM-PAEs

An application is described as a flow graph where the nodes define the function to be performed (e.g. addition, multiplication). Each node is then mapped to one ALU while the graph's edges define the connections. Data flows through the network of operators. The event network can steer the data flow based on calculation results. This enables conditional execution and while loops. The strength of the XPP Array originates from the combination of parallel array processing with fast run-time reconfiguration mechanisms [7, 8]. PAEs can be configured while neighbouring PAEs are processing data. Entire algorithms can be configured and run independently on different parts of the array. A configuration typically requires only a few thousand cycles. This is several orders of magnitude faster than reconfiguration of large FPGAs.

In the course of the 4S project the XPP Array was extended in order reduce area and thus power consumption for a given application. One strategy to achieve this is automatic cutting of small sub-graphs from the overall graph and mapping them to PAEs which execute this graph sequentially, but at higher frequency [9].

6.4.3.1 Streams

The basic communication concept of the XPP-III architecture is based on *streams*. A *data stream* is a sequence of single data packets travelling through the flow graph that describes the algorithm. In addition to data packets, state information packets ("events") are transmitted through independent event connections. Event packets containing one bit of information are used to control the execution of the processing nodes or may synchronize external devices.

The XPP communication network enables automatic synchronization of packets. An XPP object (e.g. an ALU) operates and produces an output packet only when all input data and event packets are available. The benefit of the resulting self-synchronizing network is that only the number and order of packets travelling through a graph is important. There is no need for the programmer or compiler to care about absolute timing of the pipelines during operation.

6.4.3.2 XPP Design Tools

The XPP comes with a complete tool chain [8]. The array is either programmed with a proprietary mapping description language or a vectorizing C Compiler. The C-Compiler extracts implicit parallelism form standard C-code (e.g. inner loops) and converts it to control and data flow graphs which then are mapped on the PAEs. Data flow graphs are simulated and visualized graphically.

6.4.3.3 Power Estimation

The XPP tool chain was extended by an A*pplication Power Estimator*. The power requirement of the XPP Array is highly dynamic. A PAE which is not contributing to calculating in a specific clock cycle consumes nearly no power (besides leakage and clock tree). The tool parses the simulator output per cycle and integrates the power consumption for a specific task. The power tool was calibrated with the Synopsys power compiler for a specific silicon library (e.g. ST CMOS 90 nm). The output of this tool in terms of nJ per task, described in XML format is then used by the *Spatial Mapper* (as described later). The power consumption over time can also be visualized graphically. This enables to profile the application partitioning and reconfiguration strategy in terms of power.

6.4.3.4 XPP Hardware

In the 4S project an existing processor, the XPP64-A, with 64 24-bit wide ALU PAEs and 16 RAM-PAEs was used. The chip is mounted on an evaluation board with a MIPS processor which is responsible for administrative tasks, XPP reconfiguration and communication. A video board provides a frame buffer and video output enabling test and demonstration of video algorithms. This board was linked through parallel streaming interfaces to the basic verification platform's FPGA board. The FPGA multiplexes several data streams originating from the SoC (i.e. ARM or DMA) to be processed by the XPP Array acting as coprocessor.

6.4.4 Demonstrators

Hardware and software for two demonstrator platforms have been designed. The first, the *Basic Verification Platform (BCVP)* is used as a multi-purpose platform:

- Operating System verification at an early project stage
- Test and verification of new hardware tiles on FPGA
- Interfacing to other board-level hardware tiles and legacy hardware
- Application test and benchmarking
- Early demonstration of the 4S concept.

The second demonstrator hardware, the *Highly Integrated Verification Platform,* replaces some modules of the basic platform by new chips that integrate new hardware tiles on silicon.

6.4.4.1 Basic Verification Platform

This platform is built from existing boards and chips. The specification was driven by the 4S target applications which are the radio broadcast standard Digital Radio Mondiale (DRM) and MPEG4 video decoding. Following the needs of those applications the hardware is built from:

- A SoC board based on an ATMEL SoC that integrates peripherals and two ARM processors
- An FPGA extension board with interfaces
- An Radio RF front-end based on an ATMEL RF chip
- An XPP Evaluation board with video interface which is connected to the FPGA extension board

Most evaluation and demonstration tasks were performed on this platform. The FPGA was used for the implementation and initial verification of the Montium reconfigurable cores and associated NoC routers. Another FPGA configuration provides a parallel interface to the XPP evaluation board with the XPP64A reconfigurable processor chip. The interface is capable to multiplex several independent data streams which are required for the video decoding application.

6.4.4.2 Highly Integrated Verification Platform

To show that the complexity of a heterogeneous architecture can be handled on silicon, a SoC (named *Annabelle,* ATMEL CMOS 130 nm) and analogue RF frontend (*Amelie*) have been designed and produced. The hardware platform reuses most modules of the Basic Verification Platform. However the SoC board and Radio front end have been replaced by boards for the new silicon chips.

6.5 Silicon Implementation

Implementation of a complex System on Chip on silicon was one of the project objectives. As outlined in the introduction one first needs to specify the application field. In 4S we decided to demonstrate the DRM on silicon. It became obvious that two chips need to be designed. One is a flexible RF radio front end, the second was the heterogeneous SoC. The chips integrate all functions which are required for a universal digital radio receiver. The DRM requires lots of signal processing power and flexibility.

6.5.1 Analogue RF Frontend

RF processing is a typical field where analogue hardware is required. We have developed a widely configurable analogue RF building block which connects directly to the SoC. The chip provides the analogue building block of the proposed heterogeneous tiled 4S architecture model. Though this device is not part of the overall task mapping scheme – simply because no alternative implementation exists – it is mandatory for the DRM proof of concept.

The new analogue RF chip is an advanced AM/FM/DRM frontend tuner with fast PLL integrated on a single chip. It represents a complete, automatically adjustable AM/FM/DRM front end, containing an analogue AM/DRM up/down conversion and FM down conversion system for an Intermediate Frequency (IF) of 10.7 MHz which allows an economic filter concept. The impedance driver at the IF output is designed for the A/D converter which is integrated in the new digital SoC. The chip registers are programmed through a 3-wire serial protocol originating from the SoC.

The fast tuning concept realized in this part is based on patents and allows lock times less than 1 ms for a jump over the FM band with a step width of 12.5 kHz. An automatic tuner alignment is provided by built-in D/A converters for gain and offset compensation (to support the computer-controlled alignment). The frequency range of the IC covers the AM/DRM band as well as the FM broadcasting band, the Japan band and the Weather band.

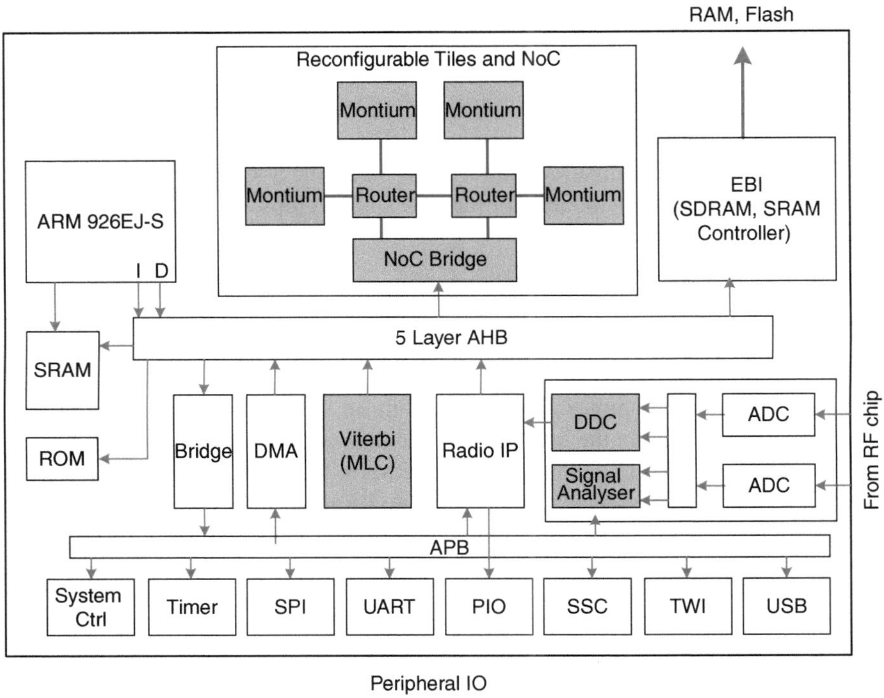

Fig. 6.5 SoC block diagram, new modules are shaded

6.5.2 Digital System on Chip

The SoC (Fig. 6.5) incorporates the complete digital signal processing chain required for a software defined radio receiver. Analogue inputs from the RF chip are converted by two integrated A/D converters. The digital samples are fed into new signal analysers providing quick information about the signal quality and to two Digital Down Converters (DDC) which provide the digital input to be processed by the processor tiles. The mentioned digital modules have been designed in the course of the project. The Viterbi decoder is also a new ASIC tile that unburdens the processors and provides best performance/power ratio for this task. Baseband processing is performed by the reconfigurable fabric which contains four reconfigurable Montium cores that are connected to the NoC. The reconfigurable fabric is connected to the AHB bus and serves as a slave to the AMBA system. In fact, the reconfigurable fabric acts as a reconfigurable co-processor for the ARM926 processor. Computation intensive DSP algorithms are typically offloaded from the ARM926 processor and processed on the coarse-grained reconfigurable Montium cores inside the reconfigurable fabric.

6.5.3 Runtime System and Tools

6.5.3.1 RTOS

Systems on chip need to perform several tasks with partially hard real-time constraints. In 4S the *eCos* real-time operating system was chosen as target operating system [1]. eCos, released under the GNU GPL license, is during design-time widely configurable to the final processor footprint and has been designed to support applications with real-time requirements. It is fully pre-emptive and provides minimal interrupt latencies, the necessary synchronization primitives, scheduling policies and interrupt handling mechanisms needed for real-time applications. eCos also provides all the functionality required for general embedded application support including device drivers, memory management, exception handling, C, math libraries, etc. The eCos' hardware abstraction layer (HAL) was adapted to support the digital SoCs, e.g. communication with the reconfigurable tiles, hard-wired accelerators and the peripherals. For booting services the *RedBoot* debug and bootstrap tool was used. Both 4S demonstrator platforms are using eCos.

6.5.3.2 OSYRES Framework

The goal of *OSYRES* is to provide an abstraction of the multi-processing platform. OSYRES [2] hides the details of the communication between functional nodes and hides the details of instantiation and allocation of functional nodes on a multi-processor platform. The OSYRES framework has been used for proof of concept and implementation of the DRM and MPEG4 applications.

OSYRES is built on top of eCos and provides the 4S-specific features. The architecture of OSYRES is optimized to support of data flow applications. The design of a data flow application differs from the traditional sequential designed application. In a *sequential* designed application, functions are called in a sequence, which is defined by the application's state machine. In a *data flow* application, functions are triggered by the availability of data or messages at their input ports. In a multiprocessor system with a run-time function allocation, the arrival of messages and thus the order at which functions are called is not predictable anymore unless they are explicitly synchronized. In OSYRES synchronization is explicitly done via messages that automatically synchronize two functions. In addition, functions may run in parallel on different processors. This requires a different way of defining the application in *Functional Nodes (FN)*.

6.5.3.3 Application Specification

An application in OSYRES is defined by a set of task graphs consisting of Functional Nodes connected by channels described in a data flow graph (see Fig. 6.6). A task

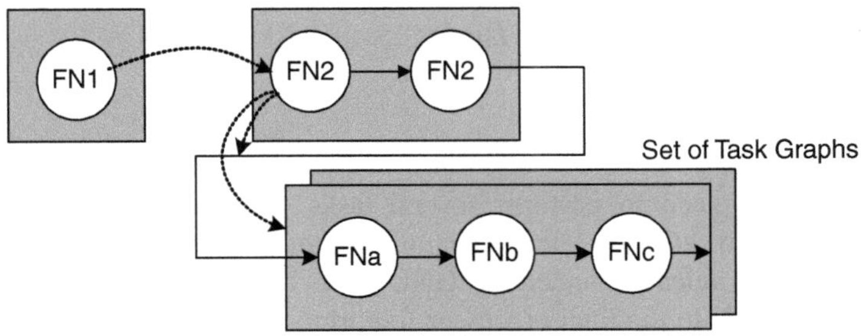

Fig. 6.6 Functional Node (FN) definition

graph is a combination of functionality, which is always instantiated as a whole. It is possible to instantiate multiple instances of the same task graph. A task graph instantiation is performed by one of the already running functional nodes.

A FN will always be instantiated on one processing device. It is a self-contained functional entity, which communicates only via its ports with other FNs. These ports form the end-points of *uni-directional channels*. The channels will be instantiated on the communication links between the processing devices such as a bus or Network on Chip (NoC).

The approach of task graphs and self-contained entities, which communicate via well-defined messages results by definition in a modular design of the overall application.

6.5.3.4 Functional Node Design Concept

An OSYRES Functional Node is defined by its functionality at its ports. There may be different Functional Node implementations for each different type of processing unit but the external behaviour must be the same for all implementations. The FN implementation itself does not take any resource of the processing unit, except for the storage of the functional implementation image. At run-time instances of functional implementations are created and it is only then, when they take resources from the processing unit. The amount of resources that is needed for an instance of a FN implementation will be defined at design time and specified in the system definition (memory, MIPS, energy) characterizing the node.

All instances of a Functional Node implementation executed by the same processor type use the same code. To allow multiple instances, each instance has its own administration, which is created at FN instantiation. This administration is the storage of all FN local data.

In addition of the operational functionality, a FN also contains special functionality to control the instantiation of new instances. Control events may be events to *Create*, *Delete*, *Start* and *Stop* a FN instance as well as channel connection events like *Add* and *Remove* ports.

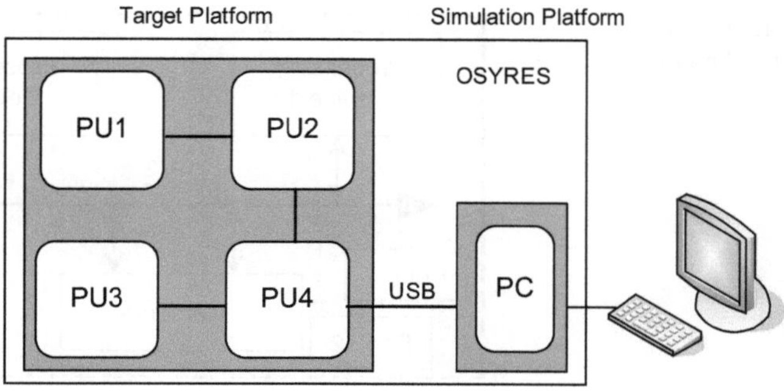

Fig. 6.7 Simulation with a PC and four Processing Units (PU)

6.5.3.5 Application Verification

As OSYRES can be implemented on different platforms, there is an implementation for the most common development or simulation environments; e.g. OSYRES for Windows using the Visual C development environment. Before the real target platform is available, one can already start simulating the application on such a common PC platform, e.g. by using simulators as MatLab for the actual processing.

On this simulation platform, applications can be verified in two steps. The first step is the verification of the application structure. In this step, the task graph structure and the messaging between FNs is verified. The second step is the verification of the functionality of the FNs individually. As the interfaces are clearly defined, only module-level test is needed. After successful verification, the overall application can be simulated first on the PC platform and in a later stage on the target platform. By combining the OSYRES platforms of the target (e.g. SoC) and the common PC platform, one unified OSYRES platform is obtained. This allows a smooth transfer from the simulation to the target platform. Starting from a running application on the simulation platform, FNs are gradually transferred to the target platform.

Figure 6.7 shows an example where the target is connected with the simulation PC via a USB cable. Starting from an application running in the simulation environment, functional processes are transferred to the target. The abstraction of communication in OSYRES assures the connections within the application, regardless where the functional processes are located.

6.5.3.6 System Definition

OSYRES provides an abstraction of the multi-processing platform. Figure 6.8 shows an example of a system platform which could be a dual ARM System on Chip, with one shared bus on which the memory and three peripherals are connected.

Each ARM processing device runs its own operating system i.e. eCos. Within this local RTOS peripherals and memory areas are allocated to one of the processing devices.

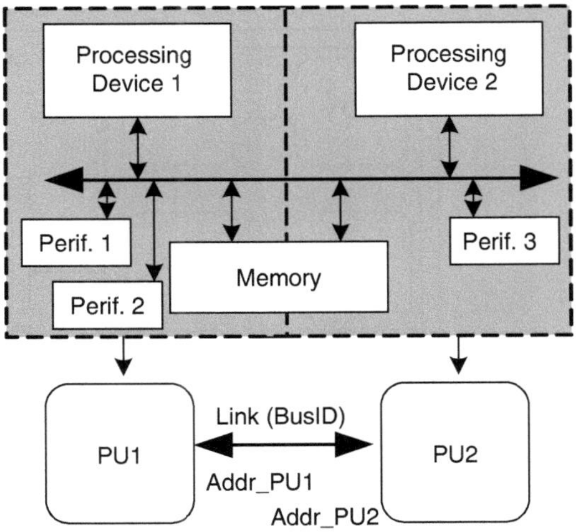

Fig. 6.8 Sample system platform with two Processing Units (PU)

Specific bus drivers are added to access the connected communication medium, in this example a shared bus.

A *Link* in OSYRES is defined as a method to access the other processing device via the communication medium. This method is implemented in a *Link driver,* which is built on top of the RTOS bus driver.

A *Processing Unit (PU)* in OSYRES is defined by the RTOS defined functionality and the OSYRES specific functionality as the communication layer, scheduler and control network as well as the previously mentioned Link drivers. In the context of the 4S demonstrator platforms a PU may be an ARM core, a Montium processor, the XPP processor or a hardwired core such as Viterbi.

The OSYRES functionality is built on top of the local RTOS. The OSYRES communication layer offers access to the platform-wide communication layer. The OSYRES control network provides the system-wide instantiation of Functional Nodes. Each PU has a local instantiation of this control network in the form of a *PU Manager*. This PU manager is responsible for the instantiation of the FNs on the PU. All PU Managers communicate with one central *Control Node*. This Control Node is located on one of the PUs and includes the run-time spatial mapping algorithm.

Reconfigurable structures as introduced in the previous sections may run FNs as well. In order for the OSYRES platform to control the reconfigurable processing devices, it needs to be adapted to control the reconfigurable structures.

These reconfigurable devices are considered as Processing Units without an operating system and will be controlled by a remote PU Manager, which runs on one of the generic processing devices like the ARM. The remote PU Manager translates control commands, like Create, Start and Stop into device specific actions (like Load, Configure and Start) towards the reconfigurable device. In addition, the remote PU manager controls the communication with the reconfigurable device, which does not support the OSYRES communication protocol. This allows the integration of OSYRES on devices which are not able to run an operating system.

6.5.4 Spatial Mapper

Common practice is to map applications to a hardware platform at design-time. In 4S we researched methods and algorithms how to perform the mapping at run-time. Run-time mapping offers a number of advantages over design-time mapping. It offers the possibility

- to adapt dynamically to available hardware resources. The available resources may vary over time due to applications running simultaneously or adaptation of the algorithms to the environment.
- to enable unforeseeable upgrades after first product release time, e.g. new applications and new or changing standards.
- to circumvent defective parts of a SoC. Larger chips mean lower yield. The yield can be improved when the mapper is able to avoid faulty sections of the chip. Also aging can lead to faulty parts that are unforeseeable at design-time.

The mapping tool *SMIT* (Spatial MappIng Tool) [11] is integrated into OSYRES and determines at runtime a near-optimal mapping of a given set of applications to the 4S heterogeneous architecture. The application is modelled as a set of communicating Functional Nodes. In the context of the 4S project "optimal mapping" means that the energy consumption of the SoC is minimized with regard to the mapping, while other constraints are still satisfied. Other optimization criteria such as user response time, memory consumption etc. could be the optimization target but were not the focus in 4S. SMIT uses the performance characteristics that are determined for each functional node implementation at design-time.

6.5.4.1 Inputs

A number of inputs are required for SMIT:

Application description
An application is assumed to be given by a set of task graphs, consisting of Functional Nodes and interconnections between processes needed for communication. This partitioning is typically done manually by an experienced designer, but future tools may perform partitioning also automatically. Additional Quality of Service requirements for the application such as required timing behaviour (e.g. throughput, latency) are also specified in the application description.

Library description
For each Functional Node of an application, one or more implementations have to be provided. A FN implementation is the implementation of a FN on a particular type of tile, e.g. object code for an ARM or configuration data for an FPGA, Montium or XPP. A FN implementation is annotated with performance figures, e.g. the amount of energy it takes to execute the function on a particular tile of the architecture or the amount of cycles it takes to execute the function on a particular tile of the architecture.

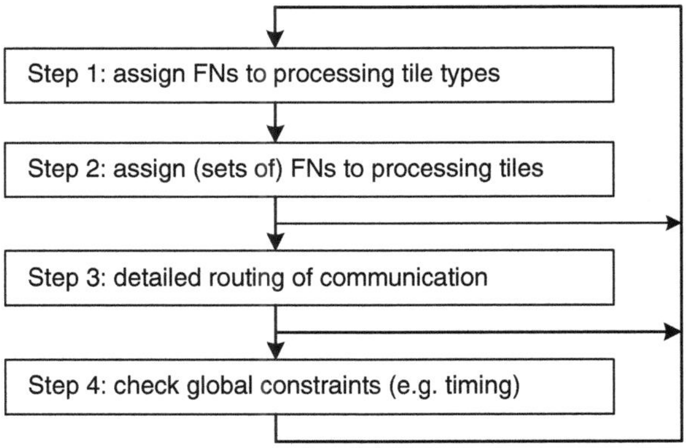

Fig. 6.9 Hierarchical mapping algorithm

Architecture description
The heterogeneous SoC architecture consists of multiple tiles of different types (e.g. ARM, reconfigurable cores) interconnected by a Network-on-Chip (NoC) or Bus. For each tile a number of characteristics have to be provided on beforehand, such as the type of the tile, the amount of available memory, the clock frequency, etc. The NoC consists of routers and links. The links are used to interconnect routers or a tile with a router. It is possible to have different links in parallel between the same source and destination. Also the NoC characteristics have to be provided, such as the topology of the network, the frequency of the clock of the network, latency per router, etc.

6.5.4.2 Mapping Approach

The mapping of an application to a multi-processor architecture is a General Assignment Problem (GAP), which is known to be NP-complete. To deal with the complexity of the mapping problem, SMIT employs a hierarchical iterative approach (Fig. 6.9). The idea is to solve the problem using multiple levels. At each level a particular decision is made that reduces the search space. Decisions of higher levels are considered to be fixed at the lower levels. On higher levels not all details are taken into account to improve the speed of evaluation.

In other words, higher levels use a higher abstraction. This hierarchical approach has the danger that decisions which seem to be promising on a higher level due to the underlying assumptions of the abstraction level, show to be bad or even lead to resulting sub problems which are infeasible. The allowed maximum runtime of the mapper routine is restricted so far that at least a *good* mapping solution, can be achieved. Good mapping guarantees that the Quality of Service requirements of the application are met. *Good* mapping is preferred over *optimal* mapping which would again consume too much processing power and energy.

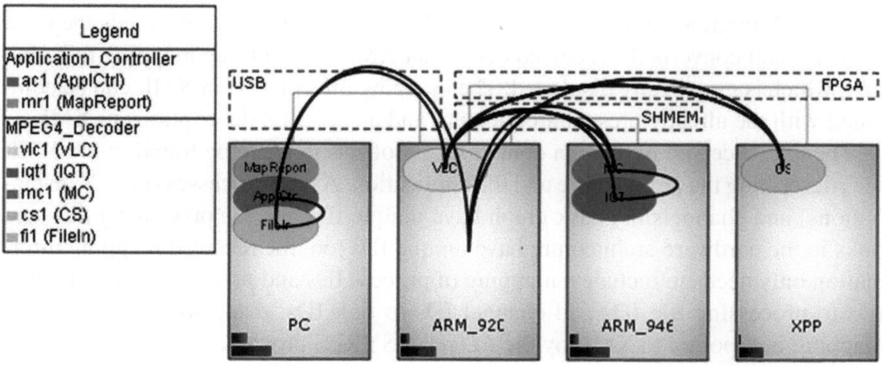

Fig. 6.10 Sample visualization of FNs

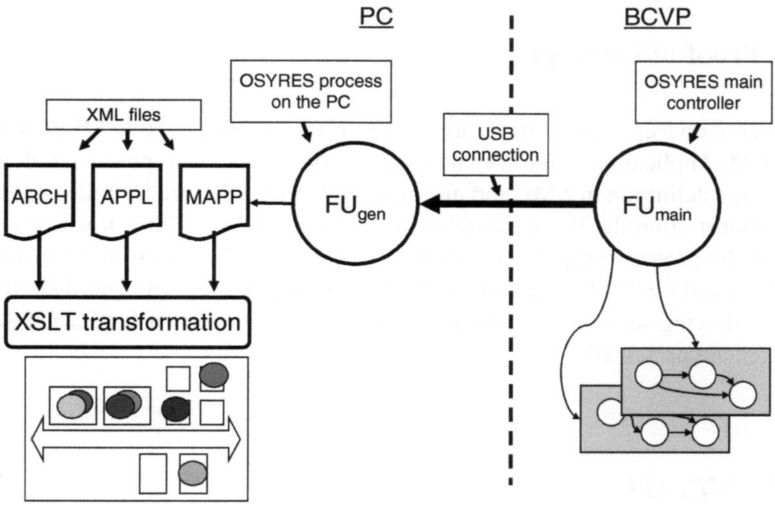

Fig. 6.11 Visualization of mapping using OSYRES

6.5.5 Mapping Visualization

For debugging and demonstration purposes, the mapping proposed by SMIT can be displayed on screen (Fig. 6.10). The visualization tool performs this task by utilizing the proposed XML tool flow in combination with the platform transparent communication layer of OSYRES (Fig. 6.11).

For visualisation two instances of OSYRES are launched: one instance is active on the target platform and the other instance is executed on a PC. Both instances communicate via an USB connection. The mapping information that SMIT provides is sent to the USB interface by a Functional Node (*FUmain*) of OSYRES.

A second Functional Node (*FUgen*) of OSYRES running on the PC reads the USB interface and converts the received data to an XML file with the mapping information, which is graphically displayed. The mapping information in XML can be combined with the already present architecture and application descriptions in XML.

The data received by FUgen contains a set of IDs that can be found in the XML files describing the architecture and the application. As the processes (i.e. FN instantiations) and channels in a task graph have unique IDs and the processing tiles and links in the hardware architecture have unique IDs too, the received mapping information only needs to include a mapping of process IDs and process implementation IDs to processing tile IDs and channel IDs to link IDs, respectively. The current mapping is exported to XML by the *FUgen* OSYRES process.

This approach demonstrates also the benefits of the OSYRES concept: it is easy for the programmer to distribute the functional nodes to several of heterogeneous processing units without the need to care about communication details.

6.6 Proof of Concept

The overall concept was verified through several scenarios targeting MPEG4 Video and DRM. Applications have been designed according to the proposed tool flow, i.e. application definition in XML and mapping to OSYRES functional nodes and links. For demonstration the Basic Verification Platform was used since it was available early in the project and provided reconfigurable cores (Montium implemented on the FPGA and the XPP evaluation board), two ARM processors and hardwired tiles. In the following section some of the scenarios which have been successfully demonstrated are presented.

6.6.1 MPEG4

The MPEG4 Video case stands for a typical video decoding application. The application was chosen in addition to DRM to demonstrate the usability of the concept on very different application fields.

6.6.1.1 MPEG4 Task Graph and Mapping Scenarios

The Task Graph in Fig. 6.12 shows the nodes of the MPEG4 decoding chain which is composed out of four Functional Nodes that have been implemented to run on several tiles (i.e. *PUs* in the OSYRES context):

- Variable Length Coding (VLC): ARM920
- Inverse Quantization (IQT): ARM920, ARM946, XPP

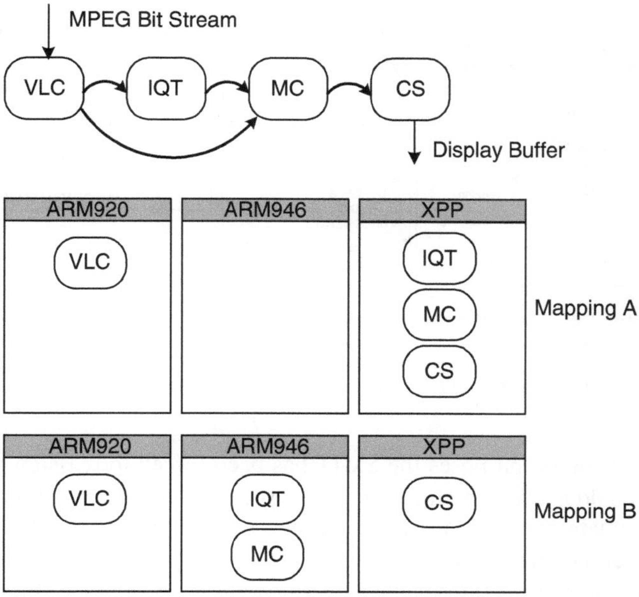

Fig. 6.12 MPEG4 Task Graph and two mapping scenarios

- Motion Compensation (MC): ARM920, ARM946, XPP
- Colour Space Conversion (CS): XPP

Each implementation provides also a performance model (XML) as input for the spatial mapper. The performance models describe the energy for each function, memory requirements, processing resource utilization and communication costs. The performance models have been generated through simulation and a new power estimation tool for the XPP. Energy for the ARM processors has been calculated using datasheet values (mW/MHz) and by profiling the number of clock cycles which are required for a specific function. Based on the Task Graph and performance models SMIT mapped the FNs to the available hardware resources. OSYRES provides the abstraction layer for FN creation and channel establishing. Besides VLC and CS, which can only be executed on specific tiles, SMIT had the choice to distribute the other tasks freely. Note that the dynamic reconfiguration properties of the XPP are used to execute the Functional Nodes sequentially.

The application's switching points have been specified at design time. It is not feasible to allow switching at any point in time in an application since the number of states to be saved and transferred to the target FN varies during execution. Furthermore it may be difficult to transfer processing states from architecture to another, because e.g. the internal data and state representation may differ. For MPEG4 best switching points are I-Frames, since those frames do not need information from previous or subsequent frames.

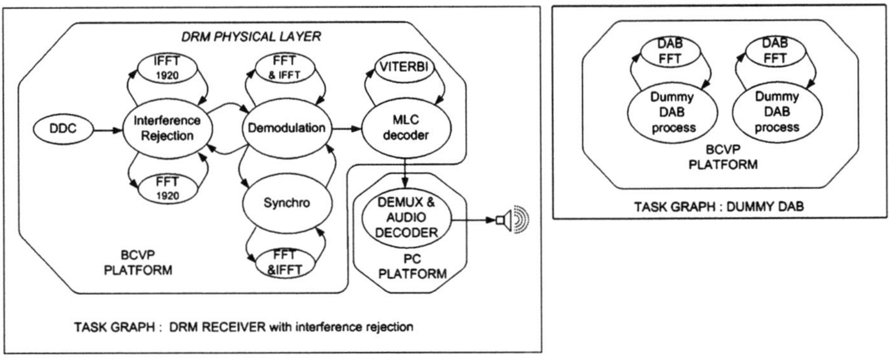

Fig. 6.13 DRM with interference rejection and DAB task graphs

For demonstration purposes the SMIT has been forced to reassign mapping by adding placeholder FNs which can only be executed on a specific tile. SMIT then had to recalculate the mapping. Note that the example in Fig. 6.12 shows only a subset of potential mappings.

A live demonstration with the basic verification platform and XPP showed that with an MPEG4 stream seamless switching between several mappings could be done in real-time without disturbing or interrupting the video output.

6.6.2 DRM

The Digital Radio Mondiale (DRM) physical layer was ported to the OSYRES environment. According to the OSYRES model, the DRM physical layer was described as task graph (Fig. 6.13). The FFT and IFFT used in the DRM interference rejection were implemented for both, the ARM and Montium. Additional features allowed to switch on/off the interference rejection and to launch a background receiver (which is used to scan for alternative frequencies). In addition a "placeholder" DAB receiver (of which only the FFT was implemented) could be added. Several usage scenarios were specified according to the typical needs of a car radio receiver. For demonstration purposes the switching between scenarios was done by keystrokes mimicking varying reception quality. The tests have been performed on the basic verification platform. Available hardware resources (i.e. OSYRES *PUs*) were the ARM processors, the hardwired down converters (DDC), the Viterbi decoder and three reconfigurable Montium tiles. For simplicity reasons audio decoding was performed on a PC linked via USB.

Figure 6.14 shows the task graph of a scenario where interference rejection is added e.g. because of reception getting worse. Interference rejection requires in addition calculating a FFT1920 and an IFFT1920. SMIT selected from all possible mapping the most power efficient resource assignment as shown. So, the most compute-intensive tasks are mapped on the Montium.

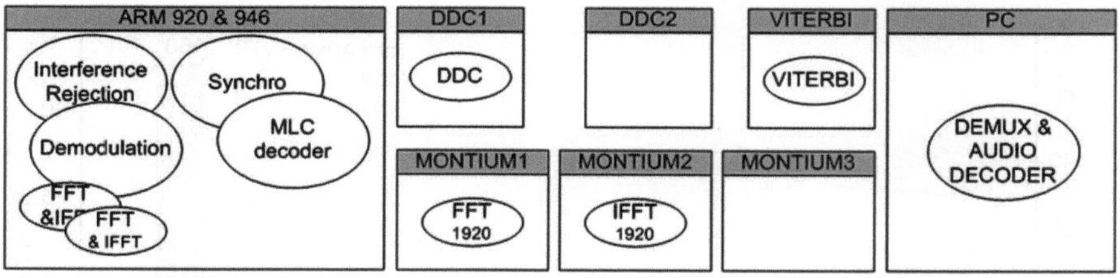

Fig. 6.14 DRM receiver with interference mapping (FFT1920 and IFFT1920)

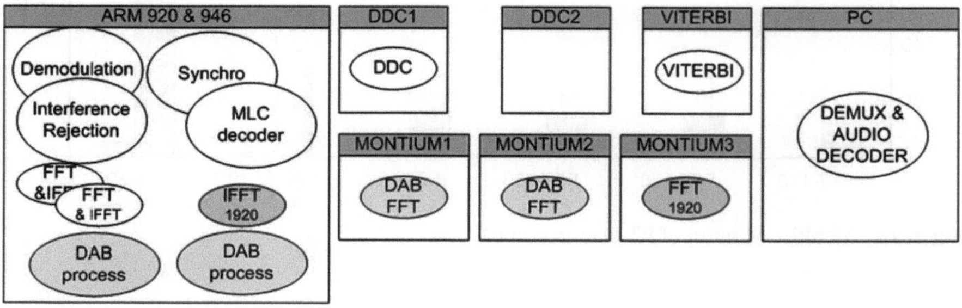

Fig. 6.15 Mapping with additional DAB FFTs

Another scenario in Fig. 6.15 starts again from a DRM receiver and an already mapped DAB receiver running in background (which delivers traffic information in a car radio). In this scenario the DAB has utilized already two Montium tiles to compute the DAB FFTs. Now interference rejection was switched on: SMIT mapped the FFT1920 to the third Montium tile and the IFFT to the ARM. In any case the mapping was performed without interruption of audio.

6.6.2.1 SMIT Mapper and OSYRES Benchmarking

Experiments on the basic platform delivered the following figures on an ARM9 running at 86 MHz while six FNs have been mapped to four PUs.

- SMIT mapping algorithm: 0.7 ms
- OSYRES task graph generation overhead: 13 ms (2 ARMs and 2 Montiums)
- Message transfer: ~0.15 ms (NoC and Montium driver).

6.6.2.2 SoC Demonstration

In order to prove that a complex heterogeneous platform can be implemented on silicon using state-of-the-art ASIC design tools, the Highly Integrated Concept Verification Platform was realized and functionally tested and an estimation of the

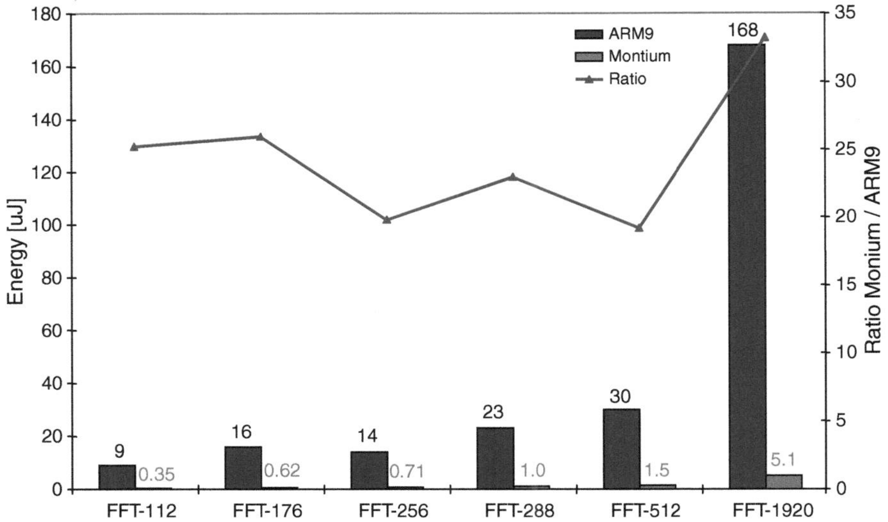

Fig. 6.16 ARM9 – Montium FFT energy comparison

dynamic power was made. The test suite demonstrated that all devices including the Montium and routers were fully functional ("first time right"). Dynamic power estimation was performed on the post-layout netlist. Fig. 6.16 shows the ARM9 and Montium comparison of energy for single FFTs. The power reduction factor is between the factor 20 and 33 in favour of reconfigurable technology compared to general purpose processors. Absolute values of dynamic power consumption of the Montium are in the range of 0.24–0.6 mW/MHz. Reconfiguration and control overhead of the reconfigurable hardware is about 10%. In addition, power reduction comes in combination with higher processing power.

6.6.3 *Outlook*

The intention of 4S was to establish a sample heterogeneous reconfigurable platform and to map tasks to the platform resources during runtime. The mapping algorithm was setup to optimize overall power consumption. With current SoC architectures that have only a small number of heterogeneous cores (< 10) the spatial mapper has only a limited set of mapping scenarios to choose from and optimisation potential is limited. The situation changes if dozens or even hundreds of cores are available. With those complex platforms it is not feasible anymore to define the mapping at design time, thus a runtime tool is mandatory to perform the mapping. The mapping algorithm does not necessarily need optimising energy consumption only. Also other criteria such as QoS may steer the mapping.

An important outcome of 4S is that task mapping can be controlled dynamically during runtime without interrupting the application. The following list gives an outlook to future scenarios which may benefit from the 4S approach:

- Devices with silicon geometries below 65 nm will suffer from severe process variations even on the same die. There will be processing cores which can only operate with reduced frequency while others achieve full frequency. Based on chip test results a performance model for each of cores can be provided and the runtime system automatically maps tasks according to requested application needs without overloading the slower cores.
- Future chips will suffer from defects and silicon lifetime constraints. A spatial mapper can adapt to the availability and current performance of cores during the lifetime of a product. This redundancy concept is completely transparent for application software.
- The 4S concept enables the most aggressive power reduction concepts such as frequency and voltage scaling and dynamic power switching. The operating system maps tasks seamlessly to the configured performance.
- Application software does not need to be tailored to a specific heterogeneous device architecture. The same application software will run on different heterogeneous platforms and hardware re-spins. For example, imagine high end and entry level mobile devices, all running the same application software that automatically distributes its tasks to the available resources.

The mapping concept is currently part of research activities within Thales. They extend the 4S mapping concept by further optimisation parameters. Lessons learned from 4S are further investigated in e.g. the *CRISP* project. The CRISP project studies heterogeneous reconfigurable many-core SoCs, runtime mapping and self-repairing reconfigurable core concepts for improved dependability. An extended version of the XPP reconfigurable processor was implemented in the MORPHEUS [18] project on silicon. Both projects are discussed in specific chapters of this book.

6.7 Conclusion

The 4S project researched innovative hardware concepts along with compile-time and run-time tools to realize a computational powerful and flexible platform for Ambient Systems. On the hardware side, we targeted the development of a dynamically reconfigurable System-On-Chip (SoC) architecture. It basically offers the required flexibility to adapt its hardware resources to diverse processing requirements that result from the demands of different applications as well as environmental changing conditions (e.g. different communication protocols or fading channels). Different to reconfigurable architectures like RAW [16], Pleiades [17], PicoChip [12], MorphoSys [14], Cell [13] or Imagine [15] which are complex multicore designs whose focus is not primarily put on power efficiency, our approach particularly targeted embedded devices where maintaining an ultra-low power profile is of great importance. Therefore we investigated and developed a *heterogeneous* SoC platform concept, composing miscellaneous hardwired as well as configurable analogue and reconfigurable digital core buildings blocks, interconnected by a Network on Chip (NoC) communication infrastructure.

We thereby were able to contribute to the research of reconfigurable architectures as we applied a holistic view on the overall system to achieve energy efficiency. Unlike in the past, various aspects have, therefore, been investigated in depth in an overlapping manner, i.e. tradeoffs in combining reconfigurable analogue and digital building blocks, design-time application characterization for energy-efficient data processing on heterogeneous processing tiles and dynamic migration of tasks to reconfigurable processing tiles to maintain or even improve quality of service and energy efficiency. With respect to the technique mentioned last, 4S also extended the run-time tools by a system-aware middleware solution, called OSYRES. Based on user demands, environmental, power and quality of service constraints its spatial mapper (SMIT) evaluates an optimal assignment of tasks to tiles and thus efficiently exploits the flexibility of computation in time and space. The decision where to place tasks is thereby not limited to the starting time of applications. It rather reacts dynamically on any changes to the mentioned criteria. Thereby, the proposed methodology, to realize applications upon a task graph model, enabled overcoming the drawback of ending and restarting an application to handle task migration on heterogeneous systems. Instead, the migration process is handled by the middleware through interaction with the running application in background, so that the user is unable to recognize the remapping process.

With all the developed methodologies, techniques and hardware concepts the outcome of 4S project had a significant impact on energy efficient embedded computing targeting applications with run-time dynamic behaviour. We have shown that heterogeneous dynamically reconfigurable architectures are very suited to realize flexible and energy efficient embedded computing platforms. This was not only done on theoretical and prototyping level but also verified based on two highly complex chips (analogue and digital) that have been realized within the project. The exploitation of the flexibility of the hardware is thereby given by a set of compile- and run-time tools that enable dynamic mapping of task to computational tiles without interruption of running applications. For the first time ever, this seamless mapping technique has been shown on a heterogeneous dynamically reconfigurable SoC architecture based on two complex applications, MPEG4 and DRM.

Besides technical related aspects, 4S also stimulated commercial interests and activities in reconfigurable computing by attracting distinguished partners of Europe's leading ICT industry. Upon that, the ideas, concepts, methods and hardware designs researched in 4S have been the basis for further European projects, e.g. CRISP [19] and MORPHEUS [18]. Finally, all major industrial and academic partners of the 4S project have perused their activities exploiting reconfigurable technology by participating in consecutive projects.

Acknowledgements The authors wish to acknowledge Jens Becker (ITIV, Karlsruhe Institute of Technology) and Jan Stoter (WMC) for review and, last but not least, all project partners, reviewers and EU project officers for their contributions that enabled to successfully fulfil the goals of the 4S project.

References

1. eCos: http://ecos.sourceware.org
2. OSYRES: http://www.ti-wmc.nl
3. Marcel D. van de Burgwal, Gerard J.M. Smit, Gerard K. Rauwerda and Paul M. Heysters, Hydra: an Energy-efficient and Reconfigurable Network Interface, Proceedings of the 2006 International Conference on Engineering of Reconfigurable Systems & Algorithms, 26–29 June 2006, Las Vegas, USA, pp. 171–177, CSREA Press, ISBN 1-60132-011-6
4. Gerard K. Rauwerda, Paul M. Heysters and Gerard J.M. Smit, Towards Software Defined Radios Using Coarse-Grained Reconfigurable Hardware, IEEE Transactions on Very Large Scale Integration (VLSI) Systems, vol. 16, no. 1, pp. 3–13, January 2008
5. P.T. Wolkotte, G.J.M. Smit, G.K. Rauwerda, L.T. Smit, An Energy-Efficient Reconfigurable Circuit Switched Network-on-Chip. In *Proceedings of the 19th IEEE International Parallel and Distributed Processing Symposium (IPDPS'05) – 12th Reconfigurable Architecture Workshop (RAW 2005)*, 4–8 Apr 2005, Denver, Colorado, USA. 155. IEEE Computer Society. ISBN 0-7695-2312-9
6. XPP-III Processor Overview (White Paper), URL: www. pactxpp.com
7. Reconfiguration on XPP-III Processors (White Paper), URL: www. pactxpp.com
8. Programming XPP-III processors (White Paper), URL: www. pactxpp.com
9. M. Weinhardt, M. Vorbach, V. Baumgarte, and F. May: Using Function Folding to Improve Silicon Efficiency of Reconfigurable Arithmetic Arrays, Proceedings of the IEEE International Conference on Field-Programmable Technology FPT'04, Brisbane, Australia, Dec. 2004
10. G.J.M. Smit, A.B.J. Kokkeler, P.T. Wolkotte, P.K.F. Hölzenspies, M.D. van de Burgwal, and P.M. Heysters. The Chameleon Architecture for Streaming DSP Applications. *EURASIP Journal on Embedded Systems*, 2007. 78082. ISSN 1687–3955
11. L. T. Smit, G. J. Smit, J. L. Hurink, H. Broersma, D. Paulusma, and P. T. Wolkotte, "Run-time assignment of tasks to multiple heterogeneous processors," in *5TH PROGRESS Symposium on Embedded Systems*. STW Technology Foundation, 2004, pp. 185–192.
12. A. Duller, G. Panesar, and D. Towner. Parallel Processing – the picoChip way! In J. Broenink and G. Hilderink, editors, *Communicating Processing Architectures 2003*, pages 125–138, 2003.
13. J. A. Kahle, M. N. Day, H. P. Hofstee, C. R. Johns, T. R. Maeurer, and D. Shippy. Introduction to the cell multiprocessor. In *IBM Journal of Research and Development*, 2005.
14. Guangming Lu, H. Singh, Ming-Hau Lee, N. Bagherzadeh, F.J. Kurdahi, E.M.C. Filho, and V. Castro-Alves. The MorphoSys dynamically reconfigurable system-on-chip. In *Evolvable Hardware, 1999. Proceedings of the First NASA/DoD Workshop on*, pages 152–160, 1999.
15. Jung Ho Ahn, W.J. Dally, B. Khailany, U.J. Kapasi, and A. Das. Evaluating the imagine stream architecture. In *Computer Architecture, 2004. Proceedings. 31st Annual International Symposium on*, pages 14–25, 2004.
16. M. Taylor et al., "The RAW microprocessor: a computational fabric for software circuits and general-purpose programs," *Micro, IEEE*, vol. 22, no. 2, pp. 25–35, 2002.
17. H. Zhang et al., "A 1-V heterogeneous reconfigurable DSP IC for wireless baseband digital signal processing," *Solid-State Circuits, IEEE Journal of*, vol. 35, no. 11, pp. 1697–1704, 2000.
18. F. Thoma et al., "MORPHEUS: heterogeneous reconfigurable computing," in *Field Programmable Logic and Applications, 2007. FPL 2007. International Conference on*, 2007, pp. 409–414.
19. CRISP project, Cutting-edge Reconfigurable ICs for Stream Processing, http://www.crisp-project.eu

Chapter 7
AETHER: Self-Adaptive Networked Entities: Autonomous Computing Elements for Future Pervasive Applications and Technologies

Christian Gamrat, Jean-Marc Philippe, Chris Jesshope, Alex Shafarenko, Labros Bisdounis, Umberto Bondi, Alberto Ferrante, Joan Cabestany, Michael Hübner, Juha Pärsinnen, Jiri Kadlec, Martin Danek, Benoit Tain, Susan Eisenbach, Michel Auguin, Jean-Philippe Diguet, Eric Lenormand, and Jean-Luc Roux

Abstract The ÆTHER project has laid the foundation of a complete new framework for designing and programming computing resources that live in changing environments and need to re-configure their objectives in a dynamic way. This chapter contributes to a strategic research agenda in the field of self-adaptive computing systems. It brings inputs to the reconfigurable hardware community and proposes directions to go for reconfigurable hardware and research on self-adaptive computing; it tries to identify some of the most promising future technologies for reconfiguration, while pointing out the main foreseen Challenges for reconfigurable hardware. This chapter presents the main solutions the ÆTHER project proposed for some of the major concerns in trying to engineer a self-adaptive computing system. The text exposes the ÆTHER vision of self-adaptation and its requirements. It describes and discusses the proposed solutions for tackling self-adaptivity at the various levels of abstractions. It exposes how the developed technologies could be put together in a real methodology and how self-adaptation could then be used in potential applications. Finally and based on lessons learned from ÆTHER, we discuss open issues and research opportunities and put those in perspective along other investigations and roadmaps.

7.1 Project Partners

1. CEA, LIST, France (coordinator)
2. University of Amsterdam, The Netherlands
3. University of Hertfordshire, UK
4. University of Karlsruhe – Karlsruhe Institute of Technology, Germany

---

C. Gamrat (✉)
CEA, LIST, Centre de Saclay - Point Courrier 94,
F-91191 Gif sur Yvette Cedex, France
e-mail: christian.gamrat@cea.fr

J.M.P. Cardoso and M. Hübner (eds.), *Reconfigurable Computing: From FPGAs to Hardware/Software Codesign*, DOI 10.1007/978-1-4614-0061-5_7,

5. Università della Svizzera italiana, Switzerland
6. Imperial College, UK
7. VTT, Finland
8. Universitat Politècnica de Catalunya, Spain
9. UTIA AV CR, Czech Republic
10. CNRS, France
11. INTRACOM, Greece
12. THALES, France
13. ACIES, France

- Project Coordinator: Christian Gamrat, CEA, LIST, France
- Start Date: 2006-01-01
- Expected End Date: 2009-06-30
- EU Program: 6th Framework Programme, FP6-2004-IST-4, FET-ACA Project No. 027611
- Global Budget: 6 M €
- Global Funding by EU: 4 M €
- Contact Author: Jean-Marc Philippe, jean-marc.philippe@cea.fr

7.2 Introduction

A few decades ago, a computer was a very big, expensive and intriguing machine for much of the people. It was thought by many as something that would probably never cross their daily life. It was a time when computers were rare and enormous machines surrounded by millions of intrigued people. Nowadays the picture has completely changed and we find people surrounded by dozens of computers, many of them sitting hidden in the most awkward places. At the same time, we have witnessed a silent change in the way most computing resources are being used: instead of having one mainframe computer used by hundreds of persons, one can see now each person using at the same time multiple computing resources. They can be found everywhere running practical applications in our mobile phones, our TV sets, our cars, our houses, in the streets of our towns, etc. All those embedded computing resources are increasingly heterogeneous and increasingly interconnected and build up very complex intricate computing mesh. Indeed, many modern applications do not rely on a single computing resource but rather on a group of various computing elements. And the situation is bound to be even more complex in the future as new applications needs and new technology emerge.

In the context of applications running on vast amount of heterogeneous and potentially volatile computing resources, the tasks of designing, programming, optimizing and managing systems are key issues. When seeking solutions for those problems, one very straightforward observation comes to mind: it would be far easier for the application designer if each part of the system could embed enough intelligence and independence so that they could locally and dynamically optimize their resources and the way they perform tasks according to the function they have to perform. In such a system, the application designer could focus on the topic he knows best: the application. There is certainly more in the idea of self-adaptive computing than the fact of helping the designer, even though this is an important reason for its use in implementing complex and heterogeneous systems. The computing system would therefore act like an army of tiny assistants (or agents) not only executing a given program but also assisting the designer to dynamically fine-tune the program according to implementation specific details. This intuitive observation has been the root idea envisioned when preparing the ÆTHER project.

The fact that current computing techniques will not be able to cope with the rising complexity of future applications and architectures has long been identified by research groups. One of the most notable efforts has been put forward by the IBM Autonomic Computing Research initiative [1]. Even before this, authors who laid down the foundation of modern computing [2] were aware that building the ideal computer system should involve a degree of autonomy in order to deal with the computing complexity of high-level tasks. High level tasks were then considered as those trying to mimic the human intelligence and cognitive processes. The rationale was then straightforward: an ideal electronic computer (electronic brain) should behave like a biological brain exhibiting a level of autonomy, self-healing and self-control. In the meantime, the fantastic evolution of microelectronics boosted by Moore's law [3] has led to multi-billion transistors chips that can implement hundreds of microprocessors in a typical computing system. At the same time, the complexity of the software layers has dramatically increased, leading to systems that become very difficult to design and operate.

Being very aware of the intricacy of the various levels of a computing system, the ÆTHER consortium embarked on the challenging path of pluri-disciplinary co-engineering three of the main aspects of self-adaptive computing: software programming, run-time system and hardware implementation. This document presents the solutions we came about to address some of the major concerns of building a self-adaptive computing system. After exposing our vision of self-adaptation and its requirements, we describe and discuss the solutions we put forward for tackling self-adaptivity at the various levels of abstraction. We then present how the proposed solutions could be put together in a real methodology and how self-adaptation could then be used in real applications. Finally and based on lessons learned from ÆTHER, we discuss open issues and research opportunities and put those in perspective along other investigations and roadmaps.

7.3 Self-Adaptation

Self-adaptation can be defined as the ability of a system to react to external changes by modifying its behavior based on a local policy. This definition introduces two important concepts:

- Obviously, such a system first needs to exhibit adaptability. Adaptability is the result of a **reactive process** modeled by a feedback loop. **Loops** (for short) are the basic objects of any adaptation process and as such shall be the basic objects to deal with. And **loops are everywhere around us**!
- Secondly there is the notion of **Self**. In the above simplistic definition, "Self" is tightly related to the "local policy". In its simplest form the "Self" of a system is entirely described by its local policy. "Self" is the local behavior of the system, it is the knowledge that needs to be maintained locally and that provides the autonomous (or local) behavior.

Therefore, Self-Adaptation is not only about adaptive loops. Self-adaptation means loops plus knowledge and this was well described by Kephart and Chess [4] of IBM Autonomic Computing research group. A lot can be said on the relevance of this simple definition on computing systems as it implies various notions:

- What does the system need to observe in order to trigger adaptation? What is an observer?
- What is the mechanism allowing the adaptation process? Is it parametric or structural?
- What kind of rule or policy should be applied to the system so that it can adapt?
- What level of supervision is needed? Can the process be completely autonomous?

In the simplest adaptive system, a closed loop allows for controlling a process based on the observation of how the process performs against a given performance target. The target or objective is generally set or computed by an external mean. Example of such adaptive systems includes the dynamic regulators that can be designed using analogue or digital techniques for specific tasks: temperature control, speed regulation, etc. Such systems are generally designed in a specific fashion for the task to be realized. Sensors and actuators play a key role in their implementation. In a self-adaptive system, the target objective can be adjusted locally based on the context and/or the observation of various parameters. Let us look at a simple example. A fairly basic application is one of a system that dynamically adapts the temperature of a device. In its simplest form, the circuit depicted in Fig. 7.1 will continuously control the temperature of the object by applying a corrective action on the actuator directly computed from the difference between the measured and target temperatures. Such a simple setup is alright if the system is in a known context, the components do not change much and finally the target can be expressed in a very accurate way like a numerical temperature value. Now what happens if something changes in the system environment so that the statically designed control loop can no longer achieve its goal?

In the simple adaptive loop, the best that can be done is probably to detect that the system is entering a non-optimal scenario by monitoring the closed loop error and identifying that the programmed behavior of the controller cannot cope with the situation. In this case the best that can be done is to raise an error signal and inform that the system is diverging from an optimal scenario. It is important to note that the optimal scenario has been defined statically at design time and is embedded into the controller behavior.

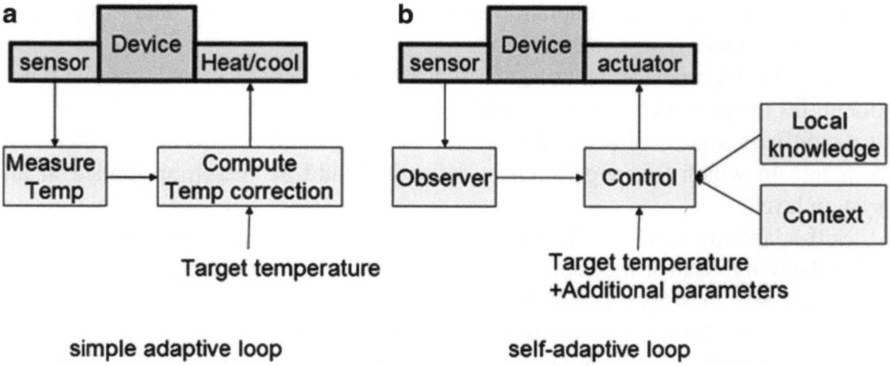

Fig. 7.1 Representation of a simple adaptive loop (**a**) compared to a self-adaptive loop. (**b**) Self-adaptation is more than an adaptation loop; it is control loop + knowledge

In the self-adaptive system, the optimal scenario (probably the most frequent one) would be part of the initial knowledge embedded into the system at design time. A knowledge management structure would constantly monitor the behavior of the loop and keep track of patterns of activity. This unit would also keep track of the related context and store initial parameters. In such an arrangement, a local knowledge base contains a set of optimal scenarios in the form of various parameters (e.g. range of values, bitstreams, algorithms, etc.) applicable for different known situations or contexts. In case a non-optimal scenario occurs, a self-adaptive system will be able to start a process for selecting a scenario (i.e. part of the current knowledge base) adapted to the current context. In a more sophisticated implementation, the system will be able to synthesize a new scenario adapted to an unknown situation and update the local knowledge base.

That's where we enter the realm of self-adaptivity. The first ability of the system should be to detect that it is diverging from a normal scenario. The second ability it must have is to take a decision on what is the best action to take when the nominal control scenario is out of bounds.

Using a programmable computer, adaptive and self-adaptive loops can be easily programmed in software using known software techniques but there is more than the mere programming of loops in the design of a self-adaptive system.

Software can be programmed in a self-adaptive way without necessarily relying on specific support mechanisms provided by the system (self-adaptation embedded in the source code); this self-adaptation mechanism can be entirely described at the software level. When software is considered, a number of possibilities for self-adaptation are available: run-time and dynamic change of the application goals (i.e., the application changes its high-level requirements for the system), adaptation based on selection of different behaviors (i.e., a different implementation of the same algorithm is selected), and intra-algorithm adaptation (i.e., some of the parameters of the considered algorithm implementations are changed at run-time). All of these self-adaptation mechanisms could be directly implemented in the software application, even though the first method requires support from the system to be effective.

This is fine as long as the adaptation process does not require any data from or any action to the underlying system.

Nevertheless, management of self-adaptivity should not be mixed with the functionalities in applications as this would put a burden on application developers and it would be a source of programming errors. *Separation of concerns*, which is a principle in which this separation is defined, should be therefore applied at all system levels.

Self-adaptivity can be both in the software and in the underlying hardware; the self-adaptivity control mechanism must provide a way to manage adaptations in hardware and in software effectively, thus allowing reaching application goals. Adaptation at different levels must be coordinated properly.

At hardware level there are two possible kinds of self-adaptation: structural (i.e., change in the functional units or in the interconnections) and on parameters (i.e., hardware parameters – such as frequency – are changed during run-time). Self-adaptive hardware can either manage adaptations internally or it can delegate (partly or entirely) this management to the software layer. For providing internal self-adaptation, the hardware needs to be able to change its configuration in a transparent way with respect to the software layer. Whenever software support is required for self-adaptivity, the hardware must notify to the software its reconfiguration capabilities. In both cases, the hardware may provide some information on the application execution and on the parameters that can be monitored and/or directly controlled by the software layer. Different kinds of hardware architectures may be utilized in a self-adaptive system. Software developers should be enabled to write applications without necessarily knowing the structure of the underlying hardware and the mechanisms used for self-adaptation. In fact, the management of all of these details would make the job of the programmer too complex, it would break portability of applications, and it would remove any convenience in using self-adaptive systems.

The task is then to find ways to engineering loops within the complete computing system: loops in the software, loops in the run-time system and loops at the processor hardware level. Moreover, the adaptation loops at the various level of the computing system shall be implemented in a way that interactions across them are not only possible but inevitable.

7.4 Self-Adaptation from the Software Engineering Point of View

What is *self-* at the software engineering level? Here we are not talking about platforms that adapt to the environment, we are talking about a program that is ***not*** engineered to respond to some changing environment (that would be merely part of the ***control*** structure of the code), but rather is ***presented*** in a way that makes it possible to achieve self-adaptation at the ***system*** level. Hence self-adaptivity in software engineering is always an indirect phenomenon, its manifestation is more efficacy than effectiveness, and in that it is an enabling factor rather than the effect itself.

How is software to be engineered to promote ***self***-adaptivity? Obviously, by trying not to ***prevent*** it by the obfuscation of the algorithmic properties with the machinery of **imperative code**. Whatever layers of adaptation lie below the top level of software engineering, these layers would need to have room for manoeuvre if they are to deploy their adaptation capabilities. But what specifically do we mean by this?

There are two main forms in which an application can be presented. First of all, it can be phrased in terms of state-transitions of an application-specific virtual machine. Indeed, that is what typically happens in software development processes in industry nowadays. A programmer creates a virtualization level by preparing program code for basic computations of an application domain, such as FFT, filters, correlators, etc., in the form of library function, and then a program is written which follows a transition diagram of the "business logic" (a video encoder would be one good example) using library functions with each transition. The problem with this form of presentation is that the notional state machine that makes those transitions is too rigid a metaphor, and so it encourages the implementer to realize it literally as a sequential state-transition process. The notional transition diagram, however, is rich in symmetry, which enables a good programmer to produce parallel code, whereby several avenues across the diagram are taken at the same time while preserving its sequential semantics. Nevertheless, doing so requires a deeper understanding of the business logic than the original design required and/or passed on. Such understanding is not readily available given the prevailing training and recruitment paradigms in industry, but even where it is, the process of designing genuinely parallel code is very complicated and error-prone.

There is another form in which programs can be presented. That form is based on the view of an application as a kind of dynamic circuit, in which information flows along streams that connect component nodes. It is **a *stream processing* view**, in which a program is conceptualized as a graph and the nodes of that graph define the *mapping* of inputs onto outputs. For example a node that multiplies input data by two before outputting it is seen as an order-preserving mapping of any x in the input stream onto the output stream with the value of the corresponding element being $2x$. The radical difference from the state transition view is that the "library function" here, i.e. multiply-by-2, is not *called* by a transition machine of any kind; instead it is activated by the arrival of the data itself. The stream processing view is inherently non-prescriptive as far as the order of actions, it is asynchronous and naturally parallel. Moreover, it is highly available to programmers of all levels of proficiency since the desirable properties (concurrency, asynchrony, decentralization) are not ***engineered*** by the developer, but are discernibly present simply in the way the application is presented. They aid self-adaptivity by allowing the ***system*** to choose its way of executing the program, by activating those nodes that have data and are not overloaded, while reducing the pressure on the nodes that are either overloaded or have insufficient resources. The programmer, the system designer and the application domain specialist are not involved in making those adaptations; rather the system itself finds its way through the configuration space at run time.

In ÆTHER we have seen it as our purpose to employ the stream-processing abstraction as a vehicle of software specification and implementation. In order to expose as much potential self-adaptivity to the lower levels as possible we came up with a simple design principle, which we call ***aggressive decomposition***. Unlike the state transition view, where splitting a state machine into several communicating state machines is potentially error-prone, the stream processing view is essentially safe: for example, splitting a node into two connected in a pipeline is as well understood as the mathematical notion of function composition. Communication overheads are not an issue either: if the graph structure becomes too fine, one can fuse a graph segment back into a single node down at the implementation level without jeopardizing the semantic correctness.

The principle of aggressive decomposition is stated as follows:

Decompose the application into a graph using the smallest meaningful nodes possible

By a *meaningful node* we assume a node whose function can be understood in terms of the application domain concepts from its name and interface definition. For example, an FFT node is meaningful, and so is matrix dot product, but multiplication by 2 may be not.

The way we propose to achieve the required decomposition is a top-down strategy. A designer in collaboration with an application domain expert produce a stream processing graph annotated with nodal functions at the nodes and data types at the arcs. Then each nodal function is considered in turn. For a node that is complex enough to be decomposable, a subgraph is produced which replaces the node, until all nodes are smallest meaningful nodes. Then an application programmer is engaged to produce nodal code for all the nodes according to the relevant application domain algorithms, while an independent *concurrency engineer* writes some coordination code (using the language proposed by ÆTHER, S-Net) which defines the graph structure and the packaging and synchronization of any data as it moves about the graph. In reality, the strategy could be partly bottom-up, partly top-down: some components may already be available and some networks can be designed without the knowledge of the bigger picture. Finally the coordination program is presented to a self-adaptive SANE platform, which dynamically merges sections of the graph into single nodes for adaptation purposes and from time to time splits merged nodes back into networks when the conditions change. We have partly implemented this vision and have shown that this style of application design is possible and natural to a range of applications. It is up to the future research to demonstrate that lower levels of the system hierarchy as envisaged by ÆTHER can use the exposed ***adaptability*** to achieve ***self-adaptation***. ÆTHER has certainly made a start in this direction.

7.5 Separation of Concerns

Dijkstra was probably the first who saw software engineering as a systematic activity based on separation of concerns [5]. Indeed, it is as true now as it was in his life time, only the concerns have become much more involved. To an extent, the software

Table 7.1 Separation of concerns between concurrency engineering and application programming

Concurrency engineer	Applications programmer
Basic understanding of component logistics	Expert knowledge of component logic
Expert knowledge of concurrency issues	Basic understanding of component composition for a given application
Focus on: coordination of components in multi/many-core systems	Focus on: algorithms, correctness, abstract complexity of components
Additional focus on: system self-adaptivity and flexibility	Additional focus on: component compatibility and generality

engineering strategy outlined above is about separation of various concerns: those of adaptivity from those of program specification, those of coordination from those of computation, etc. However, we believe that the same level of disruptiveness is required in the mechanisms of *abstraction* (another name for separation of concerns) as it is in more immediate implementation-related aspects of application development. Here the principle that ÆTHER has put forward is one of ***universal componentization***. The components in our stream processing view are, naturally, graph nodes. What is new here is that we do not allow them to have a variety of connection types and a variety of state-transition behaviors.[1] We have found that for the many types of applications components do not need to have a *persistent* state. This means that whatever internal stages the function may go through, whatever decisions may be made that cause it to change the course of computation, as soon as the result is produced and sent to the output, the state can safely be destroyed. This means that the processing node is a pure function that inputs one message and outputs zero, one or more messages and then re-initializes itself.

Moreover, we institute a SISO (single-input, single-output) principle [6, 7], whereby the component has only one input stream and produces only one output stream. On the surface this strategy seems very restrictive, but in fact it is not. The reason for it is our use of non-determinism as a phenomenon that almost replaces the multiplicity of incoming stream connection by allowing data to be arbitrarily ordered when streams are merged into one. It is yet another principle that promotes uniformity and separation of concerns. Suffice it to say that careful use of data types and some structural second-order components that we call *combinators* make non-determinism an enabling mechanism for multiply connected networks of singly connected components, as well as an additional mechanism of self-adaptivity.

Our approach draws a clear dividing line between the coordination infrastructure, fully defined by the concurrency engineer, and the computational infrastructure for which the applications programmer is responsible, and that line also divides up the variety of concerns (see Table 7.1). The concurrency engineer sees components as black boxes equipped with a certain meaningful interface. It is his concern to supply the

[1]It should be said that despite the fact that the whole application is a graph rather than a state-transition machine, the nodal functions are pieces of conventional code written by an ordinary applications programmer, hence they *do* have *a* state-transition behavior.

required data by bringing in, mixing and synchronizing various streams present in the graph in arc form so that a unit of work could be produced from a unit of message.

It is also the concurrency engineer's concern to ensure that the unit of work which produces output messages through the component interface is supported by the outgoing arcs that will deliver these messages to the nodes that need them for their units of work. The concurrency engineer has no great need to understand the intricacies of the processing scheme; if the designer has done his job properly, the same engineer can be used to process data in areas as different as multimedia compression and high-energy physics. What will change is the graph and its labelling, but not the nature of the stream coordination. The quality indicator of the concurrency engineer is his ability to appreciate the system's requirement for self-adaptation and to introduce monitoring, feedback and reconfiguration facilities on top of the basic data-processing scheme. He is empowered to do so by the high level of abstraction provided by our paradigm, which guarantees that encapsulated data cannot mix and be delivered to the wrong place in a type-correct program.

At the other end of the software production line is an applications programmer, now liberated from *all and every* communication, concurrency or synchronization concern. The programmer now assumes the availability of bite-size work, prepackaged for execution with all the necessary data and state information that the node needs, and for which the computation must result in the production of zero or more messages. Even the fate of these messages is not a concern for the programmer, only the values contained in them. Never before in the history of distributed parallel computing was the application programmer so much liberated. There is no miracle here either: the functions formerly foisted on that programmer are now passed on to a generalist (as far as the application domain) concurrency engineering expert. That is the main achievement of the ÆTHER software engineering philosophy – and the reader will notice that its feasibility is inexorably linked with the adaptivity of the platform. Without that adaptivity, the concerns of parallel execution cannot be approached by the concurrency engineer *generally*, and as a consequence, some of them would be passed up to the applications programmer thus destroying the separation of concerns.

7.6 Self-Adaptation from the System Engineering Point of View

First we make the observation that executing sequential code on a conventional core does not really expose much opportunity for adaptation of any kind. Some tradeoffs may be possible here in terms of performance vs. power requirements but it is not until we introduce the concerns of concurrency and reconfiguration that this picture becomes anything other than one dimensional. Thus we assert that the fundamental issues in adaptation from the systems engineering point of view are largely concerned with defining concurrent units of work, mapping those units onto available processing resources and providing a schedule for execution, where multiple units of work are mapped to the same resource. The latter is critical in providing an efficient utilization of a given resource when embedded in an asynchronous networked environment. The scheduling of virtual concurrency or parallel slackness [8] is the only generic mechanism available to enable an application to tolerate high latency operations, while maintaining throughput at the processing resource. Moreover, for efficiency, that scheduling should be data driven rather than based on any polling mechanism. This is a fundamental tenant of the dataflow principle, which executes a unit of work based on data availability rather than through some pre-programmed sequence [9].

Having established what we mean by adaptation, again we have to ask what the self at the system engineering level is. And again, we are **not** talking about control structures in the program that are engineered to respond to some changing environment. This would need to be predicated on the program having knowledge of its environment or mapping. Self-adaptivity in systems engineering comes about through the dynamic mapping and scheduling of defined units of work onto available resources in order to meet the computational demands of the application, while respecting any system constraints such as resource limitation, power consumption and physical location. Thus self implies a layer, component or components in the system, which given some knowledge of its environment and given some knowledge of its instantaneous or predicted computational load, reconfigures itself by mapping and scheduling the computation to best meet any constraints imposed upon it.

Clearly, one of the key separations of concern in systems engineering for self-adaptation is the **separation of the definition of units of work from their mapping and scheduling on a set of resources**. This implies providing the means to decompose programs into concurrent units of work in an abstract manner that does not make any assumptions about mapping and which moreover retains some characteristics important for such mapping. The main issue here is to be able to capture locality (i.e. tasks or units of work communicating with each other) in an abstract manner.

7.6.1 Granularity of Units of Work

One of the key issues in any form of parallel computing, which determines the efficiency of computation is the granularity of the unit of work which is mapped (distributed to another node) and/or scheduled.

From the point of view of distribution, if the cost of communication (time and/or energy) is large compared to the cost of computation at the remote place (time and or energy saved), then there is little to be gained from distribution unless the amount of computation performed there can be increased. In other words for a given cost in distribution, there is an amount of computation or grain size below which it does not make any sense to distribute the computation.

Similarly, from the point of view of managing concurrency locally, there is a cost for creating, synchronizing and scheduling a unit of work. Again, if the cost of managing concurrency (time and/or energy) is large compared to the cost of computation at the remote place (time and or energy saved), then there is little to be gained from executing the unit of work concurrently unless the amount of computation performed there can be increased.

These constraints provide quite a dilemma in being able to define a virtual machine model for self-adaptation. On the one hand we need to separate the decomposition of a program into concurrent units of work from the processing resources to which those units of work will be mapped. Yet on the other hand, as we see above, there are constraints on the minimum size of a computation. The only solution to this dilemma that maintains generality is to statically decompose the application into the finest level of granularity possible and then to dynamically aggregate units of work (sequentialize the concurrency) where necessary. Unless a fine grain approach is adopted, fine-grain architectures cannot fully exploit the concurrency available, which may only be at the instruction level in some algorithms. However, a well-defined concurrent composition can always be executed with a sequential schedule to meet the granularity constraints imposed by a particular processing resource.

This in turn raises a number of other issues, such as how to represent the code for the abstract machine and how to dynamically transform this code so as to perform this aggregation (*granularization*). These issues are best dealt with from a more concrete perspective and will be picked up once the Self-Adaptive Virtual Processor has been defined.

7.6.2 SVP – An Abstract Model of Concurrency

In the discussion above two major issues have been identified. Self-adaptation requires the abstract decomposition of a computation into a maximally concurrent representation. That computation must then be dynamically mapped to available resources in order to meet the constraints imposed by the computation itself, i.e. non-functional constraints or by its execution environment.

To achieve this we have defined the Self-adaptive Virtual processor, (SVP), which captures concurrency as hierarchical families of threads, where dependencies between threads are captured by defining synchronizing variables in the code (i-structures or dataflow synchronizers). SVP [10] is an abstract model of concurrency developed from prior work on Dynamically-scheduled RISC (DRISC) architectures [11] and refined into a general computational model in the NWO Microgrids (2005–2009, [12]) and EU ÆTHER (2006–2009, [13]) projects.

In order to justify the choices made in defining SVP, against the more general model of fine-grain dataflow, a number of patterns of concurrent execution need to be explored. The first of these is *replicated concurrency*. Here a unit of work is often a loop body, which can be executed independently a given number of times. That number may be statically known, dynamically know prior to concurrency creation or may be dynamically determined during concurrent execution. With the exception of dynamically terminated replicated concurrency, this is the low hanging fruit, exploited in most programming and execution models. These independent units of work can be distributed or aggregated quite trivially. In SVP, this is captured by the SVP create action, which creates a family of identical threads, where threads are differentiated only by their index, which is initialized on creation.

A second form of concurrent computation often exploited in concurrency models is *asynchronous functional concurrency,* where the unit of computation is a function application. Here, concurrency is obtained by executing one or more functions asynchronously with respect to the "calling" function. There are two approaches to this form of concurrency: *non-blocking,* where all parameters to the function need to be defined prior to its concurrent invocation and *blocking,* where a function is invoked regardless of whether a parameter is yet defined and where the function blocks when it requires that parameter. The latter exposes more fine-grain concurrency, as whereas the unit of scheduling in non-blocking execution is the function itself, in blocking execution it becomes a partition on the function defined by the synchronization points determined by the functions use of its parameters. This means that, in the limit, the units of work that are scheduled may be individual instructions. Dataflow computation is asynchronous functional concurrency, where all functions are defined as machine instructions, i.e. where control flow is replaced entirely with the dataflow firing rule, which required all operands of an instruction to be defined before an instruction is scheduled.

The SVP model combines control flow and dataflow for efficiency in implementation. It implements asynchronous functional concurrency in a blocking manner, where a function of arbitrary granularity is dynamically instantiated using the create action and where the function definition defines certain of its parameters as synchronizing.

In dataflow, a program is represented by a directed acyclic graph, where the nodes represent the instructions and the arcs the dependencies between those instructions. In dynamic dataflow, cycles, as induced for example by loops, are removed by the expedient of coloring or tagging the graph's edges. There are no other limitations or the form of the graph, which exposes all possible concurrency in a computation, which is both its advantage and achiles heel. The main barrier to its implementation is in only being able to expose a fraction of this dependency graph at any time due to the constraints of only being able to implement a finite set of synchronizers. The SVP approach is preferred over this approach in order to limit the amount of resources required for the capture of a program. It is much more efficient to use sequence as a synchronizer where the delay in instruction execution is known and can be statically scheduled.

The final form of concurrency supported in SVP is that of a classic pipeline, which is implemented as replicated concurrency with a linear dependency (or dependencies) between the threads created in index order. This is really an optimization on asynchronous functional concurrency, which reduces the cost of concurrency creation.

The SVP model is a hierarchical one. Any thread function in SVP may create subordinate threads to an arbitrary level. The only constraint on the depth of such recursive concurrency comes about in implementation, where a finite set of synchronizers (i-structures) must span all dependencies exposed. As dependencies in replicated concurrency are constrained to be linear in index order, it is trivial to sequence the creation of dependent threads in a single family and restrict the synchronizing resources required to those required by a single thread. The same applies to recursive creates, where synchronizers must be provided for all functions in the create tree. Providing both concurrent and sequential versions of the function are available, the run-time system may switch to sequential execution in a deep create tree at the point where the last of the synchronizers is allocated.

7.6.3 Implementations of SVP

SVP is defined in terms of its concurrency control actions and captured in an extension to the C language where the concurrency controls and synchronizing variables extend the syntax and semantics of the C language in μTC [14]. Using the create action and its corresponding synchronization point, which determines when the created function has terminated (and updated global memory), both asynchronous functional concurrency and replicated concurrency can be captured in the model. This provides an abstract representation of a deterministic concurrent program. To achieve an implementation SVP must provide a number of further actions and concepts. *Places* are introduced as identifiers for processing resources and in an implementation, a run-time system will attach a place at any appropriate point in the concurrency tree. The model also provides an asynchronous mechanism to terminate the execution of a concurrent unit of work and a mechanism to reflect on the result of such actions, so that fault tolerance can be built into the systems developed using SVP. Finally, the model is made fully generic (and non-deterministic) by the use of an exclusive place used as a mutex, which sequentializes any request to execute a function at that place.

In the ÆTHER project we have implemented SVP at a coarse level of granularity using compilers from μTC that translate this language into C++ and bind the resulting code with a library that implements the SVP actions. This implementation has been used to demonstrate a complete tool chain from S-Net programs down to the execution of functions on FPGAs, which are dynamically selected at run time for executing S-Net components.

One of the successes of the SVP approach outside ÆTHER project has been its use in the EU Apple-CORE [15] project, which is investigating the implementation of SVP at the finest level of granularity, namely in the ISA of a DRISC processor. In this implementation binary code can be used as the representation for a self-adaptive program and the various units of work can be mapped and scheduled to cluster of cores of varying size with any changes to the binary code at all. This project is also developing high-level compilers to and from this model. This work includes dedicated SVP support for the functional array processing language SAC [16, 17], as well as an automatically parallelizing compiler for legacy C code [18].

7.7 At the Hardware Level: Self-Adaptive Networked Entities

At the processor level, the key concept that supports self-adaptive properties at the hardware level is the Self-Adaptive Networked Element (**SANE**). By designing each computing element along the SANE design pattern, we guarantee its seamless integration within the ÆTHER framework. The prerequisite is that each SANE processor implements the SVP protocol for concurrency and resource management. Furthermore, SANE implements a compute-monitor-control loop, making computing elements aware of what they are currently executing. This is the basic mechanism which allows a SANE element to manage as much things as possible on a local basis. This way each and every bit of computing resource has a level of autonomy that makes it suitable to accept jobs delegated by the Run-Time system and returns reports indicating the actual cost of execution. Each SANE is responsible for meeting the conditions of any contract that it may agree to with the SVP program. The SANE design patterns can be applied to a variety of hardware targets. In the course of the ÆTHER project we have explored implementations based on standard processors and hardware reconfigurable technologies but we believe that it could also be applied to future architectures and technologies such as bio-inspired and nano-fabricated hardware.

Basically, the SANE concept has two main properties which are described in [19]. The first one is that it can react autonomously by changing its working parameters or its structure to improve its behavior (i.e. to meet the requirements of the contract it has accepted). It is represented by the closed "compute, observe, control" loop in the Fig. 7.2 that provides the SANE with embedded self-adaptation. The second property of the SANE is that it is a collaborative entity which is able to publish its capabilities and also listen to other SANEs for their capabilities. It is also able to delegate part of its work to other SANEs. This collaborative interface is based on the SVP protocol that is described in the previous section.

As shown in Fig. 7.2, the SANE has been decomposed into four main blocks. Each of the blocks represents one of the four main properties of the SANE. The Computing Engine block is meant to be a computing resource for processing data in the most flexible way. The computation process is monitored trough an observer that computes metrics about properties related to the parameters to be monitored. The goal of the observation feature is to capture self-adaptation triggering events in order to feed the adaptation controller. Based on the information about the self-adaptation triggering events, the adaptation controller takes all the required decisions to face potential problems or to optimize the overall computation and perform parametric or structural adaptation if needed. This observation-control loop enables the SANE to manage its own resources.

The SANE has the ability to collaborate with other SANEs by means of a collaboration interface. This collaboration interface enables the SANE to publish its abilities to the environment. It is also able to delegate and to receive applications or parts of them to/from other SANEs. It is based on the SVP/SEP protocol (SEP, for System Environment Process, being a resource negotiation protocol on top of SVP [20]).

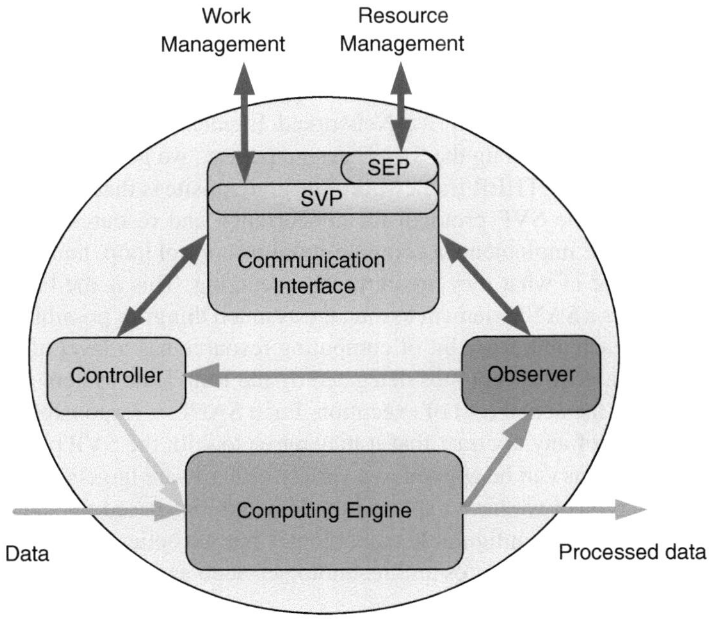

Fig. 7.2 General view of the SANE. The compute, observe, control loop represents the embedded self-adaptive behavior of the SANE whereas the communication interface represents its ability to collaborate with other SANEs

When defining a self-adaptive architecture, the designer needs to list the different events the architecture will have to cope with (e.g. failures or optimization to new applications). This step enables to provide the architecture with the right sensors/observation features to capture them. The observation feature of the SANE enables it to track self-adaptation triggering events (i.e. modifications of the environment or of the computation that will imply an adaption). From the hardware point of view, these self-adaptation triggering events can be classified in different categories, listed in Tables 7.2–7.4.

A designer should implement a SANE regarding these events in order to both implement the required sensors (to observe the events) and to provide the implementation with the required flexibility (including the inner controlling algorithms) so as to be able to react to these events.

A self-adaptive system needs observations to know about its state (self-awareness) and the state of the environment (context awareness). These observations are done through the use of sensors that take raw information and send them to the observer. The observer transforms this information into variables and metrics (i.e. complex variables or events) that can be exploited by the control to:

1. Trigger an adaptation process
2. Choose which part of the system to adapt and the type of required adaptation.

Table 7.2 Part I. Events, related measures, and possible adaptations at the hardware level (delegation and observations of the computing environment are more related to system level but may also have an influence on the local hardware resources)

Event	Related measures/observations	Possible adaptations
Deadline miss	Execution time by online profiling or workload prediction	1. Increase the processing speed by increasing the clock rate (or processing rate) 2. Change the implementation or the routing of the task to gain speed if possible (possibly switch to an interpolation implementation if exact implementation is too slow) 3. Take advantage of possible application parallelism by requesting additional local resources if available 4. Take advantage of the pervasiveness of the system by requesting computing resources in the environment to delegate the computation
Lack of battery power	Battery power given by sensors	1. Decrease clock rate or processing rate (if possible) 2. Change the implementation for a less power consuming one (including change of the physical routing links, clock or power supply gating, etc.) 3. Delegate the computation to another resource in the environment

Based on the taxonomy given in Tables 7.2–7.4, a number of sensors that can be used to take knowledge from the environment and from the system itself can be listed.

Table 7.5 shows the variety of the different variables that may be monitored and exploited by the observer. One can notice that the types of the observations are very different from each other. Thus, the interfaces between the sensors and the observer are to be considered carefully for such a system. In the end, the observer is a data transformation process that gathers the raw observations from the sensors and computes a report composed of a set of metrics. These metrics help the controller to monitor if the selected self-adaptation triggering events occur. The different adaptation scenarios described in Tables 7.2–7.4 are not possible with all the implementations. For example, changing the hardware implementation of a task is only possible with reconfigurable architectures such as FPGAs.

Table 7.3 Part II. Events, related measures, and possible adaptations at the hardware level (delegation and observations of the computing environment are more related to system level but may also have an influence on the local hardware resources)

Event	Related measures/observations	Possible adaptations
Change of the data to be processed	Observation of data type through tags Observations of the data values	1. Dynamic modification of the interfaces (clock rate, synchronization mechanisms, protocols, bus bitwidth, etc.) 2. Modification of the implementation of the function
Change of the mission (application or constraints)	No real associated measure: it depends on a control command given from the user who wants to execute a new program or another resource that delegate a task	1. Change the current application context 2. Tune the parameters of the already loaded task (soft and fast adaptation) 3. Change the task or its implementation (hard and slow adaptation)

In traditional microprocessors, there is no possibility of structural adaptation, whereas they are quite efficient for parametric adaptation (example of DVFS – Dynamic Voltage and Frequency Scaling – algorithms for frequency and voltage scaling). The adaptation of the implementation must be done at the software level. This implies that the adaptation possibilities depend on the level of flexibility provided to the computing entity by the chip designers. A completely flexible design will have the potential to deal with a larger number of situations in comparison with a design that only have one or two degrees of freedom. But the potential advantages of a completely flexible design also imply drawbacks leading the system designers to consider trade-offs.

These drawbacks are linked with both the degree of flexibility of the system and the adaptation mechanism (adaptation rules and learning abilities). The more flexible the architecture is, the more complex the adaptation engine will be (since the design and space exploration for selecting the good target configuration is larger). The adaptation mechanism is also of great importance with respect to the control complexity. This mechanism can vary from simple control mechanisms such as "if then else" statements that allow self-adaptation only between well-known situations to learning and evolvable algorithms that may be very complex and take time to converge, even if low convergence time is a practical criterion for self-adaptive systems (see Table 7.6). In order to be useful, the decision taking mechanism and the actual adaptation needs to be done quickly at a rate depending on the application processing rate.

It is obvious that the observer cannot capture all the variables to know about the complete state of the object and/or of the environment (due to the overhead of the observers and sensors). The sensors and the observer help the controller to construct and evolve (through adaptation rules) a model of the environment and the

Table 7.4 Part III. Events, related measures, and possible adaptations at the hardware level (delegation and observations of the computing environment are more related to system level but may also have an influence on the local hardware resources)

Event	Related measures/observations	Possible adaptations
Failure of a computing node or of the network (partial or global: only one part of the chip is concerned or not, transient or definitive)	Results from a set of available self-tests or remote tests (another entity does not respond to requests, etc.)	1. In case of partial failure (chip level): avoiding the faulty unit (re-routing process by self-organization, switching to a degraded mode, try to repair 2. In case of global failure: try to repair (chip level), re-routing/isolation of the faulty chip (system level) 3. In case of transient failure: error detecting/correcting code (depends on the error frequency and the deadline), etc. 4. In case of permanent failure: error-correcting code or re-routing
Change in the Environment (new resources coming/leaving, value of parameter, etc.)	Depends on the Collaboration protocol between the entities. For a publish/discovery mechanism, an entity can observe the available services offered by new resources. The system must also be aware of leaving resources.	1. Take into account new needs based on experience: reconfigure to accept new applications or tasks seamlessly (prediction) 2. Take advantage of new services offered by the environment to optimize internal process (slower the internal computations and delegate some work to a more efficient and maybe specialized entity, etc.) 3. Recover from an interrupted computation (due to a leaving resource), re-distribute the computing tasks.

Table 7.5 Examples of variables to be observed related to self-adaptation triggering events and examples of the related measurement sensors

Observed variables	Examples of related sensors
Task speed	Timers
Battery power	Power supply current sensors, in-battery sensors etc.
Data type	Sensors related to meta-data extraction and identification
Mission change	Identifier of pre-defined scenarios, critical parameter triggering pre-defined mission switch (the mission semantics is generally statically defined at a higher abstraction level and not sensed as such)
Failure	BIST sensors, on-chip noise measurement sensors [21], etc.
Environment	Sensors related to the physical environment (such as temperature, light, etc.)
	Sensors related to the computing environment (sensing new resources and their capabilities, timers for timeout measurements, etc.)
State of the chip	Temperature sensors, current sensors, etc. [22–24]

Table 7.6 Different types of self-adaptation control algorithms and related characteristics.

Control structure	Time to converge	Potential complexity	Ability to deal with unknown situations by self-adaptation
Control "if then else"	Fast	Low	Low
Evolutionary and learning algorithms	Potentially slow	High	Potentially high

chip. This model can be explicit or implicit. For example, the model can be explicit if the chip and its environment are formally described as sets of variables and equations. The model is implicitly described in the "if-then-else" mechanism since this statement contains the necessary knowledge to take an adaptation decision based on the gathered observations.

One can notice that one of the major problems of self-adaptation is the overhead of the observation-control loop compared to the computation process. Taking into account one variable describing the system, its environment or a self-adaptation triggering event implies to include the related sensors, interfaces and logic in the observer but also the related control logic in the controller. If the number of variables is high, it can slow down the decision taking mechanism. One solution could be to distribute the management of the parameters to different control loops, which may be difficult since the different chip parameters are not independent from each other. ÆTHER considered and studied different implementations of the SANE concept and its specific parts especially the Computing Engine.

7.7.1 SANE Implementation, the Case of Reconfigurable Hardware

At first, FPGA technology seems a rather natural candidate when it comes to implementing an architecture that changes over time. The capacity of new FPGAs to be dynamically and partially reconfigured (even internally for certain Xilinx FPGA families with the ICAP interface) is seen as a nice self-adaptation enabler for modifying the internal structure of the architecture that is placed on the device [25].

The fact that FPGAs are in essence fine grained reconfigurable architecture is both a very interesting property and a real problem when designing self-adaptive architecture. This paradox is linked to the above-presented trade-off between the degrees of freedom of an architecture and the required control complexity to derive the new configuration. The high degrees of freedom provided by FPGAs makes them good candidates to implement SANEs since almost every modification of the placed architecture can be envisioned. Therefore, the optimal architecture (i.e. the one that best fit to the current environment) could be obtained in principle. The problem resides in the fact that for approaching this optimal architecture, a lot of different knobs must be controlled (i.e. a new configuration composed of millions of bits should be computed). With a standard approach to hardware design, the steps necessary to compute a new configuration (synthesis and place/route) are very time consuming; they require tenths of minutes of computation and gigabytes of memory on the latest workstations. As a result, and as of today, this cannot be reasonably done dynamically at run-time.

The problem is alleviated on state of the art dynamic and partial reconfigurable FPGA designs by performing the generation of pre-defined partial configurations. Thus, the system just needs to load the pre-generated bitstreams at runtime. It means that the system simply choose between available configurations using for example a simple "if-then-else" control structure. These configurations may be hardware tasks (e.g. bitstreams) or software tasks (e.g. different implementations of an application or firmwares). This possibility offers basic self-adaptation properties such as loading the right configuration depending on demands from applications or choosing the best configuration by comparing the results and behaviors of multiple implementations. Additionally, having hardware and software flexibility enables to choose between fast adaptation (usually provided by software reconfiguration) and slower but deeper structural hardware reconfiguration.

A possibility to make progress toward the "on the fly" generation of bitstreams consists in considering coarser-grain architectures. Since a SANE needs to be able to rapidly adapt to the potential modifications of its environment, it must be quickly reconfigured. Therefore, the generation and mapping of the bitstreams required by the application needs to be automatic and very fast. These properties can be provided by coarse grain reconfigurable architectures. Such architectures have larger, more complex basic hardware cells requiring less memory bits for their configuration. One drawback is that these kinds of architecture are less flexible than fine grain reconfigurable architectures.

Another potential problem related to the use of FPGAs is that a partial configuration is only generated for a particular place on the device. The same VHDL or Verilog code (even the same netlist) will have to be placed and routed for every location on the chip. For example, with a system composed of a Xilinx MicroBlaze with four dynamic and partial reconfigurable areas as presented in Fig. 7.3, there are four places for hardware accelerators [26]. Even if the places have the same shape, four different bitstreams need to be generated per accelerator (i.e. per VHDL or Verilog description) to be able to place one hardware accelerator in

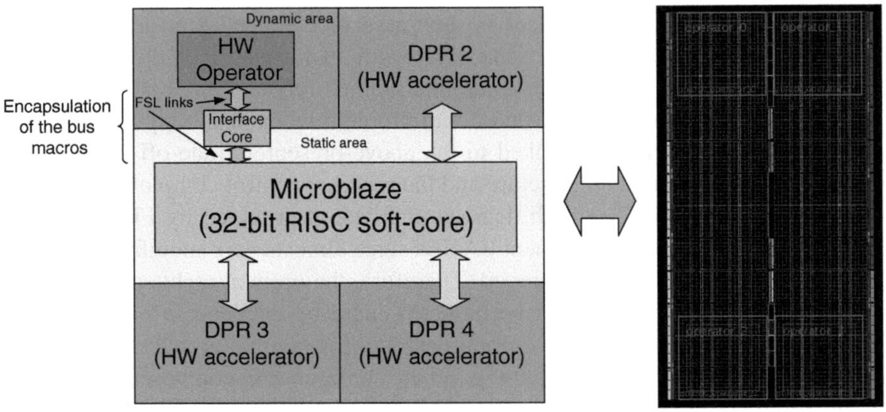

Fig. 7.3 Schematic of a partially reconfigurable chip composed of four independent reconfigurable areas (DPR) under the control of a local RISC controller (Microblaze). This is one example of a SANE hardware implementation using state of the art fine-grained reconfigurable chips. Floor plan view on the right

Table 7.7 Different types of relocation depending on the targeted hardware

Type of relocation	Targeted hardware	Required transformations
Static synthesis, placement and routing	Exactly the same as the old place	Rerouting of the I/O of the module, all the internal structure of the relocated module remains the same
Static synthesis, dynamic placement and routing	Same resources but placed differently	The placement and routing of the module must be computed again (including the I/O of the module).
Dynamic synthesis, placement and routing	Different resources (i.e. optimized or spare resources)	The synthesis phase must be done again (and consequently, all the other phases)

each area. This implies a waste of memory resources for storing all the required bitstreams. Even if this problem can be alleviated using the collaboration interface of the SANE (i.e. the ability to ask for configurations to the network if the needed configuration is not found in local memory), **re-locatable hardware** remains one of the great challenges to be tackled to design self-reconfigurable hardware architectures.

7.7.2 SANE and Future Technologies

The relocation ability can have different levels depending on the structure of the targeted hardware place (see Table 7.7).

In order to efficiently perform the relocation of hardware implemented tasks, two prerequisites are needed at least:

- First, the task should be described in a sufficient abstract language (abstract when compared to traditional hardware netlists) so that the required relocation be possible. In particular it should not contain any absolute locations for the routing information ("parameterizable" IPs).
- Secondly, a built-in distributed self-placement and self-routing mechanisms should be available. The hardware substrate can then take the responsibility to place and route the different tasks in a distributed way, without the need of a centralized place and route supervisor.

So far, the steps and algorithms implied by those two prerequisites are very computationally intensive. They would require embedding powerful processors within the hardware substrate, a very costly solution that is probably not the way to go. It is therefore a very promising but challenging opportunity to investigate solutions that could make relocatable hardware possible in future generations of FPGA by possibly using 3D-stacking and routing techniques.

7.8 The ÆTHER Computing Framework: Implementation and Applications

The technologies described and discussed above are just ways (the ÆTHER ways) to try and give solutions to the generic problem of implementing adaptation loops at the various level of a computing system (see Fig. 7.4). If defining and carefully crafting technologies adapted to the various levels is a good start, it is far from enough. The big challenge is thus to make those implementations of the common design pattern for self-adaptation work together. That is exactly what the ÆTHER computing framework is about.

7.8.1 Design Workflow

The ÆTHER framework aims at implementing a design flow and supporting tools for future dynamic, self-adaptive applications down to their implementation on distributed processor architectures (standard multicores or μGrids [27]) or dynamically reconfigurable hardware [28] that implement the Self Adaptive Network Entity (SANE) concept. The framework is based on two so-called "box" languages, Single Assignment C (SAC) and Microthreaded C (μTC), one coordination language, S-Net, that supports adaptability, and on the definition of the SANE Virtual Processor (SVP) interface between architecture-naive application capture and the implementation on specific architectures (see Fig. 7.5).

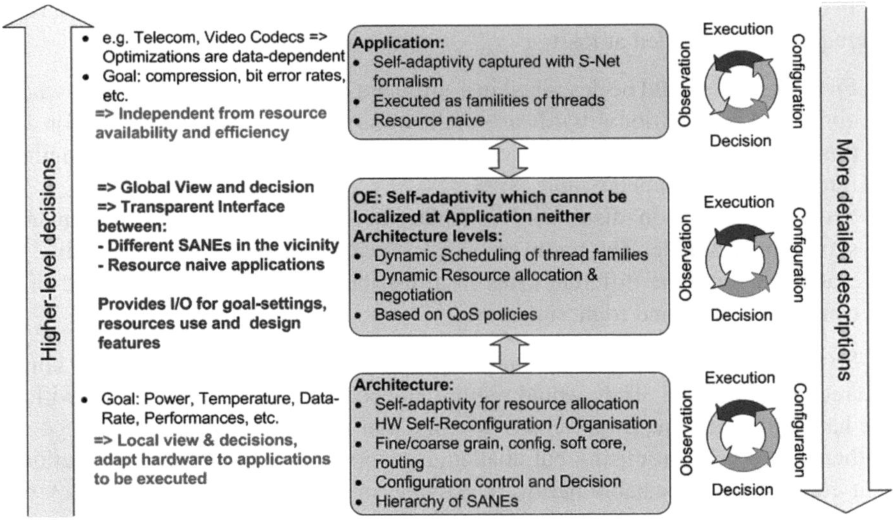

Fig. 7.4 Self-adaptation and its potential applications at different levels of the ÆTHER computing framework. OE stands for Operating Environment: it is the distributed runtime manager of the system. It allocates and schedules families of threads based both on QoS policies and resources availability

Different systems already use self-adaptivity to achieve different specific non functional goals such as self-repairing and self-organization. Self-adaptation has been used for a long time in different fields, such as, for example, networking and distributed computing. In networking, self-adaptivity is used at router level to provide quality of service (QoS) to the communications. Self-adaptivity can be used, in such a context, to recover from QoS violations. Other uses of self-adaptivity are in the context of distributed and grid systems; in this case self-adaptivity is used to provide QoS to the applications, to improve stability and fairness, and to manage resources efficiently. In complex networking environments self-adaptivity is presently being proposed for dealing with the heterogeneity in the behavior of network apparatus and to enable configuration-less insertion of new network elements.

Self-adaptivity may be also used to provide fault-tolerance capabilities as it naturally provides a way to deal with this issue by means of system reconfigurations. Applications designers might be provided with mechanisms to specify fault-tolerance requirements of applications; systems might be designed to automatically provide these capabilities by means of self-adaptation. In this context, self-adaptive systems might automatically put in place predefined self-adaptation patterns to provide fault tolerance at different levels.

The applications of self-adaptive computing are very diverse ranging from mobile and pervasive computing to any field that involves huge assemblies of computing resources on or off chip. In all of those applications fields, the computing

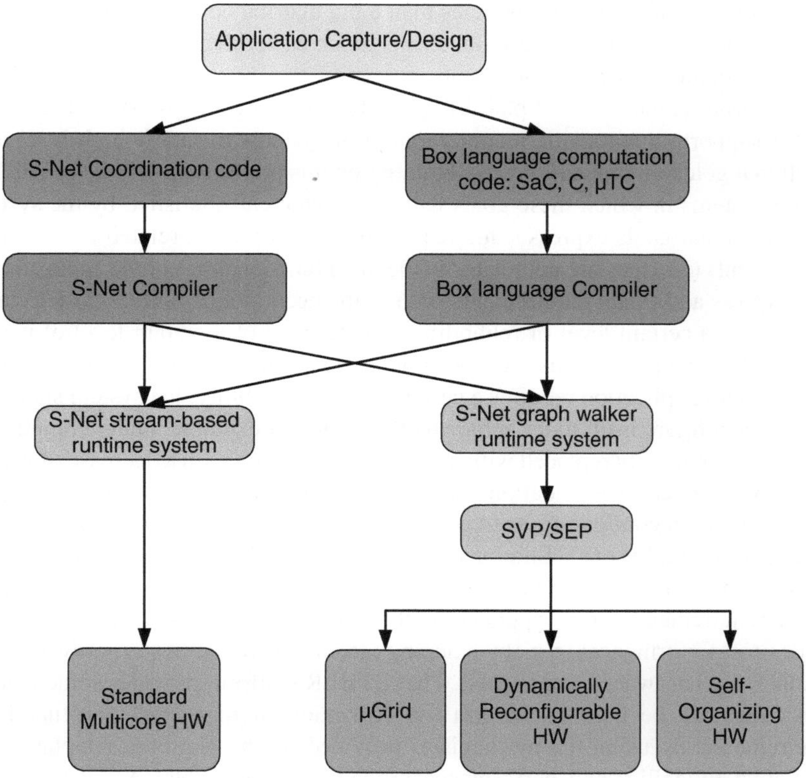

Fig. 7.5 The AETHER framework

system is exposed to potential changes in its environment that may occur as a result of external or internal events such as incoming new devices, missing resources, power shortage, reduced connectivity, component failure, etc. Self-adaptivity, coupled with proper adaptation policies, helps dealing with such events, regardless if they can be foreseen at design time (e.g., the migration of the system in different environments) or not (e.g., a part of the system that becomes faulty). The ÆTHER platform already provides all the mechanisms that are necessary to handle most of these situations: it allows the system to change its behavior by mapping different functionalities (e.g., different network protocols) on hardware blocks or by using their software implementations; furthermore, the platform provides the capability to change system parameters (e.g., the clock of hardware blocks), to control performances of the different functionalities (e.g., by moving its implementation from software to a specific hardware block and/or by parallelizing some of its parts). Even the concept of self-adaptive security at application level relies on the aforementioned mechanisms to keep the security level of applications constant, regardless the changes in the environment and in the system. Proper self-adaptation policies need to be designed for different systems: a small embedded system will have different self-adaptation policies than a big distributed computational system. Though, the description of these policies is much easier than designing the system to cope with all known possible events.

Furthermore, the ÆTHER platform provides a first meaningful step in the direction of supporting automatic reconfiguration of systems to satisfy high-level non-functional goals of applications. By building on this platform, it will be possible to design systems in which these goals will be automatically satisfied by the system. Non-functional goals express some requirements that are not related to functional requirements (i.e., they are not related to the main functionalities of the applications); they express additional requirements such as the need of a certain level of security, the need of a certain level of reliability, and the need of a certain level of performances. For example, the user or the application programmer may want to specify that a certain application requires a certain level of reliability; the system should be able to reconfigure itself to try achieving this goal. For example, faulty components may be automatically replaced with non-faulty ones, even with a decrease of performance when graceful degradation is accepted. In other cases, the system may require more complex reconfigurations (e.g., by instantiating parallel slow components to replace a fast faulty one). Similar measures can be taken for supporting security requirements. In the same way, the programmer may want specify some performance requirements for the application; the system will self-reconfigure to provide the desired performances (e.g., by mapping certain functionalities to hardware or by parallelizing hardware components). The ÆTHER platform provides some mechanisms that can be used to support the utilization of high-level non-functional requirements: by using the mechanisms provided by the platform, adaptations to cope with reliability and performance problems are available. A mechanism to translate these high-level goals into proper control algorithms for self-adaptivity has still to be developed and remains one of the key areas where further research needs to be done.

One promising field of application of the ÆTHER technologies is to introduce self-healing or self-repairing behavior in computing resources in order to provide graceful degradation of performances under a variety of threats (e.g. power shortage, missing resources, sensor or general hardware failure).

7.8.2 Implementation and Applications

To illustrate the ÆTHER framework and its supporting concepts and technologies (see Fig. 7.5) which were described in previous pages, a coordinated set of demonstrations were implemented during the project.

Among the different demonstrations, two will be described in the following paragraphs. The first one shows how an industrial-level radar Moving Target Indication (MTI) application can be captured and expressed in S-Net with boxes implemented in the functional array programming language SAC (Single Assignment C).

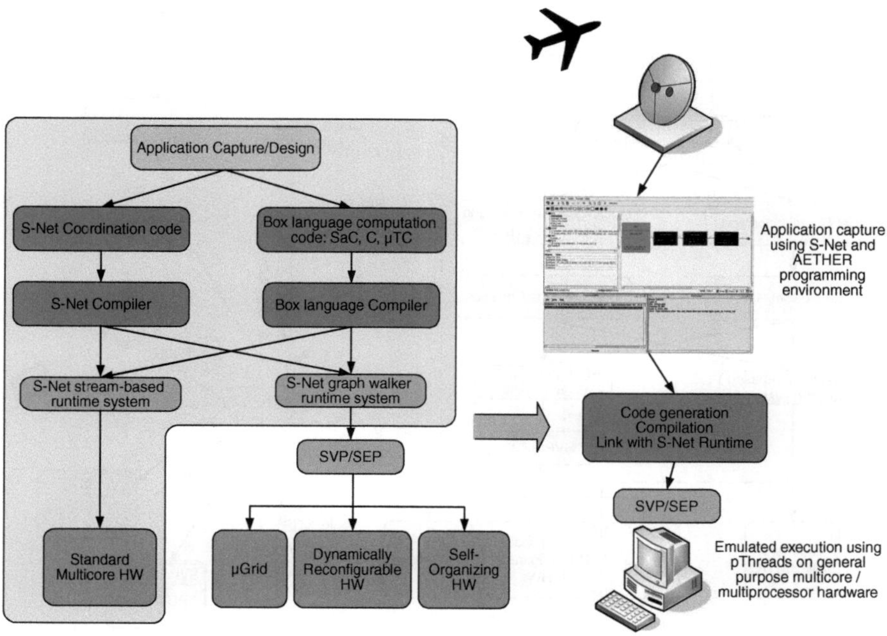

Fig. 7.6 Demonstration of a radar application, from S-Net to standard multicore hardware (emphasis on SNet and its programming environment)

The design of coordination code is supported by the S-Net graphical development environment, which is based on the MetaEdit + toolkit (see Fig. 7.6).

This demonstration shows how the graphical representation of an S-Net graph can automatically be converted into a textual representation. The S-Net compiler started from within the IDE validates the soundness of the S-Net description via type inference and type checking. The resulting code is then automatically linked with the stream-based S-Net runtime system for parallel execution on standard general-purpose multicore and multiprocessor hardware.

The goal of this first demonstration was to show the suitability of the S-Net coordination language to capture non-trivial, industrial-level applications. Its aim was also to demonstrate the entire S-Net tool chain:

- a graphical development environment for network construction
- a compiler for validation of soundness based on a type system with record subtyping
- a runtime system for automatic management of parallel execution on standard general-purpose multicore and multiprocessor hardware.

The goal of the second demonstration is to illustrate by a simple example the entire ÆTHER design flow from application capture to execution on a dynamically reconfigurable platform (see Fig. 7.7). A single application (Optical Character Recognition) is captured as an S-Net network of C-coded processing boxes.

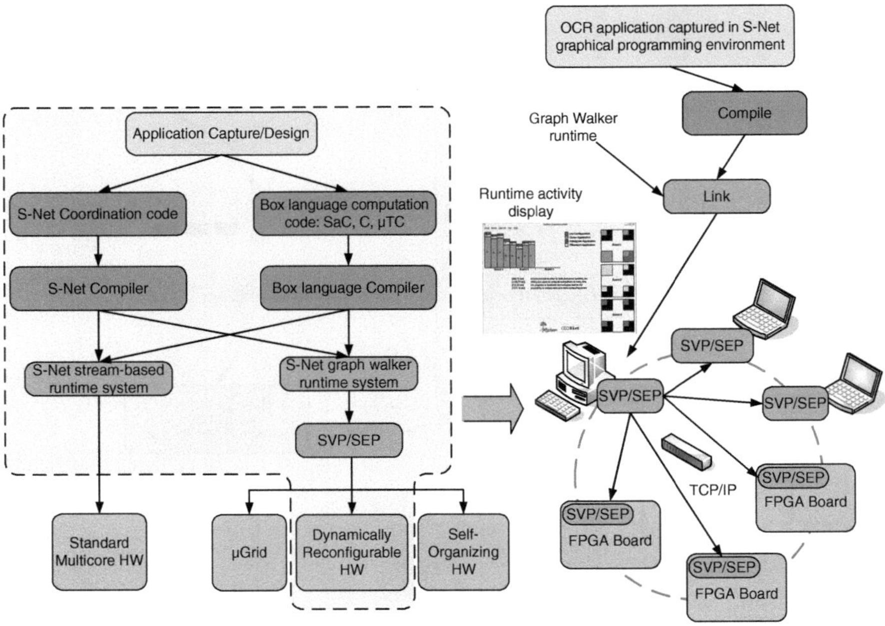

Fig. 7.7 Demonstration of the OCR application, from S-Net to reconfigurable hardware (emphasis on the execution of an S-Net-coded application on dynamically reconfigurable hardware, including the use of hardware accelerators)

Coordination design is also supported by the S-Net graphical development environment, which is based on MetaEdit + based S-Net environment. The box code and the S-Net network will be respectively compiled by standard C and S-Net compilers, and linked with the S-Net graph walker runtime which will interpret the S-Net network and interface with SVP/SEP capabilities on both a Linux platform (PC) and an embedded Linux for the Xilinx MicroBlaze processor within the FPGAs.

The application is then run on an hardware platform composed of a PC connected to three reconfigurable FPGA boards via an Ethernet network. On each of those boards, a SANE implementation contains an FPGA capable of running a character recognition task in software (on the MicroBlaze CPU) and/or in hardware, in one or several of its four dynamically reconfigurable areas. For the purpose of demonstration, the activities of MicroBlaze embedded microprocessors, PC and hardware SANEs are displayed on a monitoring PC. Reconfigurations are shown, first by having all computations done by the PC alone and then dynamically moved to hardware SANEs when they join the system, or redistributed when they leave: this will also show the capacity of the SANE to adapt themselves to execute the delegated task the best they can (by on-the-fly partial reconfiguration and dynamic retrieving of partial bitstreams over the network).

This demonstration gave an overview of the ÆTHER framework, showing application capture with S-Net, the compile chains, and SVP/SEP operations on reconfigurable SANEs. It illustrated how a programmer can build an application independently from any specific target. Furthermore and thanks to the implicit parallelism offered within S-Net, the demonstration illustrates how the Graph Walker exploiting the SVP/SEP protocol, can dynamically map tasks to reconfigurable resources and run them according to the specific properties of the target.

7.9 Open Issues and Research Opportunities

A number of documents and roadmaps have been edited that strives to identify some of the research opportunities for autonomous computing. They give us another complementary perspective on the ÆTHER research.

7.9.1 The AgentLink Roadmap

One such example is the AgentLink coordinate action that produced a roadmap report [29] on **agent based computing technologies**. Many of the topics addressed by the agent computing community are very relevant to the idea of self-adaptive computing. As the authors highlights in their conclusions, "for agent technologies, the objectives are to create systems situated in dynamic and open environments, able to adapt to these environments and capable of incorporating autonomous and self-interested components." Agent based technologies deals with computational entities designed in such a way that interaction among them is central. As a consequence agents need to "socialize" with their counterparts by exchanging information and have their own autonomous way of handling events. In this respect the AgentLink community has identified three categories of technologies that are of particular relevance for the agent system design:

- The **organisation-level** category deals with assemblies of agent entities, their structure and how they self-organize and handle a particular mission in a collective way. The ÆTHER research did not really propose any new concepts in this category and it is certainly an open topic.
- The **interaction-level** category deals with communication between agents at the language and protocol level. In a way the S-NET, and run-time protocols of ÆTHER fall into this category although with a computing approach mainly focused on managing concurrency and computing resources.
- The **Agent-level** category includes ways to implement autonomous behaviors and self-x features at the entity level. If the SANE concept naturally falls into this category it exhibits one major difference: the ÆTHER entity focused on a hardware based approach to provide autonomy at the SANE level when the agent community refers to techniques such as machine learning and artificial intelligence which are typically rooted in software engineering.

Several consequences can be drawn from this analysis. The first one is that future research in computer architecture oriented toward self-x should tackle the organization-level category described above. By studying the behavior of autonomic entities and their interactions within the context of the full ecosystem, apprehend them like collective assembly right from the start rather than aggregations of individual entities.

The second observation is in exploring more disruptive ways to implement the interactions and negotiations between entities. Finally, a mutual cross-feed between the AgentLink approach to autonomic computing, derived from software engineering, with the computer architecture engineering followed in ÆTHER should be beneficial in future research on self-x.

7.9.2 Self-Adaptive Computing and the ITRS Roadmap

The International Technology Roadmap for Semiconductors (ITRS) is an industry driven document that publishes technological visions for the developments of the semiconductor industry. The roadmap was historically focused toward silicon based technologies (essentially CMOS), but in the last few updates has made a move toward technologies beyond-CMOS in an effort to forecast their potentials as replacement technologies. This is the role of the Emerging Research Device (ERD) chapter of the ITRS. In the 2008 Document update [30], the ITRS exposes the trends and challenges in hardware and system design for the period 2008–2022:

- On the technology trends, the cell logic gate size is expected to increase almost linearly from the current values (2009) of 1 μm^2 (for MPU) and 0.01 μm^2 (for DRAM) down to 0.07 μm^2 (for MPU) and 0.001 μm^2 (for DRAM) in 2022. That's about an order of magnitude in the next 10 years which is likely to yield the same increase in processor core number for example. It is foreseen that the number of processing engines for portable consumer devices (cell phones, PDA, GPS) will jump from about 60 in 2009 up to about 1,400 in 2022. **Scaling according to Moore's law is thus continuing** for the next 10–15 years.
- On system chip trends the ITRS notes that the "More-than-Moore" approach will get a boost in the next decade. This design philosophy states that an increase of functional density can be obtained at the chip level by **allowing non-digital functionalities onboard or above the chips** (e.g. analogue circuits, power control, sensors, actuators, passive components). This is clearly a trend toward more heterogeneous Microsystems on chip. The updated ITRS roadmap also mentions the requirement for System on Chip reconfigurability to provide flexible, reconfigurable communication structures.
- On design trends, the ITRS update highlights several important challenges that will be crucial to tackle in order to design future chips. Among those: **Design productivity** needs to be enhanced in order to keep design and verification costs, **Power management** is getting critical and dynamic power and thermal control will be needed, **Reliability** is getting problematic below 45 nm because of atomic-scale effects and single-event upsets (soft errors). It is clear that runtime control, reconfiguration and healing techniques will be required.

- On the Emerging Research Device chapter, the ITRS 2008 updates notes that the most likely beyond CMOS technology would be **Carbon Nanoelectronics** (e.g. Carbon nanotubes, Graphene). If there are indications that such technologies would be able to yield adaptive components in the form of Carbon Nanotube transistors [31] most probably hybridized with CMOS processes [32], it is certainly very hypothetical at the moment of this writing.

From the ITRS roadmap we can see that the trend toward more complex computing systems is really continuing for at least the next decade. Heterogeneity of future systems will be growing, with several non-digital functions embedded within the chips and most likely included within the adaptation loops. For example power/thermal controllers embedded within microprocessors will be built with various digital and analogue components and will be based on adaptive loops that should be programmed in a way transparent to the application designer.

7.9.3 ÆTHER Research and Grand Challenges

In 2009, the European Commission ISTAG FET Working group produced a document that gathers a number of particularly important research grand challenges for the society and the knowledge in the future [33]. Based on the work gathered in several proceeding reports in the period 2003–2008, the report has identified five candidates for future research. Although all challenges are very multi-disciplinary in the topics addressed we can see among them that a good share is very relevant to the idea of self-adaptive and autonomic computing:

- **Understanding Life (Life Science)** – In this challenge we see opportunities in the Neuro-ICT challenges, Personalized ICT for health aims at providing systems for real-time, autonomous and personalized health-care. The **Neuro-ICT** challenge by aiming at understanding the functioning of the brain through emulation or simulation would greatly enhance the understanding of learning processes and in consequence **help implementing machine learning algorithms and adaptation behaviors** within self-x and autonomous computing engines.
- **Managing Complex Systems (Modeling & Simulation)** – In this challenge, the report explicitly mentions that advances in simulation techniques and most probably multi-scale simulation will benefit the field of "autonomous systems that perceive the environment, understand the situation, draw conclusions, act in an appropriate manner and cooperate with other systems". It is also indicated that such research will benefit the field of investigating **non-von Neumann principles.** Indeed the von-Neumann paradigm seems to show its limits with a strong binding with imperative programming models. Alternatives such as neuro-inspired principles have been proposed but so far have been used at very small scale. Going to the next level with those principles will requires **modeling and simulation of realistically complex systems** most probably in a mixed-technology and heterogeneous context. The same is true for investigating the collective dynamics

behaviors of large assemblies of future interacting self-x systems and their relation with their environment as emphasized in this research challenge: "**concepts for autonomous systems modeling**: Research in this area could also be directed to autonomous systems that perceive the environment, understand the situation, draw conclusions, act in an appropriate manner and cooperate with other systems."

- **Future Information Processing Technologies** – This challenge addresses future computing technologies such as quantum devices, novel processing devices, nano-computing and future data storage. If any research in those areas will definitely be key enablers for the development of self-adaptive computing, the report explicitly mentions **Self-repairing and Self-evolving computational devices** as a major research challenge. Indeed and as the report accurately says: "The research efforts in this area only scratched the surface of this problem, which will become increasingly more important as transistors shrink and computational devices are embedded in everyday objects or sent to outer space or inaccessible areas of this planet, not to speak within the human body." We could not agree more as we effectively learned while performing the ÆTHER research, that the problem of self-x and autonomous computing is tremendously complex and requires much more than what has been done so far.
- **Future problem solving technologies** – This challenge aims at studying completely new programming paradigms able to cope with the complexity of both future computing systems and future applications. The ÆTHER research very quickly identified that self-adaptive systems would require programming paradigms that need to be **non-imperative and includes ways to express semantics and non-functional features**: this is clearly a critical research topic that needs to be addressed. Conversely the report highlights: "It would be much more effective, if computers problems recognized problems to be solved from their observation of nature and the environment directly. This is a system automation problem, with sensory inputs and actuator outputs to the environment, monitoring it and determining that a new problem may arise, expressing/transforming it into a solvable formulation, solving it and acting back on environment with the solution." This way, the self-adaptive machine system becomes a central actor of the programming paradigm instead of being just a listener of a human centric language. The problem solving challenge also emphasized the importance of research on **trust, reliability and security challenges** which are indeed critical when deploying autonomous computing elements.
- **Robot Companions for Citizens** – It was probably one of the first applications in history in which the concept of a machine with autonomous behavior had been so clearly expressed. So it is today, where robotics, whether human-like or for the industry offer a reservoir of opportunities and challenges for self-adaptive computation. The research challenge in robotic is therefore of prime importance for the field of self-adaptive computing. In particular, the report indicates that research on robots brains, human-robot interaction and robot adaptive bodies will be critical. Included in this last challenge we read: "It is necessary to develop

electronic architectures that can support the requirements for adaptation, distribution, and rich sensory perception, yet at the same time, be compatible with novel body structures." Again we note that the concepts of self-adaptation and self-x are presents everywhere in this challenge. Finally the problem of ethics is evoked in the case of robots, but it goes far beyond robotics and can become a general issue when designing systems with autonomous behaviors be them robots or not.

Research in self-adaptive computing are key to almost all of the five research grand challenges identified in the context of the Future and Emerging Technologies flagship projects. Additionally it can be seen that those grand challenges are very inter-related with one providing enabling technologies for the others. The case of self–x computing systems is interesting. We note that some challenges will bring the necessary understanding tools and technologies to ultimately design and use them: understanding life, managing complex systems, future information processing technologies, future problem solving. While for others challenges they constitute enabling technologies: robotics, problem solving.

The ubiquity of the self-adaptive computing research fields in all of the grand challenges identified stresses its importance in the future of ICT regardless of the research direction that will be actually taken. The research on self-adaptive computing could have been a grand challenge of its own since it embeds all of the characteristics of the above-mentioned topics: novelty, ambition and interdisciplinary research.

7.9.4 ÆTHER Lessons and Open Issues

The ÆTHER technologies described in this chapter represent a first attempt at trying to implement the required elements of a self-adaptive computing environment. From designing adaptive loops in hardware architectures, specifying runtime system and their protocols up to defining a language construct that tries to represent the interaction of adaptive loops ÆTHER spans the whole chain of techniques and disciplines. The holistic and interdisciplinary approach of the program which has been one of the strong points of the research had the disadvantage of leaving the exploration of some of the most advanced topics to a somewhat shallow level. For example, most of the ÆTHER technologies are just about loops, concurrency and resource management taken from their standard meaning. However as we highlighted at the beginning of this text, self-adaptation is more than just loops, it is more than just the aggregation of known concepts. Self-adaptation is loops plus knowledge, the very expression of self-awareness. This is one of the main issues that remain to be tackled in the future: **embedded knowledge management**.

On the run-time part, we have seen the importance of concurrency management for managing self-adaptation at runtime. The distributing of tasks either sequentially and/or concurrently is key in allowing tasks adaptation between processors. We have shown that it was possible to devise a good enough system for managing concurrency with the proposed SVP. We have shown that it is possible to implement the required virtual processor and specific instructions even with quite different processor architectures (RISC, FPGA). However, with the growing variety and number of heterogeneous processing unit on a chip as highlighted by the ITRS roadmap, it will become increasingly difficult to devise a concurrency control system in all possible cases. And that is probably one of the main issues that will remain: **managing runtime concurrency with very heterogeneous processor architecture**.

Software ideally should provide application designers with sufficient tools and constructs to express their problems within their own field of expertise and their semantics without worrying about the specific details of the underlying computing architecture. This is separation of concerns seen at large. Through the S-NET language we have laid down the basic mechanisms to do this but there is still some work to do in order to **include non-functional features**.

Reconfigurable hardware (FPGA) has long been considered as the ideal enabling technology in order to implement dynamically adaptive hardware circuits. Indeed, the example implementations that we have developed using state of the art runtime reconfigurable devices show that it is technically possible. However this hides all of the engineering effort that need to be done specifically for each application design. And this design effort is way too important when compared with comparable software design times. Again, the ITRS roadmap highlighting this design productivity shows that although current reconfigurable technology could be used to implement run-time adaptive hardware in specific cases. It is nowhere near for general use, unless a dramatic change and breakthrough in Reconfigurable circuits's concepts is made. One clear path for such an advance would be the availability of fully **relocatable hardware circuits within a reconfigurable logic mesh**.

7.10 Conclusion

In its current state the ÆTHER project has laid the foundation of a complete framework for designing and programming computing resources that live in changing environments and need to re-configure their objectives in a dynamic way. Key concepts and technologies have been developed: S-NET, SVP, SANE but a lot of investigation and opportunities for further research remains. In many ways ÆTHER has just scratched the surface of a whole new universe. The research on self-adaptive computing has been mostly focused on investigating solutions that derives from the mainstream computing ecosystem. In many ways the solutions we have investigated, although workable in their present states, show the limits of the "standard" computing technologies when trying to develop self-x/autonomic systems.

A more disruptive approach to self-adaptivity spanning from the language level down to the processing machine implementation should be explored. In order to achieve true self-x capabilities we believe it is necessary to explore and adopt bio-inspired mechanisms, such as neural or genetic models, mechanical and architectural paradigms, such as in bionics or swarm architectures, and to move towards a comprehensive approach that captures the very principles behind the success of biological systems: biology shows everyday that it is a very successful adaptive paradigm derived from ecological, evolutional and, more generally, informational considerations.

References

1. "IBM Research Autonomic Computing." http://www.research.ibm.com/autonomic/
2. J. Von Neumann, *Theory of Self-Reproducing Automata*, University of Illinois Press, 1966.
3. G.E. Moore, "Cramming more components onto integrated circuits," *Electronics*, vol. 38, 1965.
4. J. Kephart and D. Chess, "The vision of autonomic computing," *Computer*, vol. 36, 2003, pp. 41–50.
5. Edsger W. Dijkstra, "On the role of scientific thought", in Dijkstra, Edsger W., Selected writings on Computing: A Personal Perspective, New York, NY, USA: Springer-Verlag New York, Inc., pp. 60–66, 1982
6. Clemens Grelck, Sven-Bodo Scholz, Alex Shafarenko, "Asynchronous Stream Processing with S-Net", International Journal of Parallel Programming 38(1), pp. 38-67, Springer-Verlag, Dordrecht, Netherlands, 2010.
7. S-Net website, http://www.snet-home.org/
8. David B. Skillicorn, *Foundations of parallel programming*, ISBN-13: 9780521455114, Cambridge University Press, Cambridge, England, 1994.
9. John A. Sharp (Ed.), *Data flow computing*, ISBN:0-89391-654-4, Ablex Publishing Corp., Norwood, NJ, USA, 1992.
10. C.R. Jesshope. A model for the design and programming of multi-cores, in L. Grandinetti, editor, *High Performance Computing and Grids in Action*, **volume 16** of Advances in Parallel Computing, pages 37–55. IOS Press, 2008.
11. A. Bolychevsky, C.R. Jesshope, and V.B. Muchnick. Dynamic scheduling in RISC architectures. IEE Trans. E, Computers and Digital Techniques (**143**):309–317, 1996.
12. Chris R. Jesshope. MICROGRIDS: Foundations for massively parallel on-chip architectures using microthreading. http://www.nwo.nl/nwohome.nsf/pages/NWOP_6DSBSV.
13. ÆTHER: Self-adaptive embedded technologies for pervasive computing architectures. http://www.aether-ist.org.
14. C. R. Jesshope, "μTC an intermediate language for programming chip multiprocessors," in Proceedings of the Pacific Computer Systems Architecture Conference (ACSAC'06), LNCS 4186, 2006, pp. 147–160.
15. Apple-CORE: Architecture paradigms and programming languages for efficient programming of multiple cores. http://www.apple-core.info/.
16. Clemens Grelck and Sven-Bodo Scholz. SAC: A functional array language for efficient multithreaded execution. *International Journal of Parallel Programming*, **34**(4):383–427, 2006.
17. Clemens Grelck and Sven-Bodo Scholz. SAC: off-the-shelf support for data-parallelism on multicores. In *DAMP'07*: Proceedings of the 2007 workshop on Declarative aspects of multicore programming, pages 25–33, New York, NY, USA, 2007. ACM.
18. Dimitris Saougkos, Despina Evgenidou, and George Manis. Specifying loop transformations for C2μTC source-to-source compiler. In 14th *Workshop on Compilers for Parallel Computing (CPC'09)*, Zurich, Switzerland. IBM Research Center, 2009.
19. ÆTHER Deliverable D1.1.1, First research report on SANE hardware architecture, issued 31/12/2006.
20. Chris Jesshope, Jean-Marc Philippe, and Michiel Tol, "An Architecture and Protocol for the Management of Resources in Ubiquitous and Heterogeneous Systems Based on the SVP

Model of Concurrency", In Proceedings of the 8th international workshop on Embedded Computer Systems: Architectures, Modeling, and Simulation (SAMOS '08), Mladen Berekovic, Nikitas Dimopoulos, and Stephan Wong (Eds.). Springer-Verlag, Berlin, Heidelberg, 218–228.

21. L. Li, V. Narayanan, M. Kandemir, and M. J. Irwin, "Adaptive Error Protection for Energy Efficiency", In the Proceedings of the International Conference on Computer Aided Design (ICCAD'03), November, 2003.
22. S. Lopez-Buedo, J. Garrido, and E. I. Boemo, "Dynamically Inserting, Operating, and Eliminating Thermal Sensors of FPGA-Based Systems", *IEEE Transactions on Components and Packaging Technologies*, Vol. 25, No. 4, December 2002
23. S. Mondal, R. Mukherjee, and S.O. Memik, "Fine-grain thermal profiling and sensor insertion for FPGAs", Proceedings of the IEEE International Symposium on Circuits and Systems (ISCAS'06), pp. 4387-4390, 2006
24. Xilinx, "Virtex-5 FPGA System Monitor", November 7, 2008 http://www.xilinx.com/support/documentation/user_guides/ug192.pdf
25. K. Paulsson, M. Hübner, J. Becker, J.-M. Philippe, C. Gamrat, "On-Line Routing of Reconfigurable Functions for Future Self-Adaptive Systems – Investigations within the ÆTHER Project," International Conference on Field Programmable Logic and Applications (FPL 2007), pp.415-422, 27-29 Aug. 2007.
26. Jean-Marc Philippe, Benoit Tain, and Christian Gamrat, "A self-reconfigurable FPGA-based platform for prototyping future pervasive systems", In Proceedings of the 9th international conference on Evolvable systems: from biology to hardware (ICES'10), Gianluca Tempesti, Andy M. Tyrrell, and Julian F. Miller (Eds.). Springer-Verlag, Berlin, Heidelberg, 262–273.
27. L. Zhang and C. Jesshope, "On-Chip COMA Cache-coherence Protocol for Microgrids of Microthreaded Cores", Eds. Bouge et. al., Proc Euro Par 2007 Workshops, LNCS Volume 4854, Springer, pp 38-48, 2007.
28. Martin Danek, Jean-Marc Philippe, Petr Honzik, Christian Gamrat and Roman Bartosinski, "Self-Adaptive Networked Entities for Building Pervasive Computing Architectures", Evolvable Systems: From Biology to Hardware, Lecture Notes in Computer Science, 2008, Volume 5216/2008, 94–105
29. M. Luck, P. McBurney, O. Shehory, S. Willmott and The AgentLink Community, "Agent Technology Roadmap, a roadmap for agent based computing", September 2005
30. The International Technology Roadmap for Semiconductor, ITRS Update 2008, http://www.itrs.net/, 2009
31. G. Agnus et al., "Two-Terminal Carbon Nanotube Programmable Devices for Adaptive Architectures," Advanced Materials, vol. 22, no. 6, pp. 702-706, 2010.
32. Akinwande, D. et al. "Monolithic Integration of CMOS VLSI and Carbon Nanotubes for Hybrid Nanotechnology Applications." Nanotechnology, IEEE Transactions on 7, 636–639 (2008).
33. The Information Society Technologies Advisory Group, "European Challenges and Flagships 2020 and beyond", July 2009.

Chapter 8
ANDRES – Analysis and Design of Run-Time Reconfigurable, Heterogeneous Systems

Kim Grüttner, Philipp A. Hartmann, Andreas Herrholz, and Frank Oppenheimer

Abstract The main objective of the here presented ANDRES project is the development of a seamless design flow for adaptive heterogeneous embedded systems (AHES). Based on a domain-independent formal foundation we combine domain-specific modelling languages and libraries into an integrated framework. This framework allows efficiently using and exploiting adaptivity in embedded systems. The design flow is completed by a methodology for performance analysis and tools for the automatic synthesis of adaptive hardware/software systems.

8.1 Project Partners

1. OFFIS – Institute for Information Technology, Germany (coordinator)
2. Vienna University of Technology (TU Vienna), Austria
3. Universidad de Cantabria, Spain
4. Kungliga Tekniska Högskolan (KTH), Stockholm, Sweden
5. Thales Communications S.A., France
6. Diseno de Sistemas en Silicio S.A. (DS2), Spain

- Project Coordinator: Frank Oppenheimer, OFFIS – Institute for Information Technology, Germany
- Start Date: 2006-06-01
- End Date: 2009-09-30
- EU Program: 6th Framework Program, FP6-IST-5, STREP Project No. 033511
- Global Budget: 2,9 M€
- Global Funding by EU: 2,3 M€
- Contact Author: Kim Grüttner, Email: Kim.Gruettner@offis.de

K. Grüttner (✉)
OFFIS – Institute for Information Technology,
Oldenburg, Germany
e-mail: Kim.Gruettner@offis.de

J.M.P. Cardoso and M. Hübner (eds.), *Reconfigurable Computing: From FPGAs to Hardware/Software Codesign*, DOI 10.1007/978-1-4614-0061-5_8,

8.2 Introduction

For many years EDA research and industry has concentrated most of their effort in efficient methods and tools for a better hardware, software or analogue components. But with today's highly integrated embedded systems we observe an increasing need for methods and tools that address the integration of heterogeneous components. Moreover, new technologies, such as run-time reconfigurable FPGAs offer a degree of flexibility in hardware components that was yet only known for software. This new combination of high hardware-typical efficiency in terms of performance and power and the ability to adapt to changing operational conditions is very promising. Especially complex applications e.g. in mobile telecom devices ask for high processing power but also need the flexibility to cope with changing operational modes. Although such implementations would now become technologically feasible, most of the current design methodologies do not support adaptivity as such.

There are many domain-specific design languages based on efficient Models of Computations (MoCs), up to now, there is no methodology that allows seamlessly specifying, simulating, synthesizing and verifying adaptive heterogeneous embedded systems (AHES). Actual heterogeneous system specification beyond the mixture of different languages is a recent research area. Theoretical foundations can be found in several meta-modelling frameworks [1, 2]. These frameworks are essential when defining a coherent, heterogeneous specification methodology. Ptolemy represents the first modelling and simulation framework able to integrate any MoC [3]. While Ptolemy is a powerful modelling and specification framework, embedded system design and implementation has been considered to be an open problem [4]. Ptolemy supports a "mutation" concept which allows circuit changes during runtime. However, it still lacks a synthesis strategy and an appropriate model of computation for FPGAs [5]. Similar approaches to Ptolemy, but based on C++ are represented by UmoC [6], or SystemC-AMS [7]. SystemC-AMS is a mostly European effort to extend SystemC to support various models of computation for the modelling of heterogeneous systems. The overall goal of the ANDRES project has been to resolve these issues by developing an integrated design flow for AHES. The core of this design flow is based on the industry-driven open-source modelling language SystemC. It is actively maintained, improved, and supported by the Open SystemC Initiative (OSCI) through companies like ARM, Cadence, Forte Design Systems, Intel, Mentor Graphics, NXP, STMicroelectronics, and Synopsys. As one key result ANDRES provides a SystemC-based modelling framework for designing embedded hardware/software systems on a high level of abstraction emphasizing in particular the integration of adaptivity. This modelling framework is complemented with concepts and tools for automatic hardware/software synthesis from adaptive system models. At the end of the project, ANDRES provided a complete design flow for AHES that covers specification, modelling and implementation. The design flow has been evaluated by using state-of-the-art OFDM based communication systems as use-cases. ANDRES was organized as a specific targeted research project (STREP), co-funded by the European Commission within the Sixth Framework Program.

It has started in June 2006 and lasted 40 months. The project consortium consisted of four research partners: OFFIS, Vienna University of Technology, KTH Stockholm and University of Cantabria, and two industrial partners: Thales Communications and DS2. Additional and future information about ANDRES can be found on the project website [8].

The rest of the chapter is structured as follows: We will first present the main result of the ANDRES project – the adaptive heterogeneous embedded systems (AHES) framework – including a formal approach for specification of adaptivity. The application of the AHES modelling framework will be shown by the example design of an adaptive *Binary Amplitude Shift Keying* (BASK) transceiver. Finally we are going to give an introduction to our vehicle for performance analysis in the context of adaptive systems based on the BASK transceiver model. This closes the gap between the SystemC-based AHES framework and the formal specification of adaptivity.

8.3 AHES Modelling Framework

One of the main goals of the ANDRES project has been the development of a unified design environment for modelling and simulation of adaptive, heterogeneous systems. The formal foundation of the AHES framework is **ForSyDe** (Formal System Design, [12]), with the objective to move system design to a higher level of abstraction and to bridge the abstraction gap by transformational design refinement.

The AHES framework (see Fig. 8.1) is based on SystemC and integrates three domain-specific libraries covering modelling elements for software, reconfigurable digital and analogue hardware. Additionally, the framework includes elements for easily connecting models from different domains to create executable system-level models. In addition to this support for heterogeneous models, special facilities to explicitly express adaptivity within abstract as well as domain specific models are provided. Figure 8.2 gives an overview of the AHES framework usage. Horizontally it allows the modelling and interconnection of different Models of Computation, as shown by the ASK transceiver example on the left side. Vertically it allows manual refinement and automatic synthesis to a heterogeneous hardware/software and analogue-/mixed-signal system, as shown on the right side.

The following three domain-specific libraries are the main components of the library:

HetSC enables the use of different Models of Computation running on top of the discrete-event kernel of SystemC. Its main target is safe specification of parallel software systems, but it can also be used to create early untimed models of components which may later be refined to software or hardware.

SystemC-AMS enables the specification of analogue/mixed-signal systems using different Models of Computation, including timed synchronous dataflow (TDF), linear electrical networks (LEN) and linear differential equations. In particular,

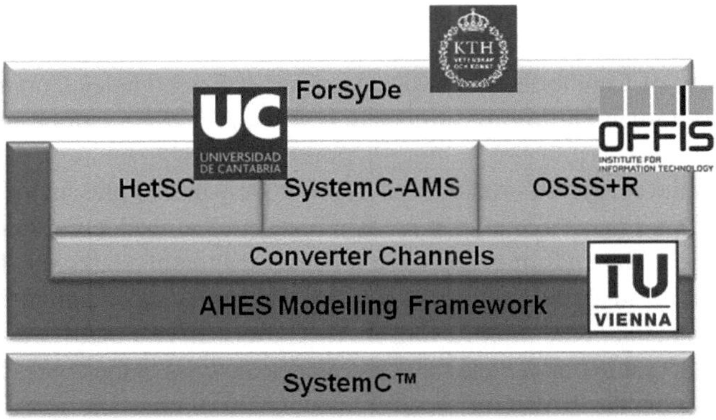

Fig. 8.1 Overview of ANDRES AHES framework

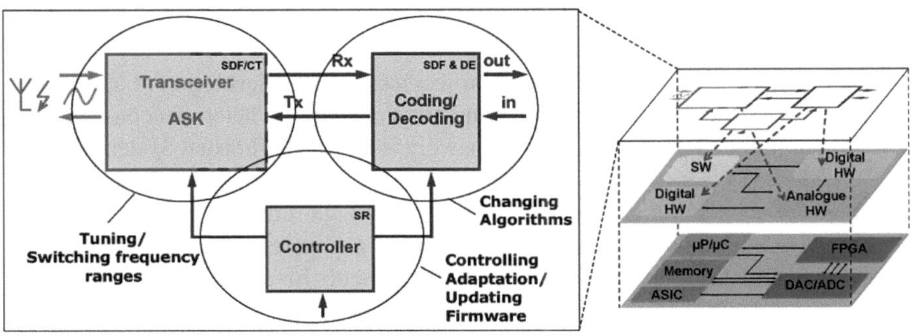

Fig. 8.2 Overview of ANDRES AHES framework usage with adaptive ASK transceiver example

the framework includes a set of SystemC-AMS Building Blocks for designing signal processing communication systems.

OSSS enables object-oriented modelling of synthesisable hardware/software systems by using objects to express communication, encapsulation and sharing of functions and resources. **OSSS + R** is an extension of OSSS which includes facilities to model run-time reconfigurable hardware systems.

To support heterogeneous models, the framework supports cross-domain communication between subsystems from different domains, most notably the so-called **Converter Channels**, which provide convenient and well-defined communication between different modelling domains.

As shown in Fig. 8.1 the whole framework is based on SystemC and is fully compatible with the latest IEEE 1666 standard [9]. All components of the **framework**

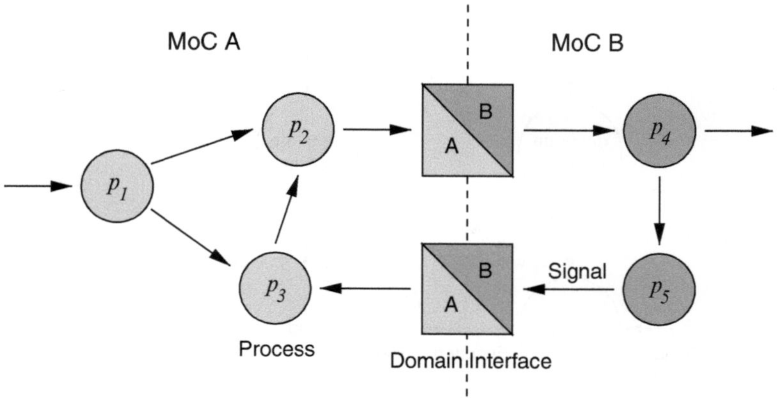

Fig. 8.3 Processes of different models of computation can communicate with each other via domain interfaces

are available under open source licenses and the full framework can be downloaded at [8]. An in depth description of the overall framework can be found in [10]. In the following, an overview over the main parts is given.

8.3.1 ForSyDe

The formal semantics of the AHES modelling framework are based on the ForSyDe [11, 12] framework developed at KTH. ANDRES extended ForSyDe by integration of adaptivity into the modelling framework. This is done by the concept of an *adaptive process*, which changes its behaviour depending on special input signals from the environment. The values carried by these input signals can be data values, but also functions or complete processes. Thus adaptation can be modelled at a varying degree of complexity.

8.3.1.1 Model of Computation

ForSyDe uses a formally defined, hierarchical heterogeneous Model of Computation (MoC), which is illustrated in Fig. 8.3. Processes communicate via signals. There are so-called domain interfaces to formally define the interaction between processes of different computational models. The following MoCs are supported: *untimed model*, *synchronous model*, *discrete time model*, and *continuous time model*.

Since the industrial designer uses SystemC as a modelling language, the AHES framework defines modelling elements and guidelines, which guarantee the SystemC models to be compliant with the ForSyDe modelling formalism. Thus methods developed in the formal ForSyDe domain, e.g. for property analysis, verification or transformation, can also be applied to the AHES SystemC models.

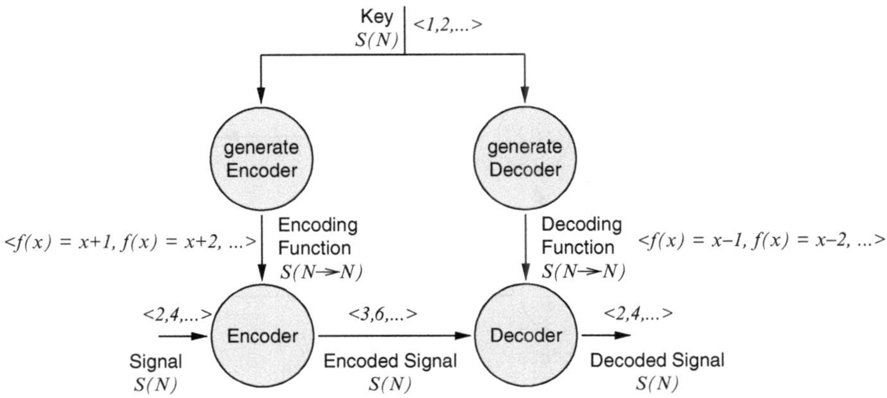

Fig. 8.4 The encoder/decoder is a typical example for adaptivity

8.3.1.2 Modelling of Adaptivity

Adaptivity is modelled by means of formally defined adaptive processes. The functionality that the adaptive process computes can be changed from the environment depending on the value of an input signal.

Figure 8.4 shows a simple, but typical example for the modelling of adaptivity. The processes *Encoder* and *Decoder* are both adaptive processes and are fed with signals carrying functions. These functions change the behaviour of the encoder. In the first cycle the encoding function adds one, while in the second cycle two is added to the incoming value. The same adaptivity mechanism is used for the decoder. To yield an implementation this model is refined and non-functional characteristics, like computational and reconfiguration time, are taken into account in subsequent design phases and the associated model using AHES building blocks.

8.3.2 HetSC

HetSC [10, 13, 14] is a specification methodology based on SystemC. It supports abstract specification of concurrent functionality under different, heterogeneous Models of Computation on top of the SystemC discrete-event (DE) simulation kernel. When following the HetSC modelling rules and guidelines, HetSC enables robust system specification, to avoid common issues in concurrent specification, such as non-determinism, non-continuity, deadlock, etc. Furthermore, automatic generation of embedded software and fast performance estimation is possible via related tools [15].

The core of the HetSC methodology is a first layer called *General Specification Methodology*, which consists of a fundamental set of modelling rules, like strict separation of communication (via MoC-specific channels) and computation only within a (network of) processes. On top of this *General Specification Methodology*, a further layer provides support for different specification styles (following different

MoCs) and for heterogeneous specification. Each specification style defines which SystemC primitives to use, specially which type of channels, and thus which synchronisation and data transfer semantics are provided. If the required synchronisation facilities are not part of standard SystemC, the HetSC simulation library provides corresponding primitives.

Each specification style adds a set of specific rules to the general rules. Styles with common semantic assumptions with regard to time handling are grouped in *HetSC domains*, most importantly the two most abstract domains: *untimed* and *synchronous*. Each domain supports several specification styles. There are, in the untimed domain: Kahn Process Networks (KPN), Bounded KPN, a deterministic variant of Concurrent Sequential Processes (CSP) with communication via rendezvous, etc; and in the synchronous domain: Synchronous Reactive (SR) and Clocked Synchronous (CS).This support of untimed and synchronous domains makes HetSC suitable for abstract specification and automatic generation of embedded software.

8.3.2.1 Heterogeneity in HetSC

HetSC supports heterogeneity in the sense that different styles and domains can be present and coherently connected within the same SystemC model. For this purpose, HetSC defines two modelling techniques, called Border Processes (BPs) and Border Channels (BCs). Formally, they enable the connection of sub-systems under different MoCs. In case of connecting parts belonging to different domains, time adaptation is required and performed.

The BP is more flexible than the BC mechanism, since the designer can define the MoC adaptation behaviour within a process, following the rules of the connected domains. The BC is safer and intentionally less flexible than the BP mechanism, since the adaptation semantic is already provided and fixed by the channel. As a result, the designer only needs to choose the correct BC as the glue component between parts under different MoCs.

Border Process and Border Channel mechanisms are suitable when the specification handles heterogeneity at the time domain. HetSC provides a wide set of predefined border channels for the most frequently used MoC connections, e.g. for the untimed and synchronous domains. More information can be found in [13].

8.3.2.2 Abstract Adaptivity in HetSC

Another extension of HetSC developed within the ANDRES project is the explicit support of adaptivity through several specification (and design) patterns called *HetSC Adaptive Processes* (HAPs) [13]. HAP patterns enable the early and fast specification of adaptivity, with a direct support of software implementation, and a comprehensible and systematic refinement to other adaptive structures, such as OSSS-based ones, to enable the implementation of adaptive processes as dynamically reconfigurable hardware [16].

An HAP is a generic SystemC specification pattern, directly reflecting the adaptivity meta-modelling concept introduced in Sect. 8.3.1.2, which provides a formal basis to the proposed ForSyDe specification methodology.

The HAP patterns are specific in the concurrency and communication structure, and on the type of specification facilities used for building it. HAP patterns are abstract and compact in the sense that they have been developed for the abstract untimed and synchronous domains. That is, they enable the specification of adaptation times at the least possible detail in terms of time. The communication of the HAP with the rest of the system is through channels, following the HetSC methodology.

The main feature of a HAP process structure is that at least one of its input channels is considered as an adaptation input channel. The rest of its input channels are considered as regular inputs. Then, an adaptation is considered to be a change in the internal state of the HAP which specifically changes adaptation context variables. This, in time, changes the functional relationship between the regular inputs and the output. Moreover, it can even change the HAP's interfaces.

In addition to the generic HAP patterns, a set of SystemC templates has been added to the HetSC library, covering the different types of adaptivity (parameter, mode, function adaptivity) under different specification domains (untimed, synchronous). A list of the available templates and their using guidelines can be found in [13].

8.3.3 SystemC AMS and Building Blocks Library

Since there is a growing trend for tighter integration of embedded hardware/software systems and their analogue (or physical) environment, today's system design methodologies require adequate modelling techniques to address the challenges of the analogue world. However, the SystemC simulation kernel has not been designed for the modelling and simulation of analogue, continuous-time systems and lacks the support of a refinement methodology to describe analogue behaviour from a functional level down to implementation level.

Consequently, the OSCI AMS Working Group has been formed to define the SystemC Analog/Mixed-Signal Extensions (SystemC AMS), which have been standardized recently under active contribution of members of the ANDRES consortium [17]. Being built on top of the standard SystemC simulation kernel, SystemC AMS has been included into the ANDRES framework to model AMS parts of heterogeneous systems.

Analogue systems can be characterised by transfer functions, their behaviour in linear time (e.g. transient conditions) or by the behaviour of analogue, electrical elements (resistors, capacities, inductors ...). This is similar to the characterisation of a digital system based on the behaviour down to the register transfer level (RTL). For simulation of those analogue parts, different MoCs have to be used. SystemC AMS uses the approach of timed synchronous dataflow (TDF), which triggers events at fixed, equidistant time intervals. This behaviour can be easily integrated into SystemC, since it allows plugging in different kinds of solvers for different MoCs (e.g. electrical linear networks, linear signal flow).

SystemC AMS builds on a layered structure. TDF is used as synchronisation layer between the different AMS MoCs (and plain SystemC). Above this layer there are the various solvers needed for the MoCs that are used by the user views in the system. The solvers are the core parts of the different models of computation. There are linear, non-linear and static solvers, for the various ways problems can be described. The view layer consists of two generic interfaces. The first one is the "netlist interface", which is common to all MoCs, allowing the description of a problem as interconnected modules. The "equation interface" on the other hand allows formulating problems directly as behavioural models or as function specifications by using differential algebraic equations.

For the connection with the rest of the system, several synchronisation and communication primitives are already present in SystemC AMS. Based on these facilities, extended and safer MoC conversion mechanisms are provided in the AHES framework, as briefly discussed in Sect. 2.5 and covered in-depth in [13].

8.3.3.1 Building Blocks Library

TDF modelling in SystemC AMS enables fast simulation of system-level models of analogue parts due to its static scheduling approach. However, the modelling can be a time-consuming task, if all modules have to be developed manually. Therefore, the AHES framework provides the most important building blocks of communication and radio frequency systems to ease this design process [18].

The implemented building blocks focus in particular on signal sources, modulation/demodulation, filters, measurement, and observation parts. They can be subdivided into three categories of blocks:

- *Signal sources*, like random numbers/bits, various waveform generators
- *Basic blocks for signal processing*, like simple mathematical functions, non-linearities, a wide set of basic RF building blocks (ADC, DAC, filters, ...), various modulators/demodulators, and other common utilities
- *Signal analysis units* (eye diagram, scatter plot, ...)

With such a set of frequently used, highly configurable modelling components, initial models of a wide range of RF and signal processing systems can be modelled easily. Examples, e.g. of an OFDM transceiver can be found in [18].

Some library components are augmented for specification of adaptivity and reconfigurability by the parameterization of modules according to respective hardware features. Non-ideal effects of analogue realisations are also partly implemented. All details about the building block library modules can be found in [18].

8.3.3.2 Adaptivity in AMS Systems

Since ANDRES has covered both, heterogeneity *and* adaptivity in today's designs, adaptivity has to be considered for the analogue/mixed-signal parts as well. Adaptivity in AMS (sub-) systems is usually limited to adjusting certain parameters or system modes as run-time reconfiguration of analogue circuits is currently not available.

SystemC AMS provides certain modelling elements for externally controlling analogue behaviour to some extent (e.g. by TDF controlled resistors, etc.). This provides externally controlled adaptivity even for analogue parts of the system. In the Building Block library, some parameters of some modules can be adjusted during run-time as well, following the same direction.

8.3.4 OSSS+R

OSSS+R is a SystemC-based modelling library providing high-level language constructs enabling application-driven modelling of (self-) reconfigurable hardware systems. As all OSSS+R elements have a well-defined synthesis semantics, designs can automatically be mapped to platforms supporting dynamic partial reconfiguration.

8.3.4.1 Modelling of Dynamic Hardware

Based on OSSS (**O**ldenburg **S**ystem **S**ynthesis **S**ubset), in OSSS+R (+ **R**econfiguration) object-orientation is used as an abstraction mechanism to describe the behaviour of dynamically reconfigurable hardware. The concept is based on the idea that changing functions in a hardware system largely resembles the use of polymorphism in object-oriented software design [19]. Polymorphism, as used in object-oriented programming, enables calling methods on an object, whose exact type is unknown to the caller. The only known reference to the object is its *interface*. Depending on the actual class of the object, the corresponding implementation of a method is executed. This enables changing parts of the software at run-time without modifying the static part of the code.

Considering a digital hardware system consisting of a static and a dynamically reconfigurable part, it is obvious that the interface between the two parts needs to be fixed. However, the functionality of the reconfigurable hardware may change. Hence, the key idea of OSSS+R is to model the reconfigurable area of a hardware system as an adaptive (polymorphic) object with a fixed interface. This *interface* is defined by a base class, while its possible variants belong to different subclasses. During run-time, different variants of the adaptive object can be configured and used (see Fig. 8.5).

To handle the management of different object configurations and to ensure persistence, OSSS+R introduces *Named Contexts*. A context represents all relevant information of an object, including its current type and state. From the designer's point of view, a context is used similar to a pointer in C++, however automatically instantiated infrastructure ensures that a context is enabled (i.e. configured) if it is accessed by a user process. Its state is automatically saved and restored during consecutive reconfigurations. Targeting hardware, a context can be accessed concurrently from

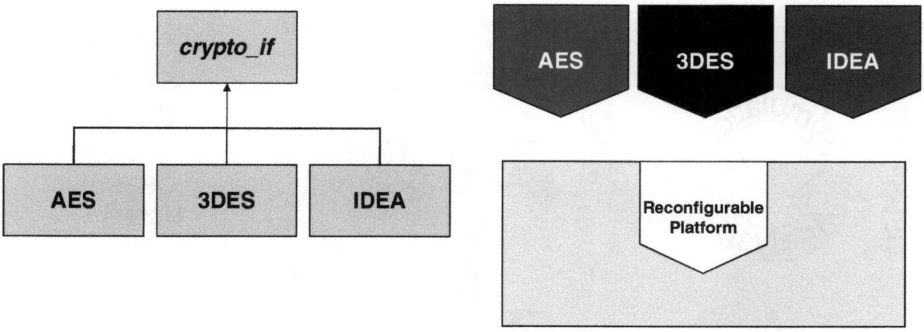

Fig. 8.5 Polymorphism and configurations

different processes. Thus, incoming request are serialised using a built-in scheduler. Contexts hide the complexity of configuration management and state preservation and enable the designer to use adaptivity transparently.

8.3.4.2 Synthesis for Dynamic Partially Reconfigurable Hardware

One of the major goals of the ANDRES project was the automated hardware synthesis for dynamic reconfigurable architectures. In the course of the project *synthesis* covered the translation of a given OSSS + R model to RT-level VHDL, which in turn serves as input for third-party backend tools, i.e. the Xilinx *Early Access Partial Reconfiguration Design Flow* [20] (see Fig. 8.6). The OSSS synthesis tool FOSSY (**F**unctional **O**ldenburg **S**ystem **SY**nthesizer) has been extended to support the language constructs for reconfigurable components introduced by OSSS + R.

The major transformation step from an OSSS + R to a pure VHDL RTL design consists of the generation of various management structures. In addition to the application and annotations (given by the designer), different arbitration mechanisms, structural information (e.g. FPGA types) etc. need to be considered. The generated infrastructure consists of a set of hierarchically organised controllers. A set of distributed controllers for each reconfigurable area handles access requests by the static design parts. Each of these access controllers uses a central reconfiguration controller per device to accomplish reconfigurations. That unit resolves conflicts between different distributed access controllers and provides an interface to the FPGA's configuration port.

The required interfaces to the reconfigurable areas can be determined during synthesis by analysing the interfaces of the corresponding *Named Contexts* that are bound to the area. This even allows the synthesis of static signal-level interfaces for unrelated method interfaces bound to a single reconfigurable area on the application layer.

All functional contents of a *Named Context* imply the generation of a separate VHDL implementation of their behaviours. In the later steps of the synthesis flow each of these functional blocks can be used for the generation of the FPGA technology dependent partial bit-streams.

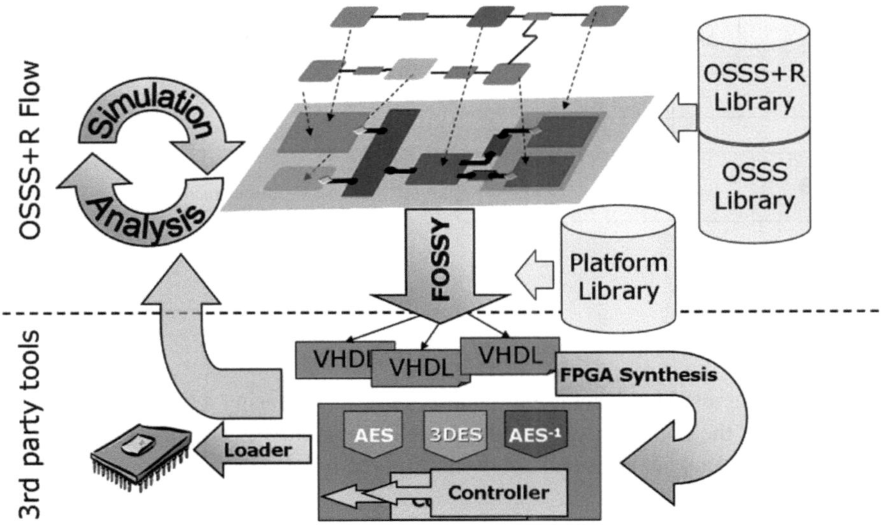

Fig. 8.6 OSSS + R synthesis flow

8.3.5 *MoC Conversion and Converter Channels*

To support truly heterogeneous models in a single design environment, connecting parts under different MoCs is an essential requirement. Apart from purely syntactical issues, time relations and data semantics have to be well-defined. In the AHES framework, several MoC conversion mechanisms are available, with different flexibility and use-cases in mind [13].

8.3.5.1 Converter Channels

The most generic cross-domain connection facilities are the so-called *Converter Channels*, which provide automatic and consistent MoC (and data-type) conversion for a wide variety of use-cases in a single modelling construct.

These channels automatically convert between different interfaces, Models of Computation and data types. Depending on the connected end-points, the correct conversion semantics are chosen automatically, which eases partial refinement of the system.

The supported MoC combinations are shown in Table 8.1, where each MoC tuple has predefined conversion semantics. SC denotes a plain SystemC signal (discrete-event MoC), SDF is the SystemC AMS SDF synchronisation layer, and FIFO corresponds to (untimed, or implicitly timed) blocking FIFO communication. Last but not least, the electrical primitives, voltage and current, (from SystemC AMS ELN) constitute their own MoCs as well.

Table 8.1 Converter channel MoC conversion capabilities [13]

From \ To	SC	SDF	FIFO	El. Voltage	El. Current
SC		✓	✓	✓	✓
SDF	✓		✓	✓	✓
FIFO	✓	✓		✓	✓
El. Voltage	✓	✓			×
El. Current	✓	✓	✓	×	

In addition to the MoC conversion capabilities, Converter Channels support certain (constrained) data-type conversions, which are frequently needed at MoC boundaries. An example for this could be a limited bit-width of an abstract A/D converter, which would require conversion from `double` to e.g. `sc_int`.

For certain corner cases, the concrete conversion semantics can be adjusted to the required behaviour as well, for example in case of buffer under-runs at the boundary between FIFO and SDF MoCs. A comprehensive description of the Converter Channel features, capabilities, and limitations can be found in [13].

8.3.5.2 Additional MoC Conversion Techniques

In addition to the generic Converter Channels, specialised MoC conversion techniques have been explored and developed within the ANDRES project. These usually address specific needs during the refinement process, which cannot be specified efficiently with the signal-like, unidirectional Converter Channels.

SystemC-AMS and HetSC: The connection of SystemC AMS and (untimed) HetSC models has been discussed in [21]. In cases when data-type conversion is not needed, the HetSC approach to MoC conversion (Border Channels and Processes) can be used as well. The main benefit of using HetSC-specific (FIFO) channels is the additional analysis capability HetSC provides. Consequently, special HetSC Border Channels have been added to the HetSC library and corresponding using guidelines required for well-defined MoC connection have been added to the HetSC methodology.

SystemC-AMS and TLM: In [22], the connection of SystemC AMS models with OSCI TLM-based models has been explored. In this area, the temporal decoupling techniques can benefit from in the internal simulator look-ahead, used within the SystemC AMS solver. This can further improve simulation performance.

HetSC and OSSS: Since the communication with ReconObject is method-based, but synthesizable OSSS models are inherently clocked models, direct communication from untimed HetSC processes to OSSS ReconObjects has not been possible before. Thus, OSSS has been extended to support method calls from untimed processes as well. In a refinement step, HetSC Border Channels can be introduced, that enable the connection between process networks and signal-level interfaces in the clocked hardware domain. This incremental refinement approach has been discussed [16].

Connection to synthesisable OSSS via (untimed) FIFOs: Synthesisable FIFO adapters have been added to OSSS to support the MoC conversion from (untimed) FIFOs to a fully synthesisable OSSS model even on the implementation level. Several specific FIFO implementations are supported (SystemC, TLM, HetSC, signal-level) both for data sources and data sinks (from the hardware perspective). These FIFO adapters further reduce development overhead during the refinement towards reconfigurable, digital hardware of certain parts of the system. Details can be found in [13] as well.

In summary, the various semantic issues regarding the connection of untimed, discretely timed, clocked and continuous models of computation have been addressed during the ANDRES project. As a result, the ANDRES framework provides well-defined, self-checking modelling elements and guidelines for bridging MoC boundaries and therefore enables true heterogeneous system specification and refinement.

8.4 Example: Adaptive BASK Transceiver

Early in the beginning of the ANDRES project an example for a complex heterogeneous adaptive system has been defined. Created by all project partners the application needed to include all relevant domains, interfaces and functional aspects that should be covered by the modelling elements and building blocks of the future modelling framework. This example system has been used during the project for demonstrating and discussing research challenges and to define interfaces between framework components. Though not necessarily a very complex application some of its components may appear in exactly the same way in real-world applications.

Figure 8.7 gives an overview of the ANDRES design flow. ForSyDe provides the formal specification model and can be used for early design partitioning, exploring the application of different MoCs and adaptation techniques to meet certain system requirements. Functional implementation is preformed using a SystemC model, utilizing the domain-specific libraries SystemC-AMS, OSSS + R, and HetSC. This executable model is refined to an architectural system representation, taking into account different implementation alternatives, either (reconfigurable) hardware, software or analogue circuits.

8.4.1 Application Description

The example application is an adaptive *Binary Amplitude Shift Keying* (BASK) transceiver which sends and receives encrypted data via a BASK modulation. The

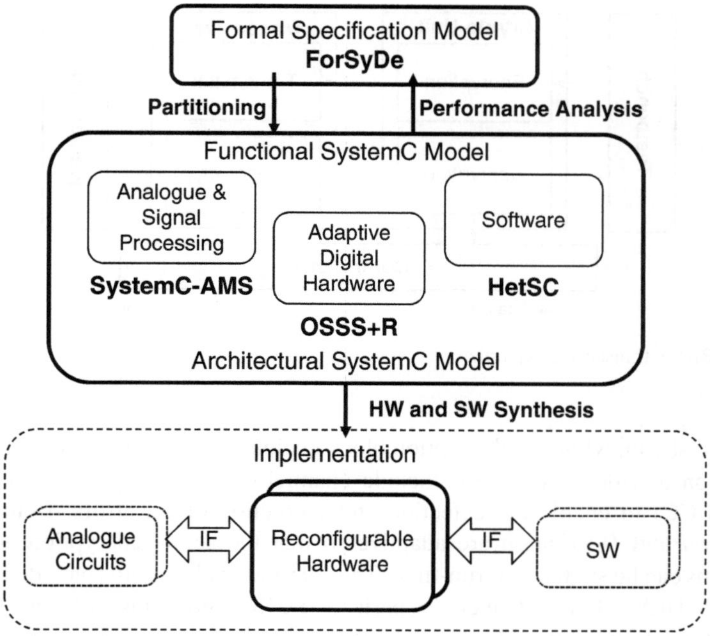

Fig. 8.7 Overview of the ANDRES design flow

transceiver can adapt to different channel conditions and transmission standards, e.g. it may change encryption algorithms or increase the transmission signal strength. Therefore it contains several components from different domains that can either be self-adaptive or configurable from a user perspective. In Fig. 8.8 the system and its components (including a testbench, an environment model, the data flow and the control flow) are shown.

The first of the three main components is the *Transceiver* consisting of a transmitter and a receiver block, representing the analogue/mixed-signal domain. The transmitter generates an analogue BASK-modulated signal from a coded binary input signal while the receiver decodes an analogue BASK-signal into a coded binary output signal. The BASK-modulation scheme was chosen as it is rather simple yet complex enough to demonstrate the interaction of the analogue domain with the digital (hardware and software) domains. In principle other more complex modulation schemes, like e.g. OFDM, would be possible using the existing AMS building blocks. The transceiver block is adaptive in a sense, that several of the modulation parameters of the analogue part can be adjusted at run-time. For example, if the channel gets noisier the gain of the output signal can be increased to keep the bit error rate low.

The second main component is the *Crypto Unit* consisting of an encryption and a decryption block. The encryption block encrypts outgoing data generating an

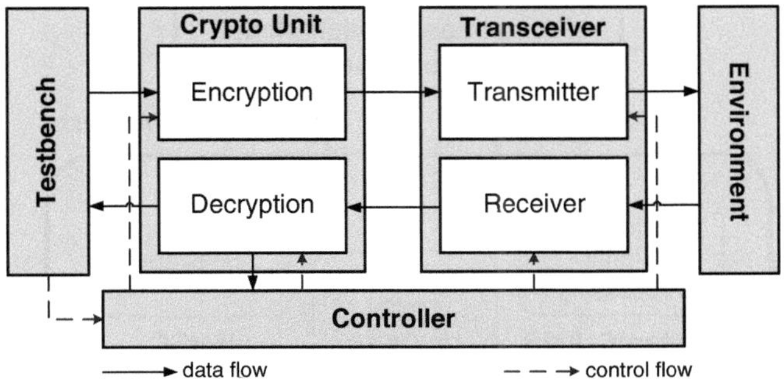

Fig. 8.8 BASK transceiver example

encrypted stream while the decryption block deciphers incoming data to plain data. Encryption and decryption also include channel coding using *Error Correction Codes* (ECC) to enable the detection and correction of bit errors. Both blocks, *Encryption* and *Decryption* are adaptive so that the actual encryption and ECC algorithms can be switched at run-time. The encryption algorithms include Blowfish and Triple DES, while ECC includes parity and CRC algorithms. All algorithms are expected to be implemented in digital hardware to be fast enough to guarantee the required throughput. Thus the *Crypto Unit* represents the run-time reconfigurable digital hardware domain.

The third and final component is the *Controller* initializing, configuring and controlling the operation of the two other components at run-time. With respect to the *Crypto Unit* the *Controller* selects the respective algorithms for encryption and monitors the current bit error rate to determine the quality of the channel. In case of an increasing error rate the Controller may choose to either switch to a more robust ECC algorithm or it may increase the signal gain of the transceiver. To reduce complexity of the system the synchronization of sender and receiver with respect to selected algorithms has been left out on purpose and is expected to be handled on upper layers. Due to its high degree of flexibility and programmability yet low requirement for computation performance, the *Controller* is to be implemented in software, thus representing the digital software domain.

While the three main components represent the actual target system to be developed, the model includes auxiliary components to model a testbench and to simulate the transmission channel.

8.4.2 Model Description

During the project several models of the BASK Transceiver have been created starting with untimed abstract models in ForSyDe, SystemC AMS and HetSC. These models have been integrated and further refined into several heterogeneous models using at least two or more of the available modelling libraries. While these demonstrated the general applicability of the modelling libraries to their respective application domains it also showed the interoperability of the libraries in a larger more complex heterogeneous system.

The initial ForSyDe representation of the model served primarily as a formal foundation for the further partitioning, analysis and refinement of the functional components into their respective domains. Based on the actual example requirements and tests for functional and structural equivalence between the ForSyDe representation and the initial abstract SystemC AMS and HetSC representations of the model could be deduced enabling the application of analysis and partitioning methods available in ForSyDe to the SystemC-based models and vice versa. Where full equivalence could not be achieved, e.g. for OSSS+R models of the *Crypto Unit*, model parameters and characteristics could be extracted and propagated back into abstracted ForSyDe representations.

In Fig. 8.9 the final model of the BASK transceiver is shown. The three main components are implemented using different libraries of the modelling framework. The *Transceiver* is implemented as a combined TDF and Continuous Time model in SystemC AMS using some of the available building blocks and some user-defined blocks. Conversion of the digital to the analogue signal and vice versa is performed by the *D2AConv* and the *AD2Conv* blocks respectively. The *Crypto Unit* is implemented as a cycle-accurate OSSS+R model using Recon-Objects to model the dynamic configuration of the different encryption algorithms. The Controller is implemented as an untimed HetSC model including two processes. The *crypto control* process monitors the bit error rate and selects the crypto algorithms of the Recon-Objects. The *power control* process also monitors the bit error rate from the *Crypto Unit* and controls the gain and threshold of the *Transceiver*. The three main components are connected via *Converter Channels*. These channels take care of converting the transferred data and the timing semantics between the different Models of Computation used in the respective domains.

8.4.3 Results

Using a standard C++ compiler with the AHES modelling framework and the OSCI SystemC 2.2.0 reference implementation the model can be compiled into an executable simulation. Figure 8.10 shows a part of the output of a simulation run. At 17,890,400 ns two bit errors are indicated. Due to these bit errors the power mode is set to *high* at 17,891,200 ns. The high error-rate results in a package being dropped by the Triple DES algorithm. The resulting signal trace is shown in Figs. 8.11 and 8.12 with the two bit errors being marked by circles at 10.6 and 11.2 ms and the signal power gain at 18 ms. More details on the example application and simulation results can be found in [23].

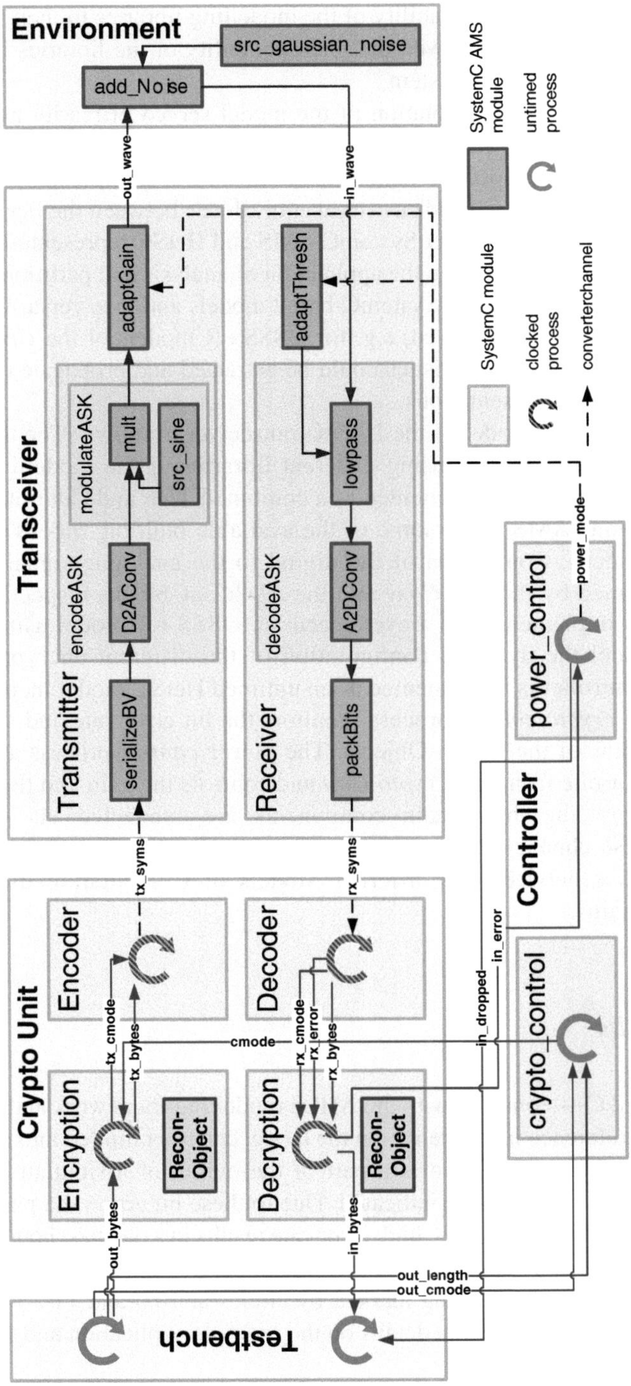

Fig. 8.9 Final model of BASK transceiver

```
...
15008 us crypto_unit@encrypt_proc(): Encrypting Triple-DES packet: 858927408
17890400 ns crypto_unit@decrypt_proc(): Errors in received bytes detected!
17891200 ns controller@do_it(): Error rate too high, switching to high power mode!
17891200 ns testbench@consume(): Packet with crypto-mode Triple-DES dropped!
32209200 ns testbench@produce(): Starting packet (Blowfish, length 2)
32209800 ns crypto_unit@encrypt_proc(): Encrypting Blowfish packet: 1128415552
34891 us crypto_unit@decrypt_proc(): Decrypted Triple-DES packet: 319951120
34891800 ns controller@do_it(): Error rate lowered down, switching to low power again.
...
```

Fig. 8.10 Simulation output

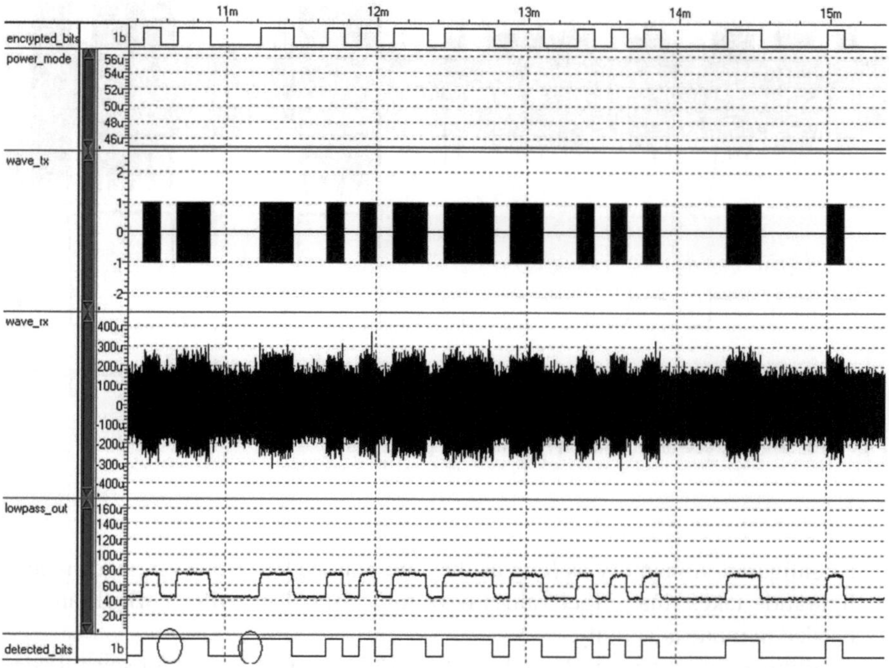

Fig. 8.11 Signal trace of simulation run showing bit errors

8.5 Performance Analysis of Adaptive Systems

Based on the AHES modelling framework the ANDRES project also worked on providing vehicles for performance analysis in the context of adaptive systems. The following problem areas have been of particular interest:

- *Trade-off between mode adaptivity and function adaptivity*. The use of function adaptivity will save many resources, since functions can be efficiently stored in

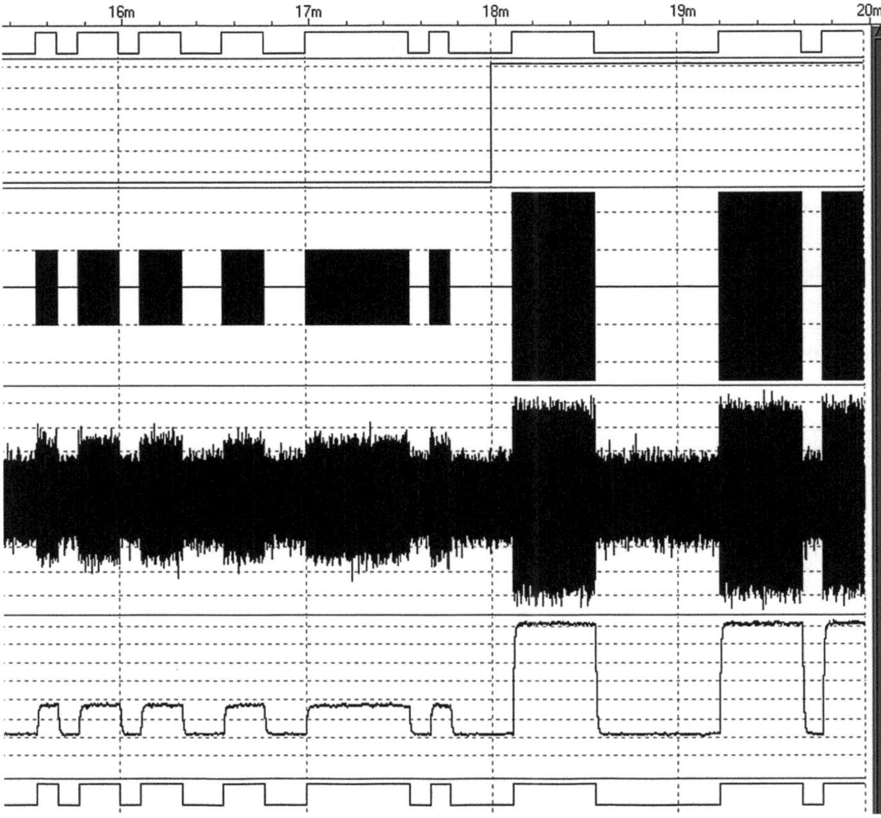

Fig. 8.12 Signal trace of simulation run showing power gain

a configuration memory as long as they are not needed. However, since reconfiguration takes time, input values have to be stored in buffers, which are not needed in mode adaptivity.

- *Usability of pre-fetching*. Reconfigurable FPGAs can have slots for several configurations. Thus, if we can predict the next function to execute, we can already load this function in advance into a second slot on the FPGA. Then a simple *switch* can activate this function, which will result in up to no delay and very small requirements on buffer size to store data.

The analysis approach chosen in ANDRES uses the formal ForSyDe representation of an adaptive application using the newly introduced adaptive processes. Though there is no direct equivalence between the ForSyDe adaptive process and a Recon-Object modelled in OSSS + R, certain characteristics and requirements of an OSSS + R model can be extracted and annotated to a ForSyDe model.

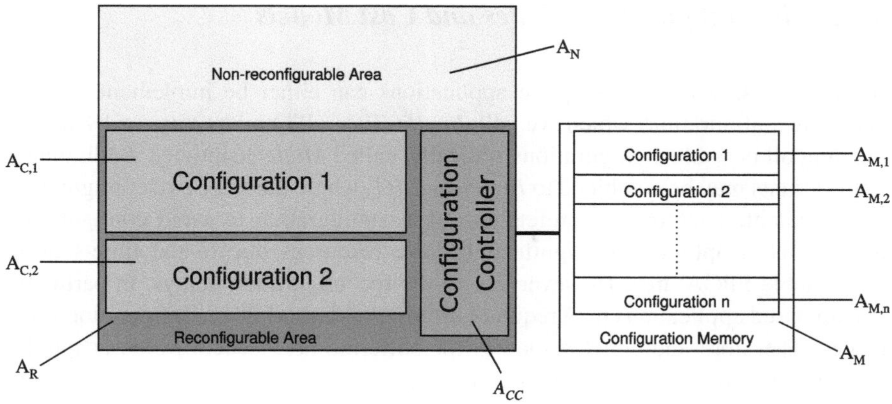

Fig. 8.13 Model of a partially reconfigurable FPGA

8.5.1 *FPGA Model*

Figure 8.13 illustrates the model of a reconfigurable FPGA as used for the performance analysis. A typical FPGA consists of logic elements (LE), memory elements (ME) and often also function specific resources. However, for the following discussion, we only assume the existence of logic and memory elements. Partially reconfigurable FPGAs can be used to design adaptive systems, since a part of the FPGA's total area can be reconfigured at run-time. For our discussion we divide the FPGA area into three areas: the *reconfigurable area* which can take different configurations at run-time, the *static area* which contains one configuration all the time and the *configuration memory* which is used to store the different configurations in a compressed format. Additionally, a partially-reconfigurable FPGA contains some infrastructure for the configuration management (initiating and scheduling reconfiguration, loading configurations), which we summarize as the *configuration controller*. Each of these areas contributes to the overall area size of an application, whereas the configuration memory consists of memory elements and the others areas consist of logic elements. The size of reconfigurable area is only as large as the largest configuration that can be configured to that area. If there are multiple configuration slots in parallel each of these slots is regarded as a separate reconfigurable area.

During reconfiguration a bit-stream is loaded from a configuration memory into one of the reconfigurable areas. The reconfiguration time on an FPGA typically lies in the area of milliseconds depending largely on the characteristics of the technology chosen for the configuration memory and the size of the configuration to be loaded into the reconfigurable area.

8.5.2 Reconfiguration Modes and Cost Models

In general using FPGAs adaptive applications can either be implemented using partial reconfiguration, which we call *Just-In-Time* (JIT) adaptivity, or by implementing all possible configurations statically, called *Mode* adaptivity. Additionally both variants may be combined to *Prefetched JIT*, where there are two configuration slots available enabling the prefetching of a configuration to avoid configuration delays. JIT adaptivity can significantly save resources due to the time-sharing of available FPGA area. However due to the reconfiguration delays, in particular stream based applications may require buffering of data while reconfiguration takes places. Therefore additional resources for buffering are needed. Mode adaptivity does not require any buffers as there is no reconfiguration delay but may require significantly more area for implementing all configurations in parallel. Including the cost for buffers, the overall area cost for a JIT adaptive application is:

$$A_{JIT} = A_{CC} + \sum_{i=1}^{n} A_{M,i} + \max(A_{C,1},\ldots,A_{C,n}) + A_{Buffer}$$

Where A_{CC} is the area cost of the configuration controller, $A_{M,i}$ is the memory cost to store configuration i, $A_{C,i}$ is the area costs for configuration i, and A_{Buffer} is the area cost of the buffers.

A mode-adaptivity implementation where all possible configurations are implemented statically has the following area costs:

$$A_{Mode} = \sum_{i=1}^{n} A_{C,i} + A_{Mux(n)} + A_{Demux(n)}$$

While there are no costs for buffers, configuration storage or a configuration controller, additional multiplexer ($A_{Mux(n)}$) and de-multiplexer resources ($A_{Mux(n)}$) are needed to switch between the available configurations. The costs for both depend on the number of configurations.

If both variants are combined to the Prefetched-JIT adaptivity, total costs are:

$$A_{Prefetch} \leq A_{CC} + \sum_{i=1}^{n} A_{M,i} + 2(\max(A_{C,1},\ldots,A_{C,n})) + A_{Mux(2)} + A_{Demux(2)}$$

Here no buffers are needed if the activation order and time of the configurations are known, i.e. a configuration can always completely be prefetched before it is used. Otherwise buffers would be needed, which in worst case would be as large as for the standard JIT variant.

8.5.3 Application Model

A minimal adaptive streaming application model is illustrated in Fig. 8.14, which serves as a base for performance analysis. Nodes denote the computation processes.

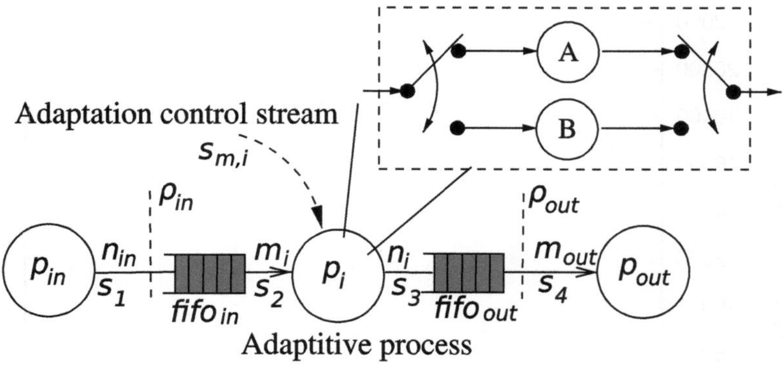

Fig. 8.14 Formal model of adaptive streaming application

Edges associated with FIFOs denote the communication channels with finite storage, which decouples the input data streams from output data streams of each communication channel (e.g. $fifo_{in}$ decouples s_1 from s_2). Processes read tokens from the input-side FIFOs, and emit the result data tokens to the output-side FIFOs at the end of the computation. The input/output token numbers are fixed and denoted as symbols at each side of the communication channels, e.g. process P_i has m_i input tokens and n_i output tokens. Meanwhile, the adaptive process P_i responds to the adaptation control stream $s_{m,i}$, and can switch between two different working modes *A* and *B*, as shown in the dashed box. While the stream source P_{in} provides a peak data rate ρ_{in} (on the communication channel cutting by the dashed-line), an average output data rate ρ_{out} needs to be guaranteed by process P_i even during its configuration.

8.5.4 Buffer Minimization and Results

Based on the architecture model of a partially reconfigurable FPGA and the application model shown in Fig. 8.14 a method for determining the maximum buffer size for a given throughput requirement and a set of configuration (with configuration sizes and times) has been developed. The method uses a hybrid approach based on integer linear programming and simulation and allows the analysis of trade-offs between JIT, Mode and JIT-Prefetching adaptation. The method has been combined with the OSSS+R modelling and synthesis techniques to allow direct parameter extraction and back-annotation from Recon-Objects to abstract application models.

An experimental result of our performance analysis using an OSSS+R model implemented on a Xilinx Virtex 4 FPGA is shown in Fig. 8.15. It shows costs for different implementations of an adaptive design using 5 configurations and varying output data rate requirements of up to 267 MBit/s. The design was implemented using JIT adaptivity (using either Block-RAM (BRAM) or Look-Up-Tables (LUT) for buffer implementations) or Mode adaptivity. The costs for Mode adaptivity are constant for all rates as no additional buffer memory is required. However for both

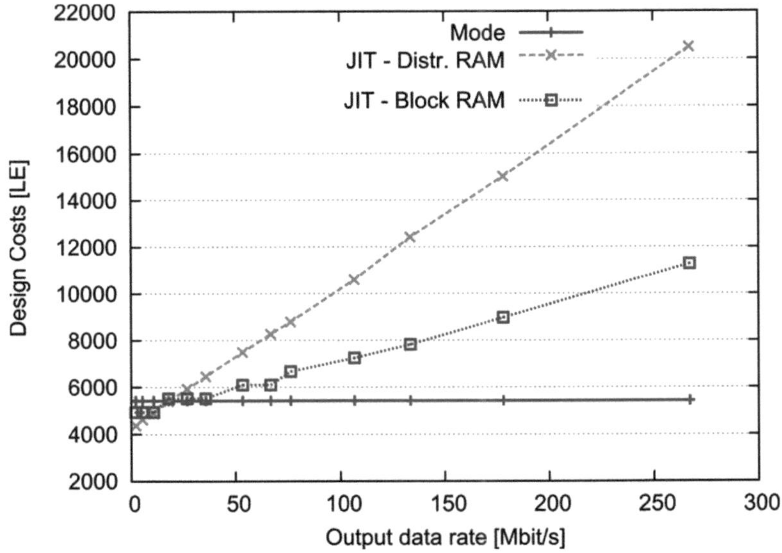

Fig. 8.15 Design costs for mode and JIT adaptivity for different output rates

JIT variants, the buffer costs increase significantly with higher output rates due to the larger buffers which are required to compensate the reconfiguration times. Though, for output rates less than 20 MBit/s both JIT variants have lower costs because the area for the buffers is smaller than the area which would be required for implementing all five configurations in parallel. The LUT RAM implementation starts to be less expensive than the BRAM variant but continues to be more costly for output data rates larger than 30 MBit/s. This is due to the fact, that the smallest unit of a BRAM that has to be allocated for a buffer is 18 kBit while for the LUT RAM it is one bit. However for larger buffers the area advantage of the BRAM quickly outweighs this disadvantage. More details on our method and experimental results can be found in [24, 25].

8.6 Conclusion

In this article we have presented the motivation, goals and main results of the ANDRES project. The project was created on the observation that the heterogeneous nature of embedded systems yielded many problems in state-of-the-art design flows. For covering different computational models, state-of-the-art methodologies incorporated different modelling domain specific languages and tools. In particular they lacked of support for adaptive architectures, especially including dynamic reconfigurable hardware. To overcome these issues, the ANDRES project has developed a SystemC modelling framework and seamless design flow for specification, implementation, simulation, and analysis of adaptive heterogeneous embedded systems. Based on the ForSyDe formalism, to express and analyse adaptivity in different MoC, three SystemC-based modelling libraries have been integrated into the AHES framework: SystemC-AMS for mixed-signal components, HetSC for software, and OSSS + R for run-time reconfigurable digital hardware. The AHES framework includes Converter Channels to connect these different libraries across different MoCs.

This modelling framework provides entry points for automatic synthesis of hardware, software, and communication infrastructure. Corresponding tools including automatic synthesis of software and runtime reconfigurable digital hardware (targeting platform FPGAs) have been implemented to assist in the implementation and analysis phase of AHES models. The AHES framework and its tools allow the efficient use and exploration of adaptivity in heterogeneous embedded systems. To support the designer in the assessment and exploration of different system performance metrics, the design flow is supplemented by a methodology for performance analysis, as presented in the adaptive BASK transceiver example.

The concepts for adaptivity, the AHES modelling framework, the synthesis tools, and performance analysis methodology have been evaluated successfully by the industrial partners Thales Communication and DS2.

In summary, the ANDRES project has established a solid foundation for modelling, synthesis, and analysis of adaptive heterogeneous embedded systems. The AHES framework is the first SystemC-based modelling library connecting the AMS, adaptive SW and adaptive HW domains. The close collaboration with industrial partners established a valuable knowledge transfer about using SystemC outside the classical HW domain.

Acknowledgements The authors would like to thank all ANDRES project partners for their valuable contributions: Andreas Schallenberg, Claus Brunzema, Anne-Marie Fouilliart, Frederic Colon, Philippe Bourgon, Fabien Colas-Bigey, Eugenio Villar, Fernando Herrera Casanueva, Axel Jantsch, Ingo Sander, Jun Zhu, Marcos Martinez, Christoph Grimm, Jan Haase, Markus Damm, Joseph Wenninger, Jiong Ou and Sarah Real.

References

1. A. Jantsch. *Modeling embedded systems and SoCs*, Morgan Kaufmann, 2004.
2. E. Lee and A. Sangiovanni-Vincentelli. *A framework for comparing models of computation.* IEEE Trans. on Computer-Aided Design of Integrated Circuits and Systems, V.17, N. 12, December 1998.
3. J. Davis et. al. *Overview of the Ptolemy project.* Technical memorandum UCB/ERL M01/11, Department of Electrical Engineering and Computer science, University of California, 2001.
4. F. Balarin, Y. Watanabe, H. Hsieh, L. Lavagno, C. Passerone and A. Sangiovanni-Vincentelli. *Metropolis: An integrated electronic system design environment.* IEEE Computer, V. 36, N. 4, April 2003.
5. D. Mathaikutty, H. Patel, S. Shukla and A. Jantsch. *UmoC++: Modeling environment for heterogeneous systems based on generic MoCs.* FERMAT Technical Report N.2005-05, Virginia Tech., 2005.

6. Alain Vachoux, Christoph Grimm, Karsten Einwich. *SystemC Extensions for Heterogeneous and Mixed Discrete/Continuous Systems*. In: International Symposion on Circuits and Systems 2005 (ISCAS '05), Kobe, Japan. IEEE, May 2005.
7. Mohamed A. Salem. *Ptolemy-Oriented Structural, Reconfigurable, and Heterogeneous Hardware Design, Verification and Synthesis*. The Sixth Biennial Ptolemy Conference, University of California Berkeley, Mentor Graphics, 2005.
8. The ANDRES project website http://andres.offis.de
9. IEEE Standards Association ("IEEE-SA") Standards Board. *IEEE Std 1666–2005 Open SystemC Language Reference Manual*, 2005.
10. J. Wenninger, M. Damm, J. Haase, J. Ou, K. Grüttner, P. A. Hartmann, A. Herrholz, F. Herrera, I. Sander, and J. Zhu. *Overall Modelling Framework for AHES*. Technical Report ANDRES/TUV/P/D1.6b/1.0, ANDRES project deliverable, August 2009.
11. I. Sander, *The ForSyDe Standard Library*, Royal Technical School, KTH, 2003.
12. I. Sander, & A. Jantsch, *System Modeling and Transformational Design Refinement in ForSyDe*. In IEEE Transactions on Computer-Aided Design of Integrated Circuits and Systems, 2004, 23, 17–32.
13. F. Herrera, S. Real, and E. Villar. *Modelling of Software. Final Library Elements*. Technical Report ANDRES/UC/P/D1.2b/1.2, ANDRES project deliverable, January 2009.
14. F. Herrera and E. Villar. *A framework for heterogeneous specification and design of electronic embedded systems in SystemC*. In Transaction on Design Automation of Electronic Systems, volume 12. ACM, August 2007.
15. HetSC project website http://www.teisa.unican.es/HetSC/
16. F. Herrera, E. Villar, and P.A. Hartmann. *Specification of Adaptive HW/SW Systems in SystemC*. In Proceedings of the Forum on Specification, Verification and Design Languages (FDL), pp. 61–66, September 2008.
17. OSCI AMS Working Group. *Standard SystemC AMS extensions Language Reference Manual v1.0*. May 2010. http://www.systemc.org
18. J. Ou, and J. Haase. *Modelling of signal processing architectures – Final library elements*. Technical Report ANDRES/TUV/P/D1.4b/1.0, ANDRES project deliverable, December 2008.
19. A. Schallenberg, F. Oppenheimer, W. Nebel, *OSSS+R: Modelling and Simulating Self-Reconfigurable Systems Field-Programmable Logic*, Proceedings - 2006 International Conference on Field Programmable Logic and Applications, 2006, 177–182.
20. Xilinx Inc., *Early Access Partial Reconfiguration User Guide* (UG208) 2006.
21. F. Herrera, E. Villar, C. Grimm, M. Damm, and J. Haase. *Heterogeneous Specification with HetSC and SystemC-AMS. Widening the support of MoCs in SystemC*. Embedded Systems Specification and Design Languages, LNEE 10:107–121. May 2008.
22. M. Damm, C. Grimm, J. Haase, A. Herrholz, W. Nebel. *Connecting SystemC-AMS models with OSCI TLM 2.0 models using temporal decoupling*. In Proceedings of the Forum on Specification, Verification and Design Languages (FDL), pp. 25–30, September 2008.
23. J. Wenninger, M. Damm, J. Haase, J. Ou, K. Grüttner, P. A. Hartmann, A. Herrholz, F. Herrera, I. Sander, and J. Zhu. *Overall Modelling Framework for AHES*. Technical Report ANDRES/TUV/P/D1.6b/1.0, ANDRES project deliverable, August 2009.
24. Ingo Sander and Jun Zhu. *Overall methodology for partitioning and performance analysis of AHES*. Technical Report D1.1c, ANDRES project deliverable.
25. Ingo Sander, Jun Zhu, Axel Jantsch, Andreas Herrholz, Philipp A. Hartmann, and Wolfgang Nebel. *High-level estimation and trade-off analysis for adaptive real-time systems*. In Proceedings of the 16th Reconfigurable Architectures Workshop, Rome, Italy, May 2009.

Chapter 9
CRISP: Cutting Edge Reconfigurable ICs for Stream Processing

Tapani Ahonen, Timon D. ter Braak, Stephen T. Burgess, Richard Geißler, Paul M. Heysters, Heikki Hurskainen, Hans G. Kerkhoff, André B.J. Kokkeler, Jari Nurmi, Jussi Raasakka, Gerard K. Rauwerda, Gerard J.M. Smit, Kim Sunesen, Henk van Zonneveld, Bart Vermeulen, and Xiao Zhang

Abstract The Cutting edge Reconfigurable ICs for Stream Processing (CRISP) project aims to create a highly scalable and dependable reconfigurable system concept for a wide range of tomorrow's streaming DSP applications. Within CRISP, a network-on-chip based many-core stream processor with dependability infrastructure and run-time resource management is devised, implemented, and manufactured to demonstrate a coarse-grained core-level reconfigurable system with scalable computing power, flexibility, and dependability. This chapter introduces CRISP, presents the concepts, and outlines the preliminary results of a running project.

9.1 Project Partners

1. Recore Systems, The Netherlands (coordinator)
2. University of Twente, The Netherlands
3. Atmel Automotive GmbH, Germany
4. Thales Netherlands, The Netherlands
5. Tampere University of Technology, Finland
6. NXP Semiconductors, The Netherlands

- Project Coordinator: Paul M. Heysters, Recore Systems, The Netherlands
- Start Date: 2008-01-01
- End Date: 2010-12-31
- EU Program: 7th Framework Programme, FP7 ICT-2007.3.4, STREP Project No. 215881
- Global Budget: 4.4 M €

K. Sunesen (✉)
Recore Systems, Enschede, The Netherlands
e-mail: Kim.Sunesen@recoresystems.com

J.M.P. Cardoso and M. Hübner (eds.), *Reconfigurable Computing: From FPGAs to Hardware/Software Codesign*, DOI 10.1007/978-1-4614-0061-5_9,

- Global Funding by EU: 2.8 M €
- Contact Author: Kim Sunesen (Recore Systems), Email: Kim.Sunesen@recoresystems.com

9.2 Introduction

Streaming applications have high market potentials and drive demands for reconfigurable platform chips [1, 2]. It is becoming increasingly more difficult to predict which applications are going to be successful in the future. For this reason, application providers are increasingly interested in programmable platform chips, which enable to anticipate on expected market trends and offer flexibility, if the market develops differently [3].

The Cutting edge Reconfigurable ICs for Stream Processing (CRISP) project develops a scalable and dependable reconfigurable multi-core system concept that can be used for a wide range of streaming applications in the consumer, automotive, medical and defense markets. The envisioned platform solution includes a massive multi-core processing architecture in combination with innovative design-time and run-time tools. CRISP addresses optimal utilization, efficient programming and dependability of reconfigurable many-cores for streaming applications. The main objective is to create a General Stream Processor (GSP) for tomorrow's streaming applications. There are many challenges in developing massive multi-core platform solutions. CRISP addresses the fundamental underlying problems. The concerns driving this project are indirectly attributable to the miniaturizations of semiconductor technology and translate into three essential questions:

How can the intrinsic processing potential of a massive multi-core architecture be exploited optimally for a wide range of streaming applications?

How can multi-core systems be programmed efficiently?

How can large multi-core integrated circuits using deep submicron semiconductor processes be made reliable and self-repairing?

The CRISP project uses a holistic and pragmatic applied research approach to find answers to these questions. The project is organized around four central themes:

Streaming applications – Targeted applications range from low-end consumer electronics and automotive applications to demanding high-end medical and defense applications.

General Stream Processor (GSP) – The GSP is a dynamically reconfigurable many-core platform for streaming applications. The scalable architecture targets flexibility, high performance, low power and a small footprint.

Run-time mapping – Computational resources in a many-core can be efficiently utilized by resource allocation at run-time. This also enables upgrading, bug fixing and hardware fault diagnosis and repair at run-time.

Dependability – Dependability and yield of deep-submicron chips are improved using new techniques for static and dynamic detection and localization of faults and (dynamically) circumventing faulty hardware.

Results of CRISP will include concrete hardware manufacturing as well as software development of tools, systems, and applications. This chapter dedicates a section to each of these themes. While the CRISP project is still running at the time of writing, preliminary results are achieved for all themes. The CRISP approach demonstrates the synergy of the four themes.

9.2.1 Related FP Projects

The CRISP project can be seen as a successor of the 4S ("Smart chipS for Smart Surroundings") FP6 project [4–6]. In 4S, reconfigurable computing proved to deliver high performance while being energy efficient, flexible, programmable, and run-time adaptable. More details can be found in the 4S project chapter of this book [4]. The results of 4S also revealed new research topics concerning scalability of multi-core systems and dependability of deep submicron technologies. On this background, three 4S project partners Atmel, Recore Systems, and University of Twente joined with NXP Semiconductors, Thales Netherlands, and Tampere University of Technology to form the CRISP Consortium to break new grounds in scalable and dependable high-performance computing using dynamically reconfigurable many-core platforms.

9.2.2 Chapter Structure

The remainder of this chapter is organized along the themes of CRISP. The next four sections introduce and explain the concepts of streaming applications including the specific CRISP demonstrators, the CRISP General Stream Processor, run-time resource management, and dependability. Before concluding, we present preliminary results of CRISP and take a look towards the future.

9.3 Streaming Applications

CRISP research focuses on efficient embedded hardware and software solutions for stream processing used in a wide range of streaming applications including baseband processing for wireless communications, multimedia, sensor, medical image, and intelligent antenna signal processing. Streaming applications can be modeled as data flow graphs with streams of data items (the edges) flowing between computation kernels (the nodes) [1, 7]. Streaming applications inherently hold locality and concurrency properties. Key characteristics include:

- High data-throughput between computing intensive kernels;
- Local processing on data items;
- Hard real-time throughput guarantees;
- Semi-static operation life-time.

The kernels and communication topology of the data flow graph, the data bandwidth between kernels, data item sizes and real-time guarantees vary with every application.

In the consumer market there are strong trends towards integrating more and more streaming applications into a single device, ever faster changing standards and algorithms demanding ever more processing power, as well as a trend of rising upfront investments for silicon manufacturing. In the medical, defense, and security markets, there is a trend towards reducing development and maintenance costs as well as using state-of-the-art technology by choosing commercial off-the-shelf components. Together these trends fuel the need for reconfigurable ICs that combine some of the flexibility known from Field Programmable Gate Arrays (FPGAs) with the speed, size, and energy advantages known from Application Specific Integrated Circuits (ASICs).

In CRISP, concrete streaming applications are worked out for digital beamforming and Global Navigation Satellite System (GNSS) reception. Digital beamforming is chosen as a typical example of a high-end application with high throughput and computing demands. Digital beamforming is used in defense and space applications but is also becoming increasingly popular in consumer electronics applications such as femto-cell and wireless access gateways. Satellite navigation systems have become omnipresent in e.g. cars, smart phones, and wrist watches. GNSS is chosen as a typical example of a low-end consumer application. The ambition of the CRISP platform is to scale from low-end to high-end applications using the same reconfigurable System-on-Chip (SoC) template. Both applications play their role in validating the success of the CRISP approach.

9.3.1 Digital Beamforming

Digital beamforming is used in an increasing range of products like sonar systems, radar systems, radio astronomy telescope systems, and base stations for wireless telecommunications. In the CRISP project, the digital beamforming application is derived from the radar field [8, 9] where requirements are demanding in terms of both data throughput and processing power. For instance, a system with 64 antenna receive channels typically has input rates of tens of Giga bytes per second and requires several Giga Multiply Accumulate (MAC) operations per channel.

Figure 9.1 depicts the functional architecture of a beamformer system. In the analog part, signals are received from multiple antennas and converted into digital signals. In the digital part, antenna processing is applied to each signal for calibration

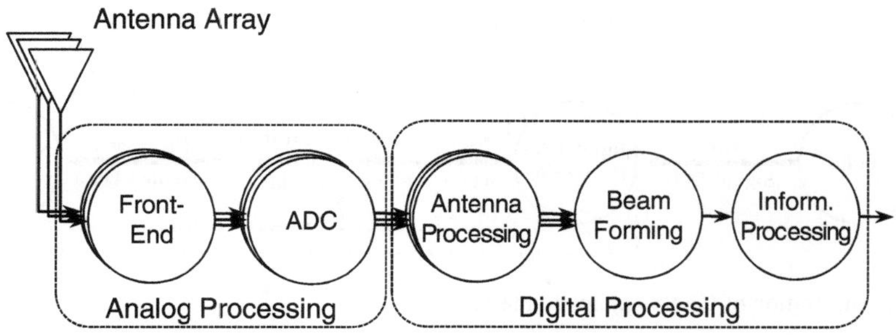

Fig. 9.1 Functional architecture of a digital beamformer system

or equalization and beamforming combines signals into beams that are further processed. Below, in the section on preliminary results further details on how the digital processing of the beamformer system is mapped on the CRISP platform.

9.3.2 Global Navigation Satellite System Application

Satellite navigation applications of today range from cheap, single-frequency receivers embedded in mobile phones to expensive, multi-frequency, centimeter-accurate scientific receivers. In the CRISP project, the GNSS application is specified and designed to support the existing U.S. based NAVSTAR Global Positioning System (GPS) [10] and the future Galileo (European satellite navigation system) [11] signals transmitted in the L1 frequency band.

Three main blocks in any GNSS receiver can be identified as (i) a radio front-end for analogue signal processing (shown as Radio in Fig. 9.2), (ii) a digital baseband processing part for navigation data decoding and signal time-of-arrival measurements (shown as Digital Baseband Processing in Fig. 9.2), and (iii) navigation calculus to determine position, velocity and time (PVT) (shown as Navigation Task in Fig. 9.2) [12].

The digital baseband processing tasks are: acquisition (search of satellites) by using long FFTs and tracking, data decoding and signal time-of-arrival measurements by means of serial correlation. These receiver domains are present in all GNSS receivers through the whole range from inexpensive mass market devices to high-end scientific receivers. In the CRISP implementation, four to eight acquisition or tracking processes are expected to be active at any point in time.

Most of the current state-of-the-art GNSS receivers are using dedicated hardware for digital signal processing (ASIC technology), but the trend is going towards Software Defined Radio [13]. The Software Defined Radio approach is motivated by the challenges faced due to the modernization efforts in satellite navigation systems. The European Galileo system together with systems from Russia (GLONASS) and China (Compass) are emerging to compete with U.S. based GPS, which is also undergoing a modernization program. The specifications of new systems are still evolving

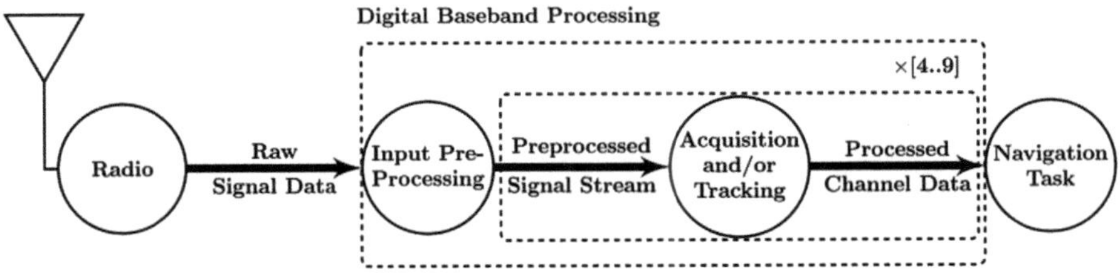

Fig. 9.2 Illustration of GNSS application tasks

and also the algorithm development in the field of multi-system reception is active. Thus, future receivers should have high computational power and flexibility for updates. A scalable platform with multiple reconfigurable processing tiles meets both of these requirements.

9.4 General Stream Processor (GSP)

The General Stream Processor (GSP) is a scalable platform template for performing virtually any streaming application, from low-end consumer applications to high-end applications. The GSP architecture is essentially a heterogeneous many-core system-on-chip (SoC) containing general purpose processor cores (e.g. ARM9™), reconfigurable cores (e.g. Xentium®), and memory tiles.

The proposed GSP architecture is designed with the principles of locality-of-reference and concurrency in mind. Hence, different levels of storage are defined in the CRISP many-core architecture; local data memories are incorporated in Xentium processing tiles and distributed on-chip memory is available in the GSP by means of the Smart Memory Tiles. Moreover, concurrent processing of computation kernels in the streaming applications can be performed on the many parallel Xentium processing tiles in the GSP. The Network-on-Chip and the distributed on-chip memories provide the communication infrastructure in the GSP to manage the data streams between the computation kernels in the streaming applications.

Within the scope of the CRISP project, a scalable GSP platform is built and subsequently demonstrated using many multi-core devices that are connected to each other to create a large many-core system-of-chips.

The GSP demonstrator is built from two kinds of chips: a *General Purpose processor Device* (GPD) and *Reconfigurable Fabric Devices* (RFD). The GSP depicted in Fig. 9.3 combines one GPD and five RFDs to build a many-core with 46 processing cores: one ARM® processor and 45 Xentium DSP cores. The Xentium DSP core is explained in more detail below. The inter-chip connection is handled by a dedicated Multi-Channel Port (MCP) chip-to-chip interface that extends the Network-on-Chip (NoC) of the RFDs across multiple chips. The MCP enables off-chip communication and is instrumental in demonstrating scalability as it allows NoC communication to extend seamlessly across chip boundaries.

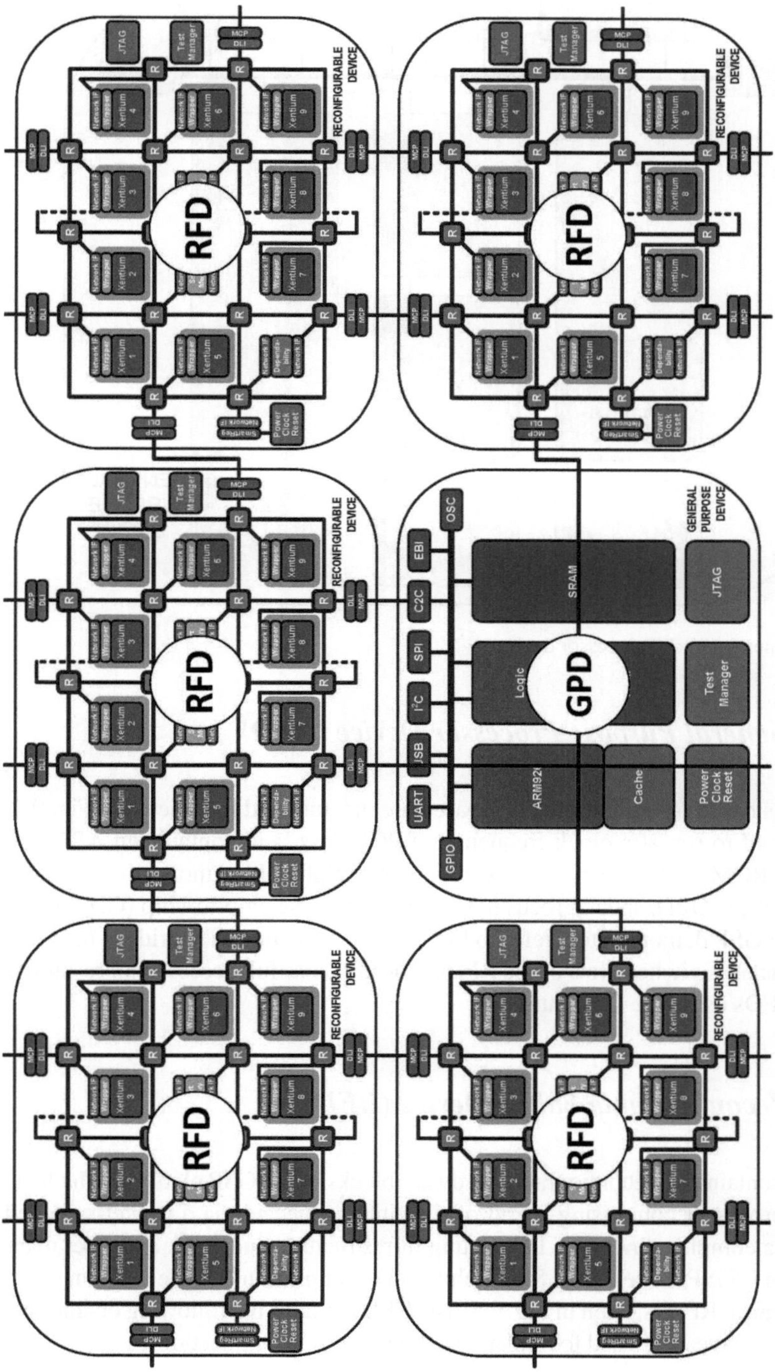

Fig. 9.3 Illustration of a GSP demonstrator connecting one GPD chip and five RFD chips in an off-chip network. This GSP connects forty-five Xentium DSP cores and one ARM® core in one network. Fig. 9.4 and Fig. 9.5 depict, respectively, the GPD and the RFD in more detail

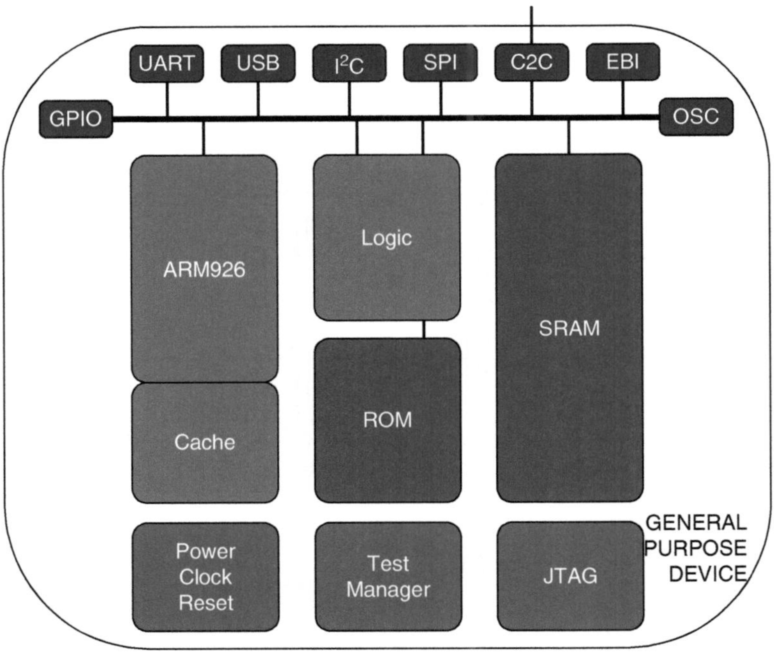

Fig. 9.4 General purpose processor device

9.4.1 General Purpose Processor Device (GPD)

The GPD is based on a traditional bus-based architecture and is depicted in Fig. 9.4. It is designed to run at a clock frequency of 200 MHz, and contains an ARM926 processor, ROM, SRAM and a wide range of peripherals. Among the common interfaces of the GPD, special focus has been given to the chip-to-chip (C2C) interface. In the GSP demonstrator (refer to Fig. 9.3) the C2C interface bridges the GPD with the packet-switched NoC of the RFDs. The GPD has full access to all resources of these RFDs using the C2C interface.

9.4.2 Reconfigurable Fabric Device (RFD)

The RFD contains the reconfigurable hardware blocks of the GSP platform. The RFD is a tiled processor comprising a grid of tiles interconnected via a packet-switched NoC. It is a complete SoC with I/O, testing infrastructure, and clock and reset management. In CRISP, a twelve-tile SoC architecture is implemented. The diagram of the twelve-instance RFD is given in Fig. 9.5. The RFD contains the following operational components that can be used to implement streaming DSP applications:

- 9 Xentium processing tiles
- 2 Smart Memory Tiles (SMT)

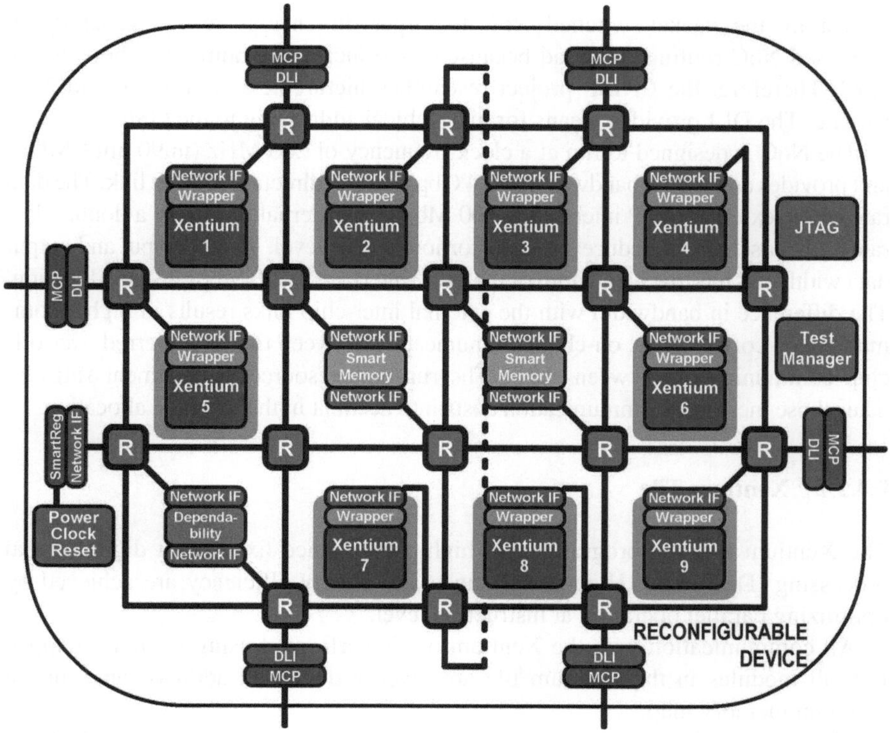

Fig. 9.5 Reconfigurable fabric device

- Dependability Manager tile (DM) + Infrastructural IP
- 6 Multi-Channel Ports (MCP) + Die Link Interfaces (DLI)
- Packet-Switched Network-on-Chip

The main communication infrastructure in the RFD is provided by the NoC. All Xentium processing tiles and memory resources are accessible via the NoC. Moreover, the NoC can be used to configure the clock manager of the RFD, and to access scan chains and memory Built-In Self-Test units of the Xentium tiles.

9.4.2.1 Scalable On-Chip and Off-Chip Communication Infrastructure

A packet-switched NoC connects the tiles in the RFD. The NoC provides means for on-chip scalability of the GSP template. Hence, larger many-core SoCs can be designed by increasing the number of routers in the NoC.

In the GSP demonstrator, off-chip scalability is addressed by extending the NoC across the RFD chip boundaries using dedicated MCP interfaces. The MCP provides a transparent off-chip interface to bridge two RFDs or to connect the RFD with the C2C interface of the GPD. The MCP interfaces enable prototyping of large many-core systems-of-chips as depicted in Fig. 9.3.

Scaling the packet-switched NoC to larger dimensions results generally in increased NoC routing overhead because of the increasing number of hops in the NoC. Therefore, the CRISP project researches hierarchical addressing and NoC routing. The DLI provides means for hierarchical addressing to the GSP.

The NoC is designed to run at a clock frequency of 200 MHz (in 90 nm CMOS) and provides an on-chip bandwidth of 6.4 Gbps in each direction of each link. The data rate of the external MCP interface is 100 Mbps per external pin using a double data rate transfer scheme to reduce signal distortion at PCB level. With an input and output data width of 8 bits, the bandwidth of the MCP interface is 800 Mbps in each direction. The difference in bandwidth with the external inter-chip links results in higher communication costs. Hence, on-chip communication between tiles is preferred over off-chip communication between chips. The run-time resource management software takes those increased communication costs into account in the resource allocation.

9.4.2.2 Xentium Tile

The Xentium tile is a programmable high-performance fixed-point digital signal processing (DSP) core. High-performance and energy-efficiency are achieved by optimizing parallel operation at instruction level.

All communication with the Xentium tile is performed using memory-mapped I/O; all modules in the Xentium tile are given a dedicated address range in the Xentium memory map.

The Xentium tile is designed to operate at a clock frequency of at least 200 MHz (in 90 nm CMOS technology, worst case conditions). If required, methodologies from simple clock gearing up to sophisticated dynamic voltage and frequency scaling (DVFS) can be implemented locally in the Xentium tile to reduce the power consumption of the processing tile. The area requirement of the tile equipped with 16 kBytes of data memory and 8 kBytes of instruction cache is ~1.8 mm^2 in 90 nm CMOS technology.

The core modules of the Xentium tile are the Xentium core, tightly coupled data memory and a NoC interface as shown in the block diagram in Fig. 9.6. Moreover, the Xentium tile contains additional logic that is used to control scan chains and memory Built-In Self-Test at run-time. The additional logic is part of the Dependability Infrastructural IP, described below in the section on Xentium tile dependability, where it is referred to as Xentium Tile Wrapper.

Xentium datapath: The Xentium datapath contains parallel execution units and register files. The different execution units can all perform 32-bit scalar and vector operations. For vector operations the operands are interpreted as 2-element vectors. The elements of these vectors are the low and high half-word (16-bit) parts of a 32-bit word. In addition several units can perform 40-bit scalar operations for improved accuracy. All operations can be executed conditionally. The Xentium datapath provides powerful processing performance:

- 800 16-bit Mega MACs per second or
- 400 32-bit Mega MACs per second or
- 400 16-bit complex Mega MACs per second

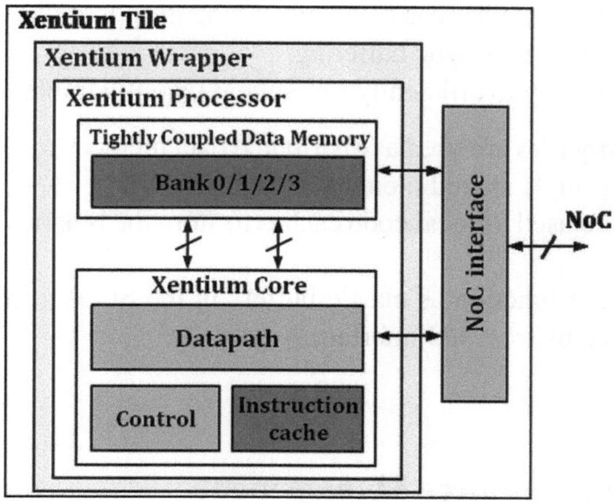

Fig. 9.6 Xentium tile

Xentium control: The control block in the Xentium core performs instruction fetching and decoding, and controls the execution units in the datapath. Instructions are fetched from Xentium-external memory (e.g. Smart Memory Tile) and are stored in the Xentium instruction cache. The programmer can indicate that a section of a Xentium program has to be prefetched by the control to ensure that the instructions of that section of the program are cached. This prevents cache misses during execution, which makes the execution time of the prefetched section of the program predictable (provided that the execution time of the data loads and stores in the section are predictable). The prefetch mechanism provides means to reconfigure the Xentium tiles in the RFD with the required DSP functions. The run-time resource management software (running on the GPD) can, for instance, issue a prefetch command in order to force one of the Xentium tiles in the RFD to reconfigure with new DSP functionality.

Xentium tightly coupled data memory: The tightly coupled data memory is organized in parallel memory banks to allow simultaneous access by different resources. The data memory can be simultaneously accessed by the Xentium core as well as by the Xentium NoC interface. By default the data memory in the Xentium tile is organized in four banks of 4 kbytes each, implemented using SRAM cells. The size of the memory banks is parameterizable at design-time.

9.4.2.3 Smart Memory Tile

In the RFD, the Smart Memory Tiles (SMT) provide shared memory that is accessible through the NoC. The SMT contains parallel memory banks with a total memory size of 64 kBytes per memory tile in the RFD.

Besides being accessible with random access, the SMT has multiple reconfigurable Address Generation Units (AGU). The AGU can generate a sequence of addresses autonomously. Multiple AGUs can be linked, giving the possibility to configure FIFO functionality or implement elastic buffering.

Memory tiles are key to efficiently implement reconfigurable many-core SoCs:

- multiple memory tiles are used as distributed data memory to reduce the required bandwidth to a single shared memory;
- memory tiles are used as local code caches to store the binaries of reconfigurable tiles in the SoC;
- memory tiles can function as elastic buffers in the SoC (e.g. to buffer input or output data streams from I/O interfaces).

9.5 Run-Time Resource Management

Due to the scalability of the GSP platform, some configuration effort is required to load an application onto the processing elements. This includes the assignment of tasks to specific Xentium cores, and configuring the routing tables of the NoC with routing information. When a programmer performs these steps (manually) at design-time, an assumption must be made on the availability of resources. This results in potential conflicts, when resources are in use by other applications, or when resources are unavailable due to hardware faults. A more extensive design-time analysis can only consider a limited number of use-cases, and event- or user-triggered actions are difficult to anticipate on. Any change in the application or in the platform that is not captured by the design-time analysis, results in incorrect or unpredictable behavior.

Even in application specific platforms, where less flexibility is required, resource allocation at run-time may have additional advantages in fault tolerance. For example, when faults are detected by a postproduction test, otherwise disabled spare parts may be used to replace the malfunctioning components; this is called static redundancy [14]. A chip then may still be usable if the number and type of faults are non-critical. This approach is used in the manufacturing process of e.g. the Cell processor, where not all eight Synergistic Processing Elements are required, resulting in a higher production yield [15]. Such scenarios only work when the application mapping to the architecture can deal with these deficiencies. On the other hand, safety critical systems often use online fault detection mechanisms. With redundant resources available, the system may continue its operation if a detected fault can be isolated. Thus, to provide support for fault tolerance, a run-time resource manager should be in place to account for the free, allocated and faulty resources in the system.

Larger systems, such as heterogeneous computing clusters, use a distributed memory model to overcome the high communication overhead of shared memory systems. Such systems often use middleware to present the system in a homogeneous manner. However, distributed memory multi-processor systems are in general heterogeneous [16], and it may be quite expensive in terms of performance and energy consumption to expose them as homogeneous systems. Keeping the property of homogeneity may even be considered impossible, due to asymmetric resource allocation for multiple applications simultaneously running in the system. Considering parallel programming, the most complex platform is a multi-user, heterogeneous cluster made up of processors of different architectures, interconnected via a heterogeneous communication network [16].

The GSP platform incorporates a distributed memory model, and besides the efficiency arguments, the Xentiums are not designed to run any middleware. Therefore, the GPD manages the resources of the connected RFDs in a centralized manner. Each application requests the amount and type of resources required for its execution. The resource manager determines whether these requests can be fulfilled. The resource manager must shield the tasks that are already present on the platform from the interference caused by newly started applications. A resource management policy [17] should enforce such conditions:

- *admission control* – an application is only allowed to start if the system can allocate, upon request, the resources the application required to meet its performance constraints;
- *guaranteed resource provisions* – the access of a running task to its allocated resources cannot be denied by any other task.

If an application cannot be added to the system without a violation of this policy, then the resources for the application will not be allocated. Hence, in that case the application is refrained from execution on the system by the run-time resource manager.

9.5.1 The Resource Allocation Problem

The resource allocation problem concerns two dimensions: the spatial and temporal dimension. Explicitly considering the (relative) location of resources within a platform may avoid inefficient resource allocations, where efficiency may be measured in the amount of resources being allocated, or in energy consumption of the platform. Orthogonal to the spatial dimension, we consider the temporal dimension. This does impose some uncertainty, as we assume that we do not know in advance which applications are started or stopped at a certain time. Therefore, optimal solutions cannot be guaranteed because it depends on future events. The resource management algorithms thus resort to finding feasible solutions, while optimizing towards secondary objectives, such as minimal energy consumption or highest performance. Additionally, in the case that multiple applications may be executed simultaneously on the platform, a reasonable platform state has to be maintained that allows for more than one resource request to be fulfilled. Given these properties and the complexity of the problem, heuristics are used to tackle the problem.

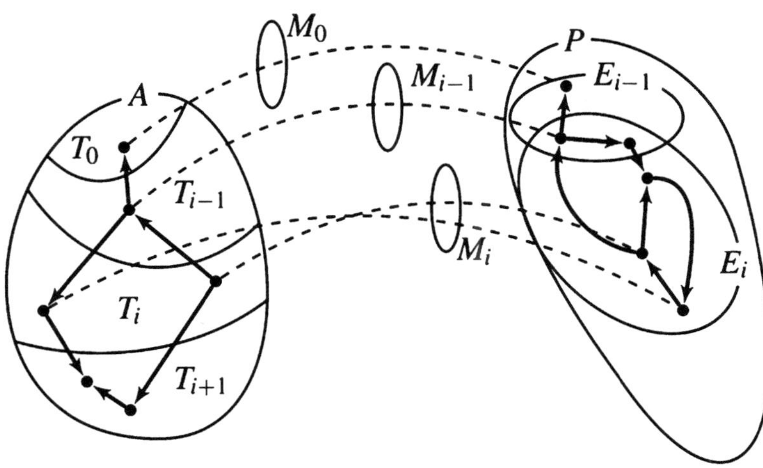

Fig. 9.7 The resource allocation problem is partitioned in multiple sub-problems, using the structure of the task graph

A graph representation of applications and platform is used to reason about the connectivity of the resource demands and provisions, respectively. In spatial resource allocation, the problem is to find specific locations to fulfill the resource requirements of the tasks T and communication channels C in an application $A=(T,C)$. A platform $P=(E,L)$ provides resources through the processing elements E, which are connected with the links $L \subseteq E \times E$. In Fig. 9.7, the graph of tasks T_i describes an application A, which needs to be mapped on a set of resources E_i composing platform P. Mapping M denotes the specific assignment of tasks to resources. Each task needs to communicate with at least one other task in the same application; otherwise, that task would compose a problem instance of its own. For each pair of communicating tasks, the communication infrastructure has to provide a communication route with enough bandwidth between their assigned processing elements. The combination of these problems may be formulated as a constraint optimization problem.

The topological aspect limits the number of possible task assignments. Therefore, exploiting the topology of the application and the topology of the platform vastly reduces the search space. The resource manager incorporates a heuristic that uses divide-and-conquer to break the resource allocation problem into sub-problems of variable size, depending on the density of the task graph.

The sub-problems that have to be solved are instances of the common generalized assignment problem, which is NP-hard [18]. Taking the communication requirements between tasks into account, for each subset of tasks T_i a subset of candidate elements E_i is selected. For every $e \in E_i$, the cost of mapping each task $t \in T_i$ is calculated. This procedure is applied to each subset of tasks T_i in application A. Thus, following the topology within application A, mapping M is incremented by each mapping M_i of tasks T_i to a set of resources E_i. In-depth details about the implementation are explained in [19]. If all tasks in the application can be mapped to a certain location in the platform and if sufficient communication resources are available, the application may be started.

One important aspect of resource allocation is to maintain clear and consistent state information of the GSP. Therefore, the resource management system software needs to know which cores and links in the SoC are available to map tasks. By using feedback from dependability software as discussed in the next sections, the resource manager receives feedback regarding broken links and erroneous cores in the GSP [20].

9.6 Dependability

The dependability of large scale multi-processor systems-on-chip is becoming an important concern, especially when the SoCs are used in mission-critical applications. Dependability is the extent to which a system can be relied upon to perform its intended functions under defined operational and environmental conditions at a given instant of time or given interval.

In the next sections, the CRISP approach for improving the dependability of the reconfigurable multi-core GSP is outlined. Dependability in the GSP focuses mainly on two aspects: making the NoC more dependable and improving the dependability of Xentium tiles. The first aspect ensures the correct operation of the NoC, which is a property that the second aspect heavily relies upon.

9.6.1 Network-on-Chip Dependability

The NoC dependability test is a GPD-hosted utility for diagnosing (i.e. detecting and locating) faults within the communication infrastructure of the GSP [21]. By using a software-only approach and solely using the existing NoC for test access, the need for specialized Design for Test structures within the NoC is eliminated. This saves silicon area and increases the ease with which the system may be deployed.

9.6.1.1 Network Fault Modeling

A model has been adopted which describes the NoC infrastructure in terms of inter-die and inter-router connections (link components) plus the switch components inside the router (intra-router routing paths), any of which may be generally referred to as a (path) component *C*. Accordingly, an arbitrary path through the network can be represented by a series of interconnected path components, C_1 to C_n.

For example, the network path illustrated in Fig. 9.8a is modeled by the five components C_1 to C_5 labeling the graph in Fig. 9.8b. Observe that C_1, C_3, and C_5 are link components whereas C_2 and C_4 denote switch components. The overall fault status of

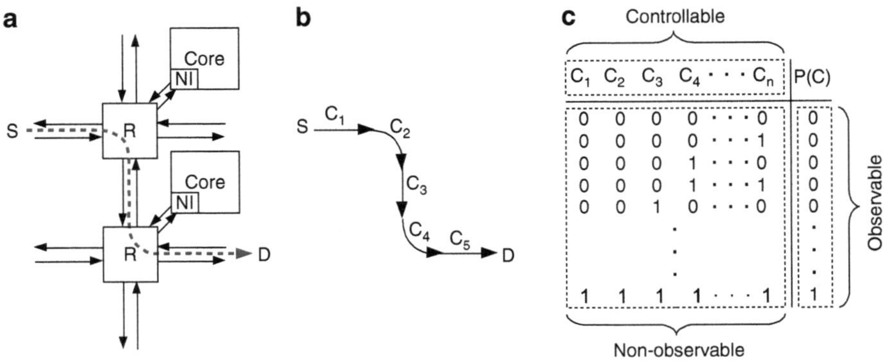

Fig. 9.8 Modeling of the NoC **(a)** Arbitrary path through the network **(b)** Path represented in terms of a connected component graph **(c)** Controllability and observability of path parameters

a path, *P(C)*, is expressed as the logical AND of each constituent path component as given in Eq. 9.1, where *C* takes on the value of '0' for faulty or '1' for good:

$$P(C) = C_1 \bullet C_2 \bullet C_3 \bullet C_4 \bullet .... \bullet C_n \tag{9.1}$$

Therefore, $P(C)=1$ denotes a fault-free path whereas $P(C)=0$ indicates that at least one path component is faulty. A faulty path component is assumed to permanently malfunction by either corrupting the packet payload or causing the packet itself to be miss-routed and/or dropped.

9.6.1.2 Network Test Concept

The task of network fault diagnosis is inherently challenging due to the fact that the path components under test are deeply embedded within a complex interconnection structure and consequently their state cannot be directly observed. However, an examination of the path function *P(C)* truth table, shown in Fig. 9.8c, suggests a promising approach. Although there are $2^n - 1$ possible fault configurations (where n is the number of components for some arbitrary route) satisfying $P(C)=0$, there is only a single state that satisfies $P(C)=1$ which is when $C_i=1$ for all *i*. The practical application of this is that whenever we can verify that a test packet has been successfully routed over some specified network path without being corrupted, it is assumed that the underlying components comprising that path are fault-free. So by routing packets through the network such that every component is part of at least one good path it is ultimately possible to render a complete network diagnosis on the *assumed faulty until proven good* basis.

To produce this outcome, we have proposed a route generation method based on an adaptation of the self-avoiding walk (SAW) [22]. The main idea is that random-like paths are computed at a source subject to the requirement that a packet may not pass through a given path component more than once for a particular route.

Note that this restriction does not necessarily preclude multiple traversals via the same router. In addition, the following constraints are applied:

- Explicitly defined start and end points;
- The physical boundary of the network topology;
- A maximum specified number of router hops.

The element of randomness afforded by SAW type routes proves to be an effective means for path discovery in the presence of unknown arbitrarily distributed faults. It is also this characteristic that makes the proposed approach highly amenable to arbitrary network topologies.

9.6.1.3 Diagnosing the NoC

Since the GPD is not embedded within the RFD(s) under test, the first phase of network diagnosis begins by testing the inter-die connectivity between the GPD and the RFD to which it is directly connected. This is accomplished by the GPD injecting one or more test packets into the RFD along a SAW type path that returns back to the GPD. Successful reception of a previously injected packet verifies the correctness of that link and enables further testing of the RFD along the fault-free path.

For testing the RFD, the previously described method is modified into a series of operations wherein the GPD writes to a designated core along one path followed by a read from that same core along a different path, both based on a SAW. Terminal ports are selected systematically in order to cover all network-to-core connections. Router-to-router connections are exercised in a random fashion as a consequence of the SAW path generation for the targeted read/write operations. Each successful read operation establishes the fault-free status of the path components along both the read and write paths used, as well as the associated network interface. Testing of RFDs indirectly linked to the GPD via one or more of the other RFDs requires additional tests aimed at verifying the usability of all inter-RFD links. The final diagnosis result includes the faults from every testable path component.

9.6.2 Xentium Tile Dependability

The dependability approach used for the Xentium tiles in the RFD adopts several key concepts from prior research to enhance the dependability of homogeneous multiprocessor architectures, including the "Know-Good-Die/Tile" concept, and concepts related to test reuse and majority-voting among identical processing tiles [23]. When we apply identical test stimuli to fault-free tiles-under-test, we expect to obtain identical test responses from these as they are all identical. If at least three tiles are tested in parallel, we can identify a faulty tile from the differences in its test responses

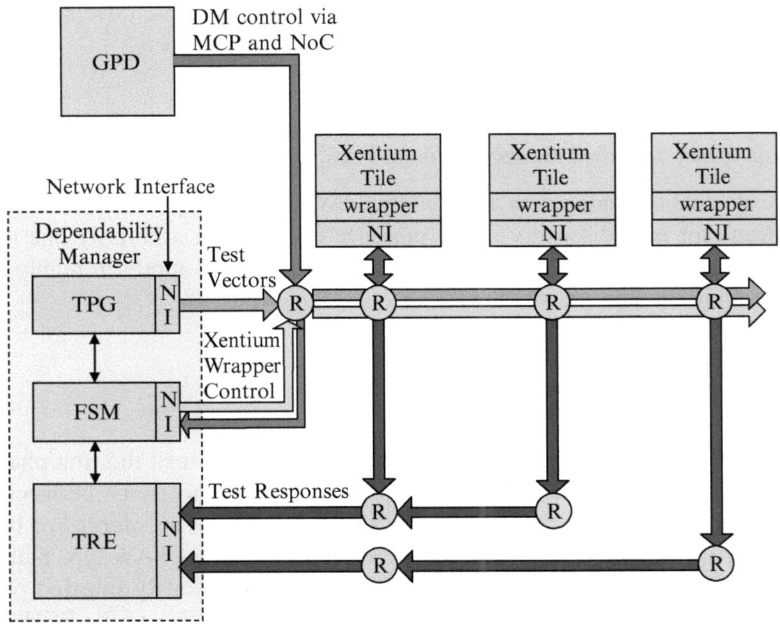

Fig. 9.9 GSP dependability architecture for the Xentium tiles

compared to the others, as long as no more than a minority of tiles-under-test (in this example one tile out of three tiles tested) becomes faulty at a time.

To permit periodic testing of tiles *at run-time* this method however requires additional, on-chip infrastructural IP that has to be designed and integrated in the RFD architecture at design-time. Our RFD dependability architecture includes a dependability wrapper around each Xentium tile, and a Dependability Manager on each chip [24]. The wrapped tiles are connected to the NoC in the same way as the original tiles were.

During dependability testing, the NoC is used as a test access mechanism to transport test stimuli to and test responses from the wrappers of the tiles-under-test. The NoC makes no difference between the transportation of functional data and dependability test data. Since the total bandwidth of the NoC therefore has to be shared between the application's functional data and the dependability infrastructure's test data, dynamically pausing and resuming of the dependability test is supported and used to ensure that the required NoC bandwidth is always available for the running applications. This *prioritization* of functional data over dependability test data ensures that the dependability tests do not impact the performance of the user application(s).

Figure 9.9 shows the RFD dependability architecture for the Xentium tiles, including its most important components: the Dependability Manager (DM), the Xentium tile wrappers around the Xentium tiles, and the GPD executing the Xentium tile dependability software.

9.6.2.1 Dependability Manager

The DM shown in Fig. 9.9 consists of a test pattern generator (TPG), a test responses evaluator (TRE) and a finite state machine (FSM) [25, 26]. The FSM controls the DM and communicates with the dependability software running on the GPD. Deterministic test patterns for the Xentium tile(s) (with 32 parallel scan-chains) have been generated at design-time using a commercially available tool. The TPG in the DM reproduces these deterministic test patterns using a linear-feedback shift register combined with a reseeding technique.

A phase-shifter packs the test stimuli into 32-bit words to suit the 32-bit wide NoC. Finally, each bit of such a 32-bit word fills one scan flip-flop in each of the 32 scan chains inside the Xentium tile. The GPD, DM and Xentium Tiles are connected to the NoC through NoC-routers (R). The silicon area of the DM is approximately one percent of the total area of the RFD.

9.6.2.2 Xentium Tile Wrapper

The Xentium Tile Wrapper allows switching the operating mode of the Xentium tile between functional mode and dependability test mode via commands issued by the DM over the NoC [24]. In functional mode, the wrapper transparently delivers data from the NoC to the functional inputs of the Xentium tile, and passes data from the functional outputs to the NoC. In dependability test mode, it delivers data from the NoC to the test inputs (scan-chain inputs and primary inputs) of the tile. A similar operation is performed at the output of the tile where scan-chain outputs and primary outputs are captured and passed to the NoC by the wrapper.

9.6.2.3 Tile Processor Dependability Software

The tile processor dependability software is an essential part of the dependability enhancement approach for the GSP. This software executes on the GPD, where it starts the background test activities. It requests sufficient RFD resources (i.e. NoC bandwidth and Xentium tiles) to use the DM together with two or three available Xentium tiles. The run-time resource manager subsequently allocates and configures the required communication routes between these components. The DM itself does not require knowledge of which cores it is testing. This is completely determined by the routes that are configured in the NoC.

The test sequence is started when the GPD writes into an FSM control register in the DM. The DM subsequently switches the wrappers of the chosen Xentium tiles to dependability test mode, preparing the Xentium tiles for receiving the test stimuli generated by the TPG of the DM. The test stimuli are subsequently multicasted via the NoC to the target Xentium tiles, and the test responses are collected and compared in the TRE of the DM.

This test sequence is halted as soon as a difference is detected in the test responses from the Xentium tiles under test, or as soon as the test is completed. The Xentium tile that generated a test response that differs from the others is identified as the faulty tile. This information is encoded into a fault status report that can be retrieved by the dependability software executing on the GPD. The faulty Xentium tile can then be isolated or removed from the usable resource table of the run-time resource management software. If no test-response differences are detected during the dependability test, all tested tiles are considered fault-free, and made available to the run-time resource management software.

This dependability test process is repeated until all Xentium tiles are tested. This way, the faulty tiles, if any, are identified and isolated from the system ensuring that the remaining Xentium processing tiles are fault-free. Depending on the dependability requirements (i.e. acceptable mean system down time / unavailability) from the end user, the dependability test activities can be performed at a desired frequency.

9.7 Preliminary Experimental Results

Reaching the ambitious objectives of the CRISP project involves software development, digital hardware design and manufacturing of two ICs and one Printed Circuit Board (PCB). The key result is prototyping the GSP platform, consisting of 45 Xentium DSP cores with run-time resource management and dependability support running both high-end and low-end streaming DSP applications. At the time of writing, the project is still running and, therefore, only preliminary results are described here. We describe, first, the results of the hardware manufacturing and, second, how the applications will be demonstrated on the GSP prototype. Demonstration of the applications on the manufactured GSP hardware unites all topics (i.e. reconfigurable multi-core SoC, low-end and high-end streaming applications, run-time resource management, and dependability) of the CRISP project.

9.7.1 Hardware Manufacturing

The CRISP project implements a pragmatic research and engineering schedule to research, conceive, implement, and manufacture two ICs and a PCB. At the point of writing, the PCB and the GPD chip are ready and the RFD dies have just left the foundry.

The PCB integrates five RFD chips and one GPD chip on a single board and prototypes the instance of the GSP architecture illustrated in Fig. 9.3, with 45 Xentium cores and one ARM® core. On the PCB, the NoCs for the RFDs are interconnected via the Multi Channel Port (MCP) interfaces. Additionally, the PCB serves as a verification platform for implementing demonstrators and integrates various I/O

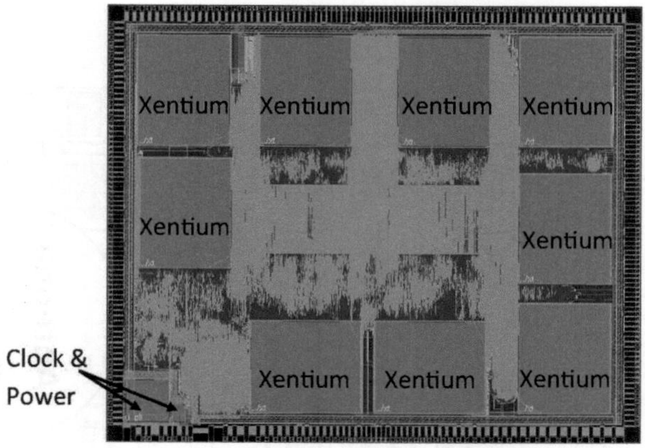

Fig. 9.10 Layout of reconfigurable fabric device (RFD) chip

interfaces and an FPGA. The FPGA is connected to all RFDs via MCP interfaces. Hence, direct access to the NoC of every RFD is provided via the FPGA.

The GPD chip is manufactured in UMC 130 nm CMOS Technology, runs at 200 MHz, and is delivered in a 400-pin BGA package. The supply voltage is 1.2V core and 3.3V I/Os. The estimated power dissipation is 400mW.

The RFD chip is manufactured in UMC 90 nm CMOS Technology. The die size is 44 mm^2 with 344 kBytes memory on-chip. The RFD will run at 200 MHz and will be delivered in a 400-pin BGA package. The supply voltage is 1.0V core and 3.3V I/Os. To manage the complexity of the RFD design, a hardmacro was created for the Xentium Tile Processor Cores. The RFD integrates nine Xentium hardmacros.

Figure 9.10 gives an overview of the chip layout. The nine big grey blocks are Hardmacros of the Xentium Processing Tile. Two Memory Tiles are located in the middle and clock and power blocks are found in the lower left corner. The logic of the dependability manager and NoC is laid out between the processing and memory tiles.

9.7.2 Beamforming Demonstrator

The CRISP beamforming demonstrator has 16 receive channels and computes 8 beams. Figure 9.11 depicts the architecture of the beamformer on the General Streaming Processor (GSP) platform. The demonstrator uses all five RFDs. Four RFDs are used for the signal processing of the 16 receive channels; each of these RFDs does antenna processing for four receive channels at an input rate of 1.28 Gbps. The beam forming function is implemented on the fifth RFD, combining the data from the 16 processed receive channels to compute eight beams at an output rate of 640 Mbps. The ongoing implementation maps the processing tasks onto 39 of the 45 DSP cores.

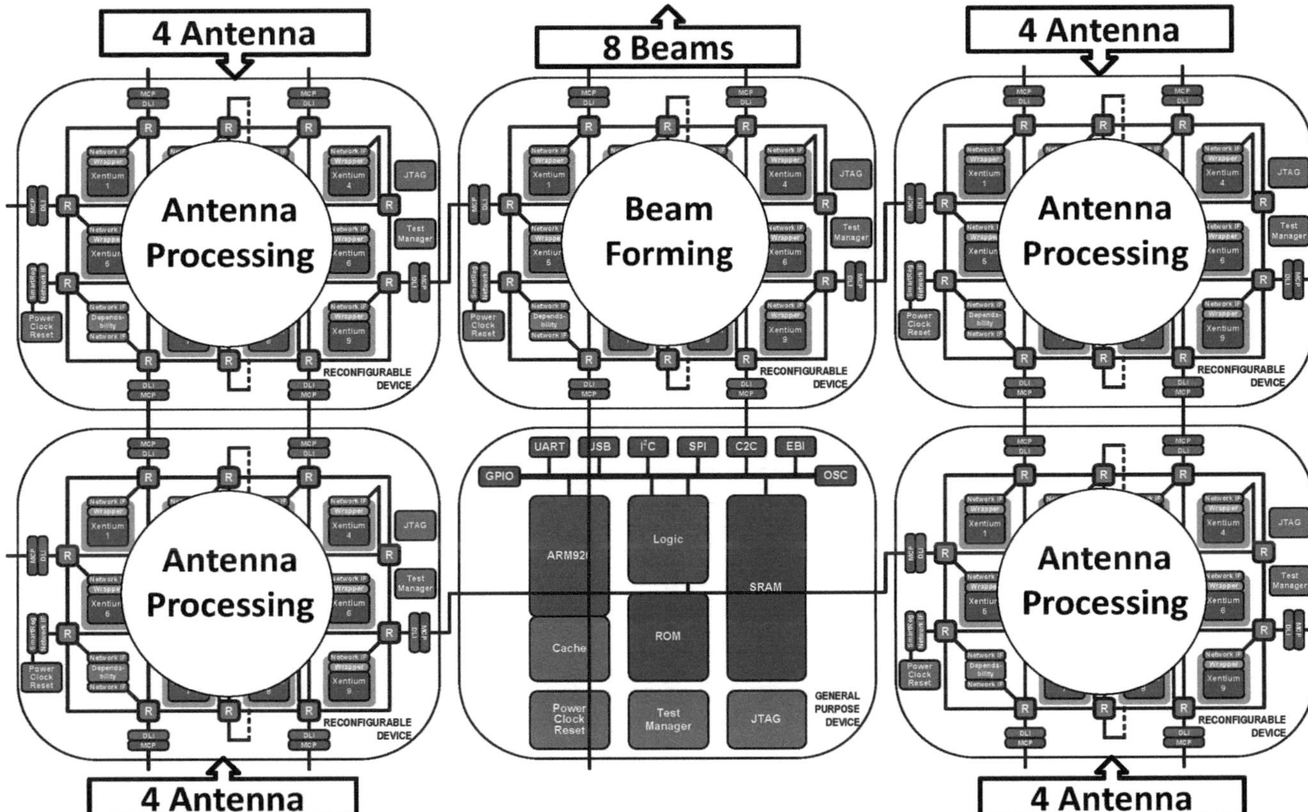

Fig. 9.11 Illustration of a mapping onto the GSP platform of the 16 channel / 8 beams digital beamforming demonstrator. Four RFDs are used for antenna processing; each processing the input of four channels at an input rate of 1.28 Gbps. One RFD is used to form eight beams from the sixteen channels at an output rate of 640 Mbps

9.7.3 GNSS Demonstrator

The GNSS application is demonstrated on the CRISP verification platform containing five RFDs using a commercial off-the-shelf radio front end. The resulting digital data stream received by the front end is fed to the verification platform and inserted in the RFD using the MCP interface.

The tasks of the GNSS application are illustrated in Fig. 9.2. All digital baseband processing can be done on one RFD. The parallel digital signal processing tasks are assigned to multiple Xentium processing tiles in the RFD and include acquisition (search of satellites using long FFTs) and tracking (data decoding and signal time-of-arrival measurements by means of serial correlation). This part of the application benefits from the parallel nature of RFD. The parallel tasks are preceded by a single task that preprocesses the input signal to ease the computational burden of the parallel tasks.

Figure 9.12 illustrates an example mapping of the GNSS application to the RFD and GPD. Three kinds of processes are mapped onto the RFD. Input preprocessing is closest to the inserting point of the input stream coming from the GNSS RF front end through the FPGA. Two acquisition processes are executed in the upper part of the RFD and four processes tracking satellite signals are executed on the right half. The outcome of the acquisition and tracking is forwarded to a single task of navigation, which is performed on the GPD. GPD forwards the navigation solution to an external PC via a serial connection. The implementation steps towards the CRISP GNSS application are explained in more detail in [27, 28]. The GNSS application is able to solve the position, velocity, and time of the receiver when four or more satellites are tracked successfully [12]. Thus, the receiver application should always (after an initial acquisition stage) have four or more cores running a tracking process to enable navigation.

9.7.4 Enabling Graceful Degradation by Reconfiguration

Dynamic resource management allows for fault-tolerance in the case that hardware faults can be circumvented, using a disjoint set of resources. However, the amount of resources required by the application does not change. In case that the platform is either seriously compromised, or its available resources are near depletion, applications may be refrained from execution. For robustness reasons, even more flexibility may thus be added to an application. By providing multiple quality-of-service levels the probability may increase that an application is allowed to start, albeit in a reduced form. This scenario is already demonstrated within the CRISP project with the GNSS application.

The next challenge is to seamlessly switch between the quality-of-service levels provided by an application. Scenarios exist where it is preferred to scale down running applications to allow additional functionality to be performed on the same

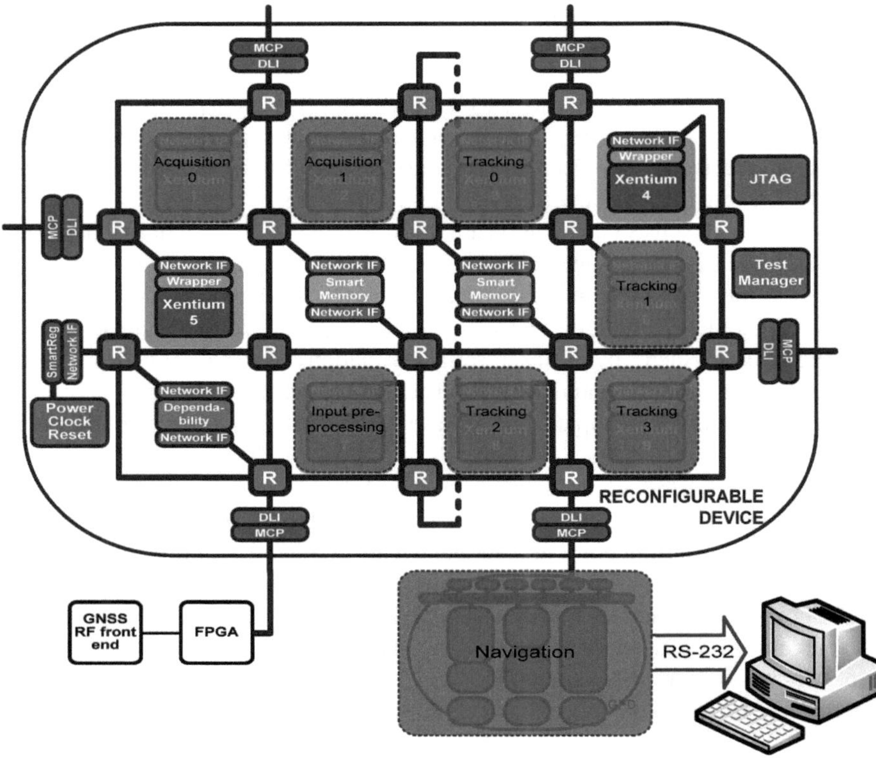

Fig. 9.12 Illustration of mapping of GNSS tasks onto the GSP using seven Xentiums on one RFD

platform. In other scenarios, hardware faults may trigger a reconfiguration of an application, such that it gracefully degrades by running in a reduced mode on a (slightly) different set of resources.

9.8 Outlook

New projects such as STARS [29] and NEST [30] have been launched to continue research along the path set out by CRISP and to pick up on CRISP results. Furthermore, CRISP motivations and goals remain highly relevant. It is expected that conventional computing architectures will be replaced by more flexible reconfigurable multi-core computing platforms and streaming applications are expected to create a huge drive and momentum for this shift.

The CRISP project anticipates on this expected architectural shift by researching streaming applications, reconfigurable IP cores, interconnect technologies, run-time tools and dependability issues. Europe is a major player in the embedded arena today and this holistic approach is necessary for European companies to achieve world leading positions in computing solutions and products for streaming applications.

CRISP aims at the important domain of streaming applications, which is expected to grow faster than other application domains with the adoption of new standards. The CRISP project will deliver a novel reconfigurable multi-core computing platform. These types of platforms are urgently needed, since hardwired logic (i.e. ASIC) is getting prohibitively expensive for an ever wider range of products. Currently, many products make use of fine-grained reconfigurable FPGAs. Their usage, however, results in severe overheads unacceptable for most mass market applications. These overheads are related to device area, power consumption, and unit costs.

Embedded systems increasingly profit from the integration of signal processing capabilities of new energy-efficient multi-core architectures with added flexibility. By combining techniques from application domain and computer engineering, embedded systems can improve the production of solutions both concerning costs and time:

Reduction of costs – by reusing IP building blocks in SoC designs, the development and verification costs of a design cycle will be reduced and design productivity be increased. Ultimately, reconfigurable multi-core platform chips can be used for a large variety of applications by reconfiguring the same chip after fabrication;

Reduction of time to market – by using energy-efficient general-purpose multi-core architectures as proposed in CRISP and abstracting from the application-specific platform, the development time of new solutions can be drastically reduced.

Moreover, the ability to reconfigure a system and to use this capability to make systems dependable is extending product lifetime. Because users can reconfigure their purchased product to increase service levels, hardware can be used for a longer period of time. Furthermore, built-in dependability techniques allow for graceful degradation. Because of these aspects, products will have a longer life cycle, become cheaper, and be less of a burden to the environment.

Dependable and reconfigurable multi-core systems will serve various application domains ranging from consumer electronics to mission critical space applications. For instance, in space applications, extending lifetime under harsh conditions, and easy upgrading are important aspects when designing dependable and reconfigurable systems. However, for consumer electronics faster time-to-market is a differentiator to win the competition. Both orthogonal application domains focus on improving dependability in combination with reconfigurable multi-core hardware, however, motivated by partly different grounds.

9.9 Conclusions

The holistic and pragmatic approach of CRISP includes hardware design and manufacturing as well as development of application, system, and development software. This approach enables research into fundamental trade-offs in the intersection of hardware and software where reconfigurable computing resides. The scalable NoC-based many-core General Stream Processor (GSP) architecture developed in CRISP delivers sufficient and scalable stream processing performance for a wide range of applications in both high- and low-end markets. The GSP is equipped with a sophisticated run-time resource management system that can dynamically change the allocation of application tasks to processing cores and data transfers to communication channels. The run-time resource management turns the GSP into a flexible and efficiently programmable coarse-grained reconfigurable computing platform. Moreover, CRISP combines the run-time management with dedicated dependability hardware and software to create a self-repairing dependable GSP. The approach is a novel way to exploit the inherent redundancy of NoC-based many-cores and to address tomorrow's predicted issues with accelerated degradation of ICs as processing geometries continue to shrink.

Acknowledgments CRISP is a multi-disciplinary endeavor. To achieve the project objectives, highly specialized knowledge is required about applications, front-end and back-end hardware design as well as development and system software. The work reported in this chapter is a team effort and the authors would like to thank the many experts from the consortium, who worked on CRISP including Sébastien Baillou, Helmut Baur, Arjan Boeijink, Werner Brugger, Christoph Büchler, Jan de Wilde, Jeroen Flierman, Michael Günther, Matthias Hofinger, Klaas Hofstra, Henk Holdijk, Philip Hölzenspies, Thomas Janz, Erik Karstens, Ben Kemper, Roy Krikke, Pieter Maier, Rene Meerman, Rik Portengen, Jordy Potman, Johannes Schäfer, Jeroen Scholten, Hans Schurer, Lodewijk Smit, Hermen Toersche, Rene van Hees, and Mark Westmijze. Further, the authors would like to thank the European Community's Seventh Framework Programme FP7/2007-2013 for supporting the CRISP project, and the project officers and project reviewers for their support and help in guiding the project towards its objectives.

References

1. Dally W. et al. Stream Processors: Programmability with Efficiency, ACM Queue, pp. 52–62, 2004.
2. U.J. Kapasi, S.R. Rixner, W.J. Dally, B. Khailany, J.H. Ahn, P. Mattson, and J.D. Owens. Programmable stream processors. In IEEE Computer 36(8), pp 54–62, 2003.
3. International Technology Roadmap for Semiconductors (ITRS), http://www.itrs.net.
4. 4S Project Chapter of this book.
5. 4S Project web pages http://www.recoresystems.com/research/.
6. G.Smit, E. Schuler, J. Becker, J. Quevremont, and W. Brugger: Overview of the 4S project. In *Proc.* 7th International Symposium on System-on-Chip (SoC'05), 2005.
7. M.J.G. Bekooij et al. Dataflow Analysis for Real-Time Embedded Multiprocessor System Design. In Dynamic and Robust Streaming between Connected CE Devices, Kluwer, 2005.
8. H. J. Visser, Array and phased array antenna basics. Chichester, West Sussex, UK: Wiley, Sep. 2005.
9. M. I. Skolnik, Introduction to Radar Systems, 3 rd ed. New York, NY, USA: McGraw-Hill, 2001.
10. GPS Interface Control Document (ICD-GPS-200D), IRN-200 C-004, U.S. Air Force, 2004.
11. "Galileo Open Service, Signal In Space Interface Control Document (OS SIS ICD)," European GNSS Supervisory Authority, draft 1, 2008.

12. M. Braasch and A. J. Van Dierendonck. GPS Receiver Architectures and Measurements. In Proc. of the IEEE, vol. 87, no. 1, pp. 48–64, 1999.
13. D. Akos. The role of Global Navigation Satellite System (GNSS) software radios in embedded systems. In GPS Solutions, 2003.
14. S.K. Eo, S. Yoo, K.M. Choi. An industrial perspective of power-aware reliable SoC design. In Proc. Asia and South Pacific Design Automation Conference (ASP-DAC), pp. 555–557. IEEE Computer Society Press, 2008.
15. J. Kurzak, A. Buttari, P. Luszczek, and J. Dongarra. The Playstation 3 for High-Performance Scientific Computing. In Computing in Science & Engineering, Vol. 10, Issue 3, IEEE Computer Society, 2008.
16. J. Dongarra, A. Lastovetsky. An overview of heterogeneous high performance and grid computing. In Engineering the Grid: Status and Perspective, 2006.
17. O.M. Moreira, M.J.G. Bekooij. Self-timed scheduling analysis for real-time applications. In: EURASIP Journal on Advances in Signal Processing, vol. 2007, pp. 24–37, 2007.
18. M.L. Fisher, R. Jaikumar, and L.N.V. Wassenhove. A multiplier adjustment method for the generalized assignment problem. In Management Science 32(9), pp 1095–1103, 1986.
19. T.D. ter Braak, P. K.F. Hölzenspies, J. Kuper, J. L. Hurink, and G. J.M. Smit. Run-time Spatial Resource Management for Real-Time Applications on Heterogeneous MPSoCs. In Proc. Conference on Design, Automation and Test in Europe (DATE), pp. 357–362, 2010.
20. T.D. ter Braak, S.T. Burgess, H. Hurskainen, H.G. Kerkhoff, B. Vermeulen, X. Zhang: On-Line Dependability Enhancement of Multiprocessor SoCs by Resource Management. In Proc.12th International Symposium on System-on-Chip (SoC'10), p103–110, 2010.
21. S. Burgess, T. Ahonen, and J. Nurmi: Software Based Approach to Fault Diagnosis for Multi-Die Networks-on-Chip. In Proc. System, Software, SoC and Silicon Debug, 2010.
22. B. Hayes. How to avoid yourself. In *American Scientist*, vol. 86, no. 4, 1998.
23. H.G. Kerkhoff, O. Kuiken, and X. Zhang: Increasing SoC Dependability via Known Good Tile NoC Testing. In Proc. Conf. on Dependable Systems and Networks (DSN'08), 2008.
24. X. Zhang, H.G. Kerkhoff, B. Vermeulen. On-Chip Scan-Based Test Strategy for a Dependable Many-Core Processor Using a NoC as a Test Access Mechanism. In Proc. 13th Euromicro Conference on Digital System Design (DSD), 2010.
25. O.J. Kuiken, X. Zhang and H.G. Kerkhoff. Built-In Self-Diagnostics for a NoC-Based Reconfigurable IC for Dependable Beamforming Applications. In Proc. IEEE Intern. Symp. on Defect and Fault Tolerance in VLSI Systems (DFT'08), 2008.
26. H.G. Kerkhoff and X. Zhang. Design of an Infrastructural IP Dependability Manager for a Dependable Reconfigurable Many-Core Processor. In Proc. DELTA'10, 2010.
27. H. Hurskainen, J. Raasakka, T. Ahonen, and J. Nurmi. Multicore Software-Defined Radio Architecture for GNSS Receiver Signal Processing. In EURASIP Journal on Embedded Systems, vol. 2009, Article ID 543720, 10 pages, 2009.
28. J. Raasakka, H. Hurskainen, T. Paakki, and J. Nurmi. Modeling Multi-Core Software GNSS Receiver with Real Time SW Receiver. In Proc. ION GNSS, 2009.
29. Sensor Technology Applied in Reconfigurable Systems for sustainable Security (STARS) Project, National Dutch Project, http://www.starsproject.nl.
30. Nederland Streaming (NEST) Project, Dutch Technology Foundation STW Project, http://caes.ewi.utwente.nl/caes/index.php/research/recently-started-projects/nest

Chapter 10
ERA – Embedded Reconfigurable Architectures

Stephan Wong, Luigi Carro, Mateus Rutzig, Debora Motta Matos, Roberto Giorgi, Nikola Puzovic, Stefanos Kaxiras, Marcelo Cintra, Giuseppe Desoli, Paolo Gai, Sally A. Mckee, and Ayal Zaks

Abstract In a scenario where the complexity and diversity of embedded systems is rising and causing extra pressure in the demand for performance at the lowest possible power budget, designers face the challenge brought by the power and memory walls in the production of embedded platforms. The focus of the ERA project is to investigate and propose new methodologies in both tools and hardware design to break through these walls, and help design the next-generation embedded systems platforms. The proposed strategy is to utilize adaptive hardware to provide the highest possible performance with limited power budgets. The envisioned adaptive platform employs a structured design approach that allows integration of varying computing elements, networking elements, and memory elements. For computing elements, ERA utilizes a mixture of commercially available off-the-shelf processor cores, industry-owned IP cores, and application-specific/dedicated cores. These are dynamically adapted regarding their composition, organization, and even instruction-set architectures, to provide the best possible performance/power trade-offs. Similarly, the choice of the most-suited network elements and topology and the adaptation of the hierarchy and organization of the memory elements can be determined at design-time or at run-time. Furthermore, the envisioned adaptive platform must be supported by and/or made visible to the application(s), run-time system, operating system, and compiler, exploiting the synergism between software and hardware. Having the complete freedom to flexibly tune the hardware elements allows for a much higher level of efficiency, riding the trade-off curve between performance and power compared to the state of the art. An additional goal of the adaptive platform is to serve as a quick prototyping platform in embedded systems design.

L. Carro (✉)
Universidade do Rio Grande do Sul, Passo Fundo, Brazil
e-mail: carro@inf.ufrgs.br

J.M.P. Cardoso and M. Hübner (eds.), *Reconfigurable Computing: From FPGAs to Hardware/Software Codesign*, DOI 10.1007/978-1-4614-0061-5_10,

10.1 Project Partners

1. Technische Universiteit Delft, The Netherlands (Coordinator)
2. Industrial Systems Institute, Greece
3. Universita' degli Studi di Siena, Italy
4. Chalmers University, Sweden
5. University of Edinburgh, UK
6. Evidence, Italy
7. ST Microelectronics, Italy
8. IBM, Israel
9. Universidade do Rio Grande do Sul, Brazil

- Project Coordinator: Stephan Wong, Technische Universiteit Delft, The Netherlands
- Start Date: 2010-01-01
- Expected End Date: 2012-12-31
- EU Program: 7th Framework Programme, FP7-ICT-2009.3.4 Embedded System Design, STREP Project No. 249059Global Budget: 4.014.511,00 €
- Global Funding by EU: 2.800.00,00 €
- Contact Author: Luigi Carro, Email: carro@inf.ufrgs.br

10.2 Introduction

The embedded systems market has become a strong focus in Europe that distinguishes itself from the more high-performance systems market in the USA, and the consumer electronics and the semiconductor market in Asia. Strong application areas in Europe are spread among different application markets like in automotive, aerospace, industrial automation, medical/healthcare, and telecommunication. A key (hidden) element in embedded systems is the embedded processor that determines most of their functionality. Traditionally, embedded applications with very specific requirements (power, size, cost, speed, or any combination thereof) were implemented in dedicated application-specific integrated circuits, leading to long design times and re-designs for new applications. In recent years, the key driver in the design of the embedded processor has been integration in order to keep (design) costs low, reduce time-to-market, and improve functionality of embedded systems. A striking example is the mobile phone, in which the number of separate chips was significantly reduced because of the utilization of a full embedded platform. The integration is still continuing on a single chip towards a more structured (on-chip) design by combining multiple IPs, and utilizing more complex and diverse programmable processor cores. Additionally, mobility is translated into limited power budgets that must satisfy all the required functionality. Several examples show that specialized hardware performs better and consumes less power than general-purpose processors. The trend towards specialization is also apparent in large –albeit specific– markets, such as the

DSP and Graphics Processing Unit (GPU) markets, where specialized architectures are fast gaining ground because their organization is very efficient for a certain class of problems. This has been the key issue of embedded systems platforms until very recently. Once again, embedded systems are nowadays moving from very specialized devices, targeting a single function, to more general-purpose platforms that must deliver high performance and low energy for a broad range of applications, which should cover all the important application domains where embedded platforms are used. Hence, the embedded processor and its platform are key to the EU development in these diverse and competitive fields.

To sustain the high-performance and low-energy behavior expected by consumers, the important dilemma which arises is whether to have dedicated specialized sub-systems in a fixed multi-core organization, or strive to do better. Creating a different accelerator for each new application that reaches the market is costly in terms of design time, and introduces a delay before its adoption by the whole community (compromising time-to-market). In contrast, adapting the processor architecture on-the-fly while an application is running would yield significant performance and power benefits, while avoiding the cost and time-to-market pitfalls. This is an old dream that as of yet has not come to fruition, but now three conditions exist that can bridge the gap between intention and reality. The first is the maturity of reconfigurable computing, since there are several different options to chose from (coarse- and fine-grain, homogeneous or heterogeneous, etc.); the second is dynamic adaptation, since this allows for fine-tuning at run-time, without user intervention; the third one is advanced compiler and operating systems technology, which have greatly evolved in the past years to cope with heterogeneous platforms and the need for on-line modifications and resource management.

In the embedded systems market, continued innovation is driving the need to improve existing functionalities and to introduce new functionalities. This is needed to differentiate products and to entice consumers/customers to purchase new or improved products. Whilst in the past product cycles span several years, the competition nowadays is cut-throat and product design cycles on average are more in the range of 1 year. This led to a more structured, platform-based design approach of embedded systems, and in turn of the embedded processors. Use of standardized IP blocks and re-use of existing designs has become common practice. Unfortunately, although a platform greatly reduces design effort and time, it precludes ultimate optimizations which are not foreseen at design time for niche markets. This way, a platform can very easily become obsolete, since new standards covering communication, audio and video processing are constantly emerging. Hence, a platform inherits all the benefits and all the inefficiency that a general-purpose processor could have when targeting specific functionalities. This scenario of changing applications in short times calls for some sort of adaptation during the lifetime of a product. Although FPGAS and eFPGAs can provide some adaptation for what concerns changing applications, their programming times preclude their adaptation during a program phase, without touching important points like memory and communication. As embedded systems must rely on limited energy supplies, most likely these optimizations must also be performed

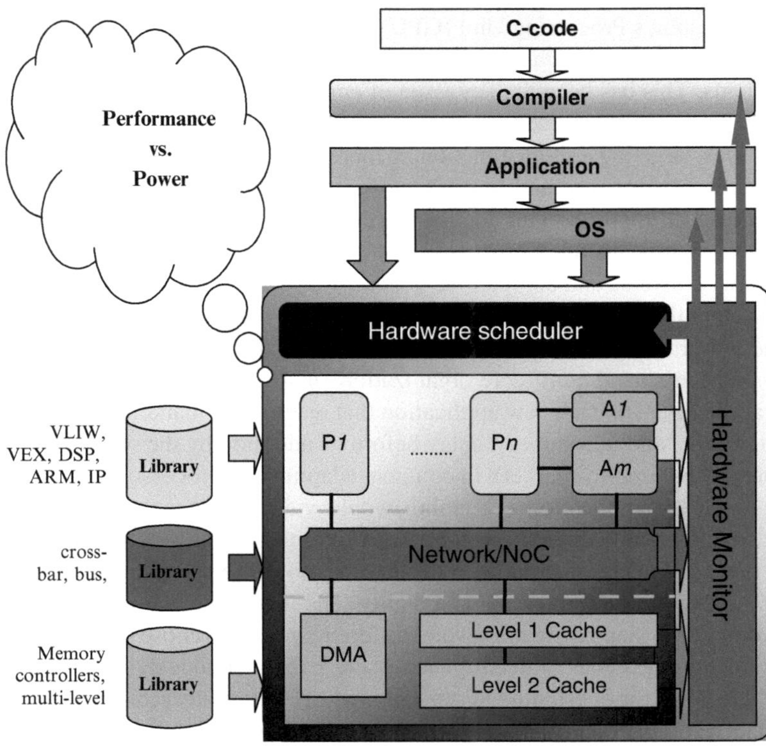

Fig. 10.1 General overview of the ERA project

at the hardware level, in order to extract the most out of the underlying technology. Due to the complexity of such a task, innovative solutions must be sought to ensure adaptability targeting optimization for different markets and for different applications within the same market. The ERA project covers exactly this problem by introducing a reconfigurable fabric able to perform adaptation at different moments (design time, application deployment, even during the lifetime of a product) to adapt to new standards without losing its power efficiency advantages, as shown in Fig. 10.1. Finally, reconfiguration will be critical for products developed for high-reliability markets (automotive, health) or for products developed in state-of-the-art processes, that are likely to have many defects, since we are reaching the limits of nature itself. In this case, without having total control of the fabrication process, one must devise clever fault-tolerant mechanisms at design time that can cope with potential low yield and aging, in order to add or maintain value to the product.

Adaptability is the key to develop embedded platforms for the new, heterogeneous and multi-applications embedded market. One of the ways to adapt is to have a reconfigurable fabric available, so that hardware changes can be done whenever necessary. Although designs made with reconfigurable hardware are gaining popularity in the embedded systems market, there are several issues that must be addressed in order to allow for the breakthrough. Some of these issues stem from the excessive power dissipation of reconfigurable devices, since their programmability takes lots of power to read form an external memory, and also their operation draws extra power due to the extra reconfigurable wiring that must be present to allow for reconfiguration itself. Moreover, one still needs a design flow that, from the final user (the embedded system designer) perspective, is simple and hides all optimization bottlenecks from view.

To cope with the reconfiguration power problem, in ERA the focus is on the development of accelerators using a coarse-grain reconfigurable fabric, composed of a reconfigurable VLIW processor, a flexible memory organization and an interconnection network that can provide better usage of power resources by distributing its routing resources online. A software stack consisting of a compiler and OS will provide the means to drive both static and dynamic reconfiguration decisions according to the application characteristics and the user objectives (in terms of power and performance).

The reason that a VLIW processor was chosen for this project is because of its excellent power/performance trade-off. However, as embedded systems target new applications, the optimal size of the VLIW is clearly an issue. Developing traditional static tools such as those available today, based on a set of benchmarks, would only produce a general-purpose machine, without the power and performance benefits one would expected to have in dedicated markets with stringent requirements. Therefore, in the ERA project we target the usage of multiple and variable issue VLIW, as well as a multidimensional VLIW, so that for each different application, and for different parts of each application, an optimized processor can be constructed on-the-fly to obtain the fastest computation (for the application) with the smallest possible power budget within the constraints of the design.

As a variable issue processor is used, extra stress is placed in the memory subsystem. This requires the adaptation of the memory as well, in order to cope with variable instructions and variable amount of parallelism available at any given time during program execution. The memory hierarchy plays a significant role in the end performance of a system and its power consumption. Our goal in the ERA project is to tailor the memory system for executing a target application to meet specific (user) objectives for performance and power. The innovation in our case is that this is going to happen in concert with the processor reconfiguration (as well as NoC reconfiguration) so as to be synergistic towards meeting the user objectives. Moreover, all modifications are to be developed during online operation. Memory reconfiguration entails a vast array of techniques: resizing on-chip memory to fit program needs, optimizing cache architectures for dynamic and static power consumption, optimizing shared-memory communication via custom cache coherence protocols, modifying the replacement policies for minimizing miss rates in caches, compressing data and removing data redundancies, managing caches shared by multiple applications, partitioning the on-chip memory for different functionalities (caches, scratchpads, loop buffers, etc.) and any combination of the above. While the reconfiguration space is vast in this project, we will combine the mechanisms that work synergistically and provide global, application-driven policies for applying these mechanisms. Our goal is to provide a malleable memory system that can be efficiently adapted to the needs of an application using both static (application information) and run-time information (hardware monitoring & feedback).

In the same manner, as the processor-cache and processor-memory bandwidth and throughput varies, so the communication needs also vary. Devising a single strategy for the communication network would mean either extra power without equivalent performance gains, or degradation in the quality of services. Therefore, in this project a reconfigurable NoC is also proposed as a manager of the communication needs.

At the software level, in the ERA project the consortium proposes a tool chain that comprises not only the application APIs needed to guide reconfiguration, but also run-time managing of the reconfiguration process via the OS. Hence, from the final user perspective, the embedded system designer himself, the adaptivity of the platform will be seen as a seamless software development process. One can envisage that the user could control the reconfiguration, since having a multitude of possibilities for tuning the system performance leads to a set of possible implementations of the same hardware accelerator, each with different performance characteristics in terms of execution time and hardware resources used. In this project, we envision the possibility of performing a coordinated system-wide dynamic reconfiguration that will be able to provide the needed hardware resources to the application in a "just-in-time" fashion, hence, providing adaptation for a wide range of application requirements. The online adaptation is seen as a user-controllable trade-off between the execution time of the software and the hardware resources – and consequently power consumption– required by the various subsystems.

Another important issue raised by the adaptation process concerns the production of optimized code. In current fixed VLIW architectures, the amount of ILP that can be exploited is fixed at design time of the architecture, and at compilation time of the application However, since the ERA project is based on a variable-issue VLIW, the compiler must be aware that run-time reconfiguration of the VLIW schedule can also be performed. The biggest challenge is performing this schedule change with minimal cost, which will involve a combination of multiple code versions and fast schedule, and code generation algorithms. In this project, we propose a compilation tool-chain that can act as a mediator between the features and requirements exposed by the applications (e.g., memory locality, memory bandwidth requirements, and processing needs, as well as execution phases, multi-threaded behavior, communication, and synchronization) and the reconfiguration potential of the architecture. In particular, different modules of the architecture will have the ability to reconfigure themselves according to the (in-hardware) perceived runtime conditions and resource usage. Our approach complements this hardware monitoring and adaptation by leveraging software development and tuning tools that are able to implement a range of optimization strategies of different complexities, aiming to induce the most suitable architectural reconfigurations that match the application requirements during run-time. Using this approach, the optimization tools can take advantage of the global view of the architecture together with the application structure and needs, in order to trigger profitable reconfiguration actions.

Most complex embedded systems use an operating system, which is responsible to load an application to the execution memory, decide how much CPU time each application will have and manage interrupt requests. In this project, we propose to modify the OS so that, knowing the task distribution in advance, manages the reconfiguration, decides when to reconfigure, and schedules the work in concert with reconfiguration decisions. This would not be possible without the usage of hardware monitors that can communicate information to the operating system level so that global reconfiguration decisions can be made.

Preliminary work shows that measuring only four statistics per core (instructions retired, main memory accesses, floating point utilization, and instruction stalls [1]) via appropriate performance monitoring counters (PMCs) and knowing how these statistics correlate with power consumption is sufficient to predict per-core power (e.g., when only whole-chip power can be measured). Such information has been leveraged for local task scheduling decisions to maintain user-specified power budgets. In this project, the adaptation based on analytic models using PMC data from real applications will be used in many forms: (i) it will be used in a profiling step to feed per-phase information back to an optimization system (which may simply choose alternate, lower power-consuming instruction sequences during critical phases, or may perform arbitrarily sophisticated power/performance optimizations, possibly utilizing hardware assists); (ii) model data will be used by hardware adaptation policies to reconfigure micro-and macro-architectural structures; (iii) the PMC data and analytic models will be adapted to perform system-wide adaptation in hardware and software.

Finally, in the context of a multicore implementation of the ERA reconfigurable platform, the OS will be called to schedule concurrent applications across multiple cores. Hardware will provide key monitoring information on application behavior (functional unit utilization, memory intensity, temperature, and the like). In this case, reconfiguration and scheduling decisions must be taken in concert since one affects the other. Simple scheduling decisions (e.g., not to run a particular application on a particularly configured core) can be determined in hardware, for instance if the core has more resources than the application can fully use, or far too few resources for the application to perform well, but deciding which applications to place where or how to configure the overall available resources to maximize the benefits for the workload as a whole will likely require a more global view of the system. The OS can keep monitored application information as part of each process state. It can then adapt an application's current core to best suit its performance and power requirements. On a system level, the OS can employ simple economic models to co-optimize reconfiguration and scheduling for power and performance across all running applications.

All these modifications are in line with the overall strategy of the industrial partners of the consortium. The usage of different accelerators is a common place in STMICRO products, but the problem of adapting the platform for different markets with different accelerator in the shortest possible time is unfortunately also present. Hence, the ERA project is targeting a real life problem, whose solution might have a major impact in the way embedded systems are designed in the EU.

10.3 Main Applications

Objective ICT-2009.3.4 "Embedded Systems Design" puts a strong focus on the development of a novel (generic) embedded systems design method that can be applied to several application areas. In the ERA project we develop a platform that can adapt itself through coarse-grain reconfigurable hardware to tailor the hardware itself for changing environments and needs of the applications running on the platform for different application markets and platform usage.

The proposed ERA platform can provide adaptability at different abstraction levels: optimization of application software at design time, OS control and optimization at run time to cope with changing conditions, and hardware adaptation at run time to efficiently tune its performance to the application or OS needs and taking into account power budgets. However, the hardware resources, thanks to their regularity, can also be used as adaptive spare parts, and hence increase the fault tolerant capabilities of the platform. This additional capacity can then be used either in highly reliability demanding markets, like the automotive one, or in consumer markets when the hardware uses the ultimate technology, which is likely to have many errors and a low yield (that should be increased). Finally, to exploit the adaptability of the proposed ERA platform, software tools will be made available to achieve this.

As for the application set, the ERA platform targets those that present the characteristics described as follows.

Heterogeneity: as an adaptive platform is the target, the goal of the ERA project itself concerns covering different heterogeneous applications. Heterogeneity in the present context refers to applications that require different hardware resources during different times. For example, a modern cell phone must deal with audio, video, and baseband processing, and for each of these tasks different accelerators are called in. The demonstrators chosen to validate the platform prototype are a good example of such heterogeneity coverage since we plan to optimize not only different applications that compose a modern cell phone, but also the operating system that is running on top of the platform. Additionally, the platform will allow for a mixture of heterogeneous components.

Predictability of non-functional properties such as performance and power consumption: one of the main goals of the ERA platform is to determine the precise performance needs of the applications that we investigate after which we will determine the "best" possible set of adaptive hardware component to achieve those goals. Monitoring of the performance and power are key elements within the project to allow adaptability at the three mentioned levels (application, OS, and hardware) to reach the required performance levels with the provided power budget.

Adaptability and self-awareness for coping with uncertainty, upgrades of components, and self configuration concepts: besides being able to be controlled by the OS (and hence based on a previously defined and coded in the OS optimization strategy), in the ERA project the platform can adapt to different code execution scenarios also in an automatic fashion. This hardware based adaptation will be performed at different blocks, covering communication and processing. The overall strategy for dynamic optimization of the communication blocks is to use performance monitors that can advise the internal control system to increase resources of a certain channel of the used NOC in order to avoid congestion, while reducing resources that are not being effectively used in other channels. For the processing blocks, the idea is to allow monitoring of the executed code, and by using dynamic strategies like hardware DLLs or binary translation change the amount of resources that are being used at a certain moment of time. This adaptation may demand extra resources from the memory subsystem that must also respond with adequate bandwidth for the new performance or power node.

10.4 Experimental Results

In this section, we will show some results that sustain the basic claim of this project, that is, the clear need for an adaptive and dynamically reconfigurable platform. Most of the initial examples have been taken from the MiBench set.

10.4.1 Processor (ILP and TLP)

Currently, there is a big discussion about the amount of ILP and TLP that should be supported by the underlying platform. In the ERA project both are supported. Variable ILP can be supported by changing the VLIW issue width, while TLP is supported by activating several heterogeneous VLIWs.

Table 10.1 shows, for different applications, how TLP and ILP change for different applications, and hence the need for a reconfigurable fabric. The benchmark set covers general-purpose, embedded and parallel applications. The second column in Table 10.1 shows the mean basic-block size, as an indicator of potential ILP. It can be observed that each application has a different number of potentially parallel instructions, and hence changing the available ILP could lead to either performance or power gains.

The last column in Table 10.1 shows the total amount of executed instructions in an in-order processor. One can see that the distribution of code among threads has a huge variability (even in the same domain), and hence to cover several applications one must also provide a different number of active processors. Moreover, when sequential code is being executed, turning off processors and their clock network might lead to power efficiency, since no acceleration is possible with simple and low-power techniques.

This scenario claims for an adaptive platform as in the ERA proposal to find the optimum spot of the power-performance curve.

Table 10.1 ILP and TLP opportunities

Benchmark	Mean BB size (#instr)	Mean thread size: 4 threads #instr	%	8 threads #instr	%	16 threads #instr	%	64 threads #instr	%	Sequential (#instr)
General purpose										
equake	4.80	1,066,842	0.11	648,980	0.06	396,256	0.04	107,967	0.01	1,007,548,408
art	6.86	642,769,700	11.26	640,642,746	11.22	640,646,160	11.22	642,795,870	11.26	5,710,455,583
apsi	14.56	77,393	0.00	69,530	0.00	65,645	0.00	59,677	0.00	6,170,539,927
Embedded										
susan_e	16.60	2,397,335	1.11	1,917,022	0.89	1,680,975	0.78	1,519,993	0.71	215,489,859
patricia	5.04	5,139,690	5.00	3,485,357	3.39	4,020,714	3.91	4,194,469	4.08	102,793,656
susan_c	17.36	1,477,499	1.68	957,759	1.09	700,196	0.80	502,259	0.57	88,061,549
susan_s	12.10	5,835,705	0.42	2,931,041	0.21	1,528,875	0.11	468,044	0.03	1,374,058,841
Parallel										
swaptions	5.92	5,542,634	1.09	2,773,691	0.54	1,389,156	0.27	350,775	0.07	509,508,947
blackscholes	4.83	14,659,706	3.13	7,339,119	1.57	3,686,241	0.79	928,638	0.20	468,936,073
md	6.51	1,994,602	2.50	1,004,970	1.26	509,355	0.64	132,462	0.17	79,625,991
jacobi	6.94	6,346,588	8.44	3,171,016	4.22	1,586,607	2.11	401,982	0.53	75,191,816
fft	8.10	2,899,996	5.04	2,006,890	3.49	1,559,014	2.71	1,199,242	2.08	57,580,983
IU	8.32	1,552,984	1.54	1,065,276	1.05	1,024,484	1.01	1,026,825	1.01	101,169,484

10.4.2 Memories

In order to optimize memory power, one must have an idea about the access pattern present in different applications. This can be achieved by observing the locality of the references of the target applications. In the initial screening of the MiBench applications we realized that the locality of references allows one to reduce the leakage power consistently in comparison to previous solutions ("improved drowsy cache" [2]).

Therefore, a detailed investigation of the behavior of the *djpeg* application was done to verify whether it contained the potential for reconfiguration based on the distribution of strides when accessing data memory. Once the access pattern (the frequency of strides) is known, it can be used to reconfigure the cache by switching off parts of it [2], or by changing its size.

In order to extract the phases that may exist in a program, the execution is divided into intervals – sections of continuous execution of a program. In our example, we used intervals with length varying from one million instructions to one billion instructions. Intervals are then grouped into phases using SimPoint tool [3] in a way that each phase contains intervals with similar behavior (similar memory access patterns in this case). SimPoint uses k-means clustering in order to split program behavior into phases. K-means clustering takes a set of points in n-dimensional space and splits them into clusters by using distance between points as a metric for similarity (points with small distance are considered similar).

To characterize each interval as a point, we use frequency vectors: The *djpeg* binary that is executing in the simulator is instrumented at translation time so that each time an access to memory occurs the difference in address between the current access and the previous access is calculated (in words), and the corresponding element of the frequency vector is incremented. For example, if current access is a load from the address $0 \times 1{,}000{,}000$ and the previous was the store to $0 \times 10{,}000{,}004$ the difference is one word and the element at position one is incremented. The maximum stride that is recorded is 1,024: the element at the position 1,023 will contain the number of memory accesses that had the stride 1,024 or bigger. At the end of each interval, the vector is saved, and all its elements are reset to zero. The set of these vectors is then passed to SimPoint. To perform the clustering, SimPoint considers each vector as a point 1,024-dimensional space and performs k-means clustering of the set of intervals. The analysis performed with the interval length of one million instructions discovered *nine different clusters* (Fig. 10.2). As the length of intervals increases, less phases are discovered - for the interval length of one billion instructions SimPoint does not discover any difference among intervals of execution.

The data shown in Fig. 10.2 clearly sustain the fact that, as there are different memory access patterns as the program is being executed, there is a requirement for variable memory bandwidth and locality policy, and thus changing these parameters one might achieve huge gains on power efficiency in the ERA platform.

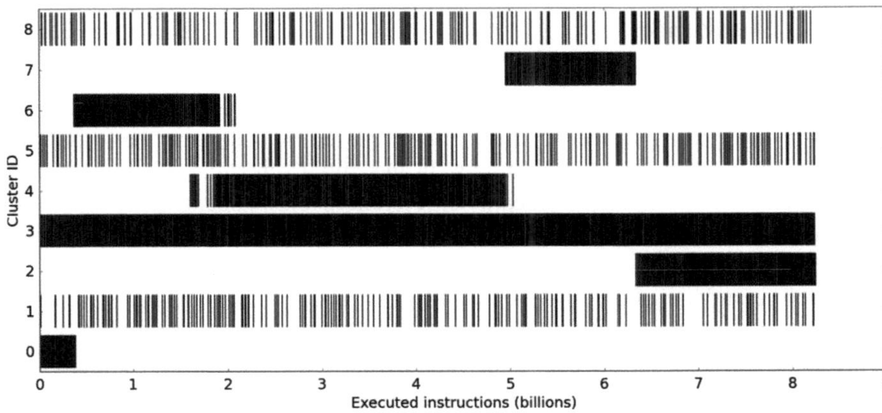

Fig. 10.2 Phases detected during execution of djpeg using the interval of one million instructions. X-axis shows the number of executed instructions while the y-axes shows the cluster that the interval belongs to (nine clusters are discovered in total)

We plan to further expand this analysis by collecting the statistics from > multiple sources and combining them to get a comprehensive view of the application behavior [4]. Data will be collected from measurements taken from emulated system (QEMU), from existing hardware and from ILP statistics generated by the compiler.

10.4.3 Network

In order to show detailed behavior of the communication patterns inside an MPSOC, we simulated four examples of real applications that can be found in Fig. 10.3. Each node represents a task, and each arc is weighted by the communication rate among tasks in Mbytes/s. In all simulations we used the SOCIN NOC [5], and used a fixed size buffer capable of storing four flits per output channel. The applications used were the MPEG4, VOPD [6], MWD (Multi-Window Display) [7] and Xbox [8], all with 12 cores, but with different communication patterns, as represented in the bandwidth of each link depicted in Fig. 10.3. In this figure, arcs in MPEG4, VOPD and MWD show rates in MB/s, while arcs in the XBOX application show rates in GB/S. A cycle-accurate traffic simulator in Java was utilized to evaluate the network hotspots and the average latency using the reconfigurable and original routers. The distribution of the cores in the NoC was specified in accordance with the communication needs of the cores, reflecting a design time choice, being based on the original application.

Figure 10.4 shows the mean efficiency of MPEG4, VOPD, MWD and Xbox when mapped to a 4×3 NoC with homogeneous buffer sizes. In this report, efficiency represents how many buffer units are being appropriately used, in accordance with

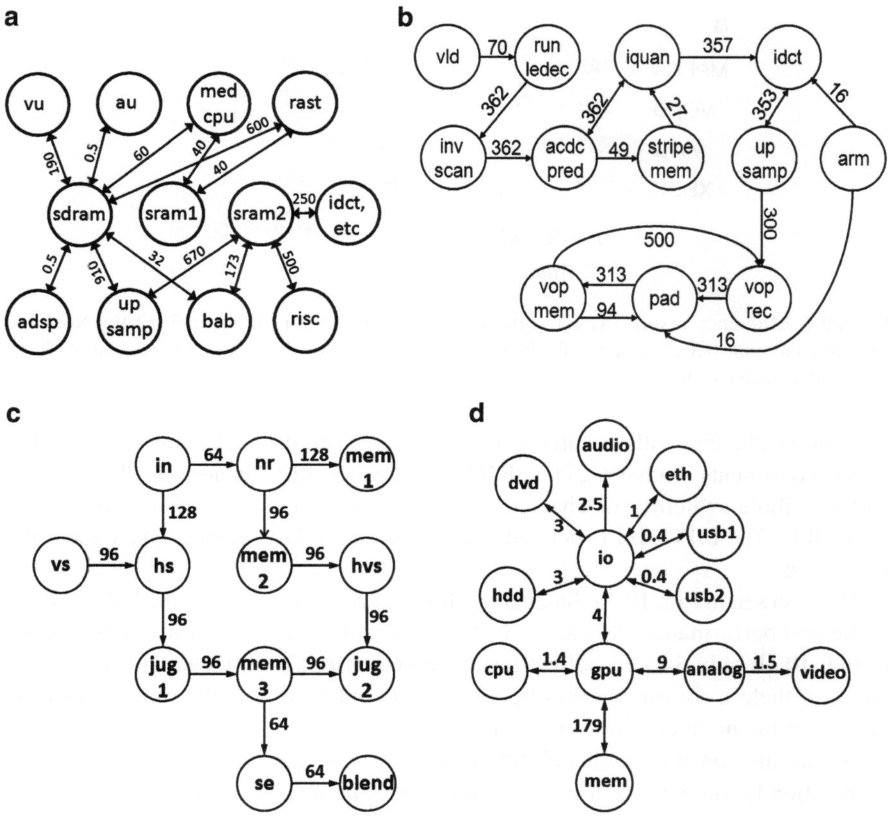

Fig. 10.3 MPEG4 (**a**), VOPD (**b**), MWD (**c**) and Xbox (**d**) task graphs

the necessity of the application. The efficiency results in Fig. 10.4 were obtained in accordance with Eq. 10.1,

$$\eta = \frac{\sum_{i=1}^{i=\#routers} \frac{\#buffers\_used\_router_i}{\#total\_buffers\_router_i}}{\#routers} \tag{10.1}$$

where the reference value has been obtained by using the best buffer distribution for a certain performance level.

Figure 10.4a presents the efficiency considering all channels of the network with the same buffer depth, and Fig. 10.4b shows the efficiency using an heterogeneous NoC, where each router might have a different number of buffers compared to others router, but all channels inside a router having the same number of buffers (achieving optimal communication throughput for a given power budget). In both cases, one can observe that the routers use excessive buffers in some channels, since not all channels present the same communication rate. In such cases, the extra buffer

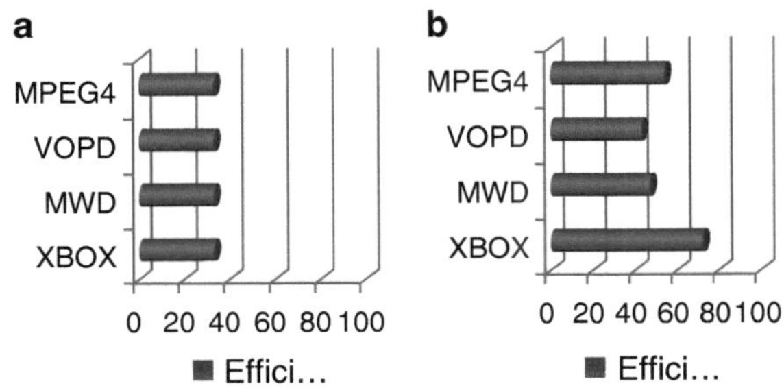

Fig. 10.4 Efficiency results (**a**) using the same buffer depth for all channels of the NoC and (**b**) using the same buffer depth for all channels of the same router, but with a different buffer depth specified to each router

units of the channel will consume unnecessary area and power. The network used in these experiments is a mesh-2D with XY-routing algorithm, handshake flow control and wormhole-switching mechanism [5]. Each input channel has a FIFO for storage of the flits. The FIFO size is defined at design time, and all channels have the same FIFO size.

One can see in Fig. 10.4a that, using a homogeneous router with the buffers sized to the best performance case, around 33% of the buffers slots are utilized. Similarly, in Fig. 10.4b only 54% of buffer slots have been used in the mean. However, they are nevertheless consuming power, but are not contributing to the reduction of the latency or the number of hotspots in the network.

Again, the concept of adaptability must be present for one to use the best communication fabric with the lowest possible power budget.

10.5 Summary of Proposed Innovations

As a means to show the effectiveness of the adaptive platform, in the ERA project commercially available FPGAs are used to demonstrate the concept. It is not our goal to devise new low-power FPGA structures, but instead propose a platform that can benefit from different types of reconfigurable hardware. The reasons for this prototyping option are as follows:

Cost-effectiveness: The NRE (non-recurring engineering) costs of FPGA designs is an order of magnitude lower than the NRE costs of ASIC designs and this is only expected to increase with newer technologies. Additionally, the costs of design tools targeting FPGAs are much lower than those targeting ASICs. This simply means that ASIC designs are increasingly only viable for mass-market embedded products and FPGA-based products will become more commonplace in the near future from a pure design cost perspective.

Time to market: The utilization of reconfigurable hardware is already commonplace as in the ASIC design trajectory; reconfigurable hardware is being utilized to perform design verification. To shorten the time to markets windows, it is increasingly becoming more economical to ship FPGAs in actual products and forego the ensuing ASIC design trajectory. This is a natural progression from hand-designing ASIC chips towards cell-based ASIC designs with the cells now being placed on reconfigurable hardware structures. This also fits perfectly with the programmability trend as the larger more complex circuit designs and processor cores can nowadays be accommodated on the latest FPGA chips.

Performance: The performance of FPGAs is increasing at a much faster rate than their ASIC counterparts due to their very regular structure. Therefore, even though FPGAs are much slower than ASICs when implemented in the same technology, it is common that in actual product development, the utilized FPGA chips are at least one generation ahead of ASICs. This considerably alleviates the speed disadvantage. Especially when one considers that FPGAs nowadays also contain many dedicated circuitry to further improve the speed of commonly used functional elements, e.g., multipliers and on-chip memory. Moreover, the increased logic density also translates into increased memory bandwidth. This could allow the industrial partners to use the prototype of the proposed platform in a commercial way much before any dedicated ASIC is ready.

Power consumption: The power consumption of FPGAs is still considerably higher than their ASIC counterparts, but methods to save power targeting FPGAs start to gain focus, and our project intends to deal with power issues from different perspectives. Finally, low-power FPGAs are starting to make their appearance.

Flexibility: Flexibility by utilizing reconfigurable hardware can be achieved in many ways and many research projects have demonstrated this. The functionality of the hardware can be changed through reconfiguration or reconfigurable hardware structures or programming the softcore(s) (or in some cases: fixed processing cores), This allows for "in-field" updates of functionality and maybe more importantly enabling the possibility for designers to address design and process variability later on in the product cycle. This further reduces the design and verification costs.

From the perspective of an integrated design manufacturer such as STMICRO, one key factor for the success of future consumer products design, is to provide a solution to the faltering ability of existing silicon design and verification methodologies to keep up with the exponential increase of chip capacity and the increased pressure on architects and designers alike to develop very complex hardware in a decreasing amount of time. As a result the number of products that will rely on a set of predefined architecture template will greatly increase. The availability of a pre-characterized parametric architecture and programmable engine platform enables faster design space exploration and minimize critical late design respins during the physical implementation and validation phases. In order to make this possible, industrial design teams would need to have access to technologies such as the one focus of this proposal; so that programmable accelerator architectures best suited to an application domain could be easily prototyped and selected from a range of well defined features whose fast-prototyping platform is already available. This would make a significant difference in terms of cost reduction for the total NRE and allow gaining a competitive advantage over designs that adopt a more traditional conservative approach to system level design.

The architecture proposed in the ERA project will allow Evidence to develop innovative solutions in the market of reconfigurable systems, allowing Evidence to acquire advanced knowledge in the usage of FPGA for signal processing and multimedia systems together with Embedded Linux microcontrollers. The framework developed as part of the ERA Project will be exploited by Evidence with the prototype study of an embedded board based on a Xilinx FPGA, as well as with applications in the signal processing area with key Evidence customers.

The technology developed for software support of reconfigurable embedded architectures in the ERA project will allow IBM to offer advanced tool-chain solutions that target changes in the underlying platform efficiently, thereby increasing the overall utilization and performance-per-power characteristics of the system. In particular, IBM will benefit from advanced and innovative compiler optimizations which leverage the capabilities of target platforms.

Additionally, we will combine all the solutions proposed in this project into a demonstrator platform that we expect will allow the industrial partners fast access to new products developed on top of it. The intended platform will serve several purposes:

Quick development platform for the industry: the clear interfaces defined in this project should allow the industrial partners to take from the platform everything they need and still incorporate their own IPs. Moreover, for low volumes even the prototype can be used as a commercially viable product, since the consortium will use available FPGA technology to validate its contribution.

Academic purposes: the ERA platform can be easily used to build different instances of embedded processing solutions and we foresee and will actively pursue the possibility of incorporating the ERA platform as a teaching tool in embedded courses or labs.

Having stated the main objective of the ERA project and the sub-objectives in order to reach it, we can now summarize the main deliverables of the ERA project as follows:

1. A hardware prototype reconfigurable multi-core system: a hardware platform prototype, encompassing a reconfigurable VLIW processor, reconfigurable memory, and reconfigurable NoC, that can be used as a demonstrator of the adaptive properties.
2. A software platform prototype, that working together with the hardware can provide adaptability in a seamless fashion for the platform user and software designer:
    - A supervisor coordinating the reconfiguration operations, providing the needed trade-off between application quality-of-service and resource usage;
    - Exploration of profile-directed compiler optimizations to increase performance and reduce power by scheduling reconfiguration instructions, and scheduling several cores;

- Portable compilation scheme capable of adapting to changes in the underlying architecture efficiently.

3. A set of applications identified that will benefit the proposed ERA platform. As the main test case of the platform, several programs that are used in portable phones will be used as our driving benchmark.

Here we list some of the main areas where we expect that the ERA project is going to advance the state of the art.

Reconfigurable processors. Most embedded devices will employ complex system-on-a-chip (SoC) platforms, based on several different sub-systems, where each one is built to optimize a specific application domain (DSP, 3D-graphics, etc.). However, there is no easy way to anticipate all the software applications for an embedded system, which means that a means for dynamic adaptation must be available.

Fine-grain reconfigurable computing where one can create many different circuits on an FPGA-like substrate available on chip, seemed a very promising approach to solve the need of dynamic adaptation. Indeed, many fine-grain reconfigurable approaches have proven to be a good option for reducing the design time of specialized functions. Many successful examples have shown good performance improvements and energy savings [9–15]. However, the fine-gain reconfigurable approach has been plagued by two major problems: virtualization of the reconfigurable hardware (especially managing its large state) for use in a multi-programmed environment (multiple concurrently-running applications) and a blurring or breaking of the hardware/software interface at the Instruction Set Architecture (ISA) level that complicates programming, compilation, toolsets and run-time systems. It is exactly in these areas where most of the effort is spent for the development of fine-grain reconfigurable systems. With the advent of multi-core processors and MPSoCs the situation became increasingly complex since the need to manage several different accelerators and at the same time minimize energy and ensure performance at the global system level became a necessity. Piperench [9], Molen [10], GARP [11] and Raw [12] are well known examples that extend the underlying instruction set. Such reconfigurable machines rely on the compiler to detect application "hotspots" and optimizing them by creating new specialized instructions. Despite the implications of its use, static code detection of potential reconfigurable accelerators is still being explored. However, when a compiler is responsible for finding possible candidates for optimization, two main problems emerge, particularly concerning the OS: besides the necessity of having the OS source code, the room for optimization is restricted to static code – dynamic library optimization is not possible, since when and how much the libraries will be used is not known before run-time.

More recently, reconfigurable approaches have been presented, where the focus was on dynamic hotspot detection. The pioneer approach is Warp [13]. This technique uses a non-trivial CAD algorithm to transform the sequences of instructions to a control flow graph. Then, it synthesizes the code and maps the circuit onto a simplified FPGA structure. As another example, CCA [14] (Configurable Computing

Accelerator) provides both dynamic and static detection. CCA is a tightly coupled coarse grain reconfigurable array, composed by simple functional units, coupled to an ARM processor. These examples of reconfigurable hardware are restricted to optimize standalone user applications, with specific hotspot optimizations and with hotspots known at design time –something that is becoming less and less frequent. In [15] one can find a coarse grain system that covers acceleration of the whole software code, using dynamic techniques and a coarse grain fabric.

The ERA project advances the state-of-the art by taking another route in reconfigurability: instead of trying to accelerate applications by reconfigurable functionality (which is hard to use) ERA relies instead on architectural reconfigurability for optimizing applications in terms of performance *and* power. The processor (as a single core or as a collection of cores on the chip) is reconfigured to match the computation needs of the application.

Memory Hierarchy Reconfiguration. Similarly to processor reconfiguration, the memory system is also reconfigured for the application. There is a significant body of work concerning the reconfiguration of the memory hierarchy, primarily for high-performance computing but also with some sparse examples for embedded computing. Similarly to processor reconfiguration, caches have been a prime target for **resizing** to save power. Although it is always better to have more cache, in many cases using much less cache can save considerable power while giving up very little in performance. There are several proposals for resizing the cache to fit program needs, most important among them: Variable L1/L2 division [16], Selective Cache Ways [17], the Accounting Cache [18], and Miss Tag Resizing for CAM-Tag Caches [19] (which is well suited for embedded architectures). These techniques differ on the memory partitioning technique used (e.g., partitioning at the memory bank/segment level, cache-way level, or even within a bank using bit-line segmentation), and the policy used to drive the partitioning. Both within and across such techniques there are significant differences in power and performance depending on the target application. The policies proposed to drive such techniques are tuned to deliver the desired results for some benchmark suite so there is no clear consensus what works and what does not for our target application domain. Furthermore, these cache resizing techniques were proposed in isolation without taking into account any possible processor reconfiguration. There are very few proposals for holistic reconfiguration of both the processor and the cache hierarchy, but even those do not take fully into account the complex interaction of the joint reconfiguration of these two components; they use instead blanket performance/power feedback policies.

Beyond resizing, critical role in the power/performance of the cache hierarchy play the techniques for reducing static power dissipation (DVFS [20] and/or decay techniques [21]) and techniques to optimize the switching activity (dynamic power) for well known cache architectures. The toolkit available for optimizing the memory hierarchy includes [20]: various power-efficient associative cache architectures (phased caches, way prediction, etc.), filter caches, loop caches, and trace caches, data compression techniques (value locality compression, zero compression, etc.), and coherence optimization techniques.

In the ERA project the reconfigurable memory architecture is based on the needs and requirements of our target application domain, i.e., embedded applications. Mechanisms for memory reconfiguration will be unified according to the synergy they show (again in relation to embedded applications). Furthermore, the management policies for such reconfiguration techniques will be application driven, rather than blanket policies that cater to the general case but can be rather inefficient for specific applications or even individual application phases. We will use the necessary hardware monitoring coupled with the application-fed requirements to construct application-driven policies that can achieve various user objectives. Of critical importance is that these policies will work in concert with the processor reconfiguration proposed in our project. Our holistic approach will be expanded to cover multiprocessor SoCs and their memory hierarchies, multiprogramming in such environments (e.g., multiple programs running concurrently and sharing the chip resources), and reconfiguration of the communication infrastructure (including cache coherence protocols and networks-on-chip) specifically for parallel applications running on multiple heterogeneous cores and/or accelerators.

In the ERA project the whole compilation tool-chain can act as the mediator between the features and requirements exposed by the applications (e.g., memory locality, memory bandwidth requirements, and processing needs, as well as execution phases, multi-threaded behavior, communication, and synchronization) and the reconfiguration potential of the architecture. In particular, different modules of the architecture will have the ability to reconfigure themselves according to the (in-hardware) perceived runtime conditions and resource usage. In the integrated approach of ERA we complement this hardware monitoring and adaptation by leveraging software development and tuning tools that are able to implement a range of optimization strategies of different complexities, aiming to induce the most suitable architectural reconfigurations that match the application requirements during runtime. Using this approach, the optimization tools can take advantage of the global view of the architecture together with the application structure and needs, in order to trigger profitable reconfiguration actions.

10.6 Conclusions

In the ERA project, we are addressing the power and memory walls that bottlenecks in the design of embedded systems to satisfy the demand for more performance. More specifically, we are focusing on the design of the embedded processor and its environment. We identified a clear trend in the design of embedded processors that is moving away from application-specific implementations to more general-purpose approaches. This is not to say that application-specific implementations are no longer important, but they are only (economically) viable when there is a large mass market to amortize the design costs or when the "best" solution is sought after (independent of costs).

In this trend, we strongly believe that reconfigurable hardware will play an important role as it is capable of providing adequate performance (for most embedded applications) while maintaining a good level of flexibility that is needed in current-day embedded systems market, in which time-to-market times are shrinking. Of course, the mentioned power and memory walls are also present in the reconfigurable embedded processor scenario, and in the ERA project, we are addressing them at the same time. We believe that reconfigurability can bring forward new solutions to allow us to scale the walls. As the embedded market is moving towards a more structured design approach, we focus in the ERA project on three main elements: computing elements, memory elements, and networking elements. For each type of element, we will propose new solutions on how to exploit reconfigurability to provide more performance, but at the same time limit power consumption and better scale memory requirements. In this chapter, we also provided some evidence why we believe that our approach will be successful, and the effective need of reconfiguration in different structures to bypass the memory and power wall. In the end, one of the goals of the ERA project concerns building a proof-of-concept demonstrator to show the benefits of our approach and which can be immediately used by the industry for prototyping purposes.

Acknowledgments We would like to acknowledge the Valorisation Centre in Delft, and in particular two persons: Linda Roos and Theresia Twickler. They are responsible for the (financial) management of the project and their contributions/efforts allowed all the researchers involved in the ERA project to properly focus on their research tasks.

References

1. Bhavishya Goel, Sally A. McKee, Roberto Gioiosa, Karan Singh, Major Bhadauria, Marco Cesati, "Portable, scalable, per-core power estimation for intelligent resource management," Greencomp, International Conference on Green Computing, pp.135–146, 2010.
2. M. Alioto, P. Bennati, R. Giorgi "Exploiting locality to Improve Leakage Reduction in Embedded Drowsy I-Caches at Same Area/Speed", In Proceedings of IEEE International Symposium on Circuits and Systems (ISCAS), pp. 37–40, May 2010.
3. G. Hamerly, E. Perelman, J. Lau, B. Calder, "Simpoint 3.0: Faster and more flexible program phase analysis", In Journal of Instruction Level Parallelism, vol 7, 2005.
4. N. Puzović, S. A. McKee, R. Eres, A. Zaks, P. Gai, S. Wong, R. Giorgi, "A Multi-Pronged Approach to Benchmark Characterization" in IEEE International Conference on Cluster Computing Workshops and Posters (CLUSTER WORKSHOPS), pp.1–4, 2010.
5. C. Zeferino, A. Susin, "SoCIN: A Parametric and Scalable Network-on-Chip" in 17th Symposium on Integrated Circuits and System (SBCCI), 2003, pp. 169–174.
6. Bertozzi, D. et al., "NoC Synthesis Flow for Customized Domain Specific Multiprocessor Systems-on-Chip", IEEE Transaction on Parallel and Distributed System, 2005, pp. 113–129.
7. K. Srinivasan and K. S. Chatha, "A Low Complexity Heuristic for Design of Custom Network-on-Chip Architectures," in Proceedings of Design, Automation and Test in Europe Conf.,vol. 1, 2006, pp. 1–6.
8. J. Andrews, N. Baker, "Xbox 360 System Architecture", IEEE Micro, vol. 26, no. 2, 2006, pp. 25–37.

9. Goldstein S. C., Schmit H., Budiu M., Cadambi S., Moe M., and Taylor R. R., "PipeRench: A Reconfigurable Architecture and Compiler". Computer 33, 4, 70–77, 2000.
10. Vassiliadis S., Wong S., Gaydadjiev G., Bertels K., Kuzmanov G., and Panainte E.M., "The MOLEN Polymorphic Processor". IEEE Transactions on Computers, 53, 11, 1363–1375, 2004.
11. Hauser J. R., and Wawrzynek, J., "Garp: a MIPS Processor with a Reconfigurable Coprocessor". In Proceedings of the 5th IEEE Symposium on FPGA-Based Custom Computing Machines (FCCM), 1997.
12. Waingold E., Taylor M., Srikrishna D., Sarkar V., Lee W., Lee V., Kim J., Frank M., Finch P., Barua R., Babb J., Amarasinghe S., and Agarwal A., "Baring It All to Software: Raw Machines". Computer 30, 9, 86–93, 1997.
13. Lysecky R., Stitt G., Vahid F., "Warp Processors". In ACM Transactions on Design Automation of Electronic Systems (TODAES), pp. 659–681, 2006.
14. Clark, N., Kudlur, M., Park, H. Mahlke, S., Flautner, K., "Application-Specific Processing on a General-Purpose Core via Transparent Instruction Set Customization". In International Symposium on Microarchitecture (MICRO-37), pp. 30–40, Dec. 2004.
15. Beck A.C.S., Rutzig Mateus B., Gaydadjiev G., Carro, L., "Transparent Reconfigurable Acceleration for Heterogeneous Embedded Applications," in Proceedings of the Design, Automation and Test Conference (DATE), pp.1208-1213, 2008.
16. D.H. Albonesi, "Dynamic IPC/clock rate optimization," in Proceedings of the 25th International Symposium on Computer Architecture (ISCA-25), 1998.
17. D.H. Albonesi, "Selective Cache Ways: On-demand Cache Resource Allocation", in Proceedings 32nd Annual IEEE/ACM International Symposium on Microarchitecture (MICRO-32), pp. 248–259, 1999.
18. S. Dropsho, et.al., "Integrating adaptive on-chip storage structures for reduced dynamic power," in Proceedings of the International Conference on Parallel Architectures and Compilation Techniques, 2002.
19. M. Zhang and K. Asanovi'c, "Highly Associative Caches for Low-power Processors", Kool Chips Workshop, 33 rd International Symposium on Microarchitecture, 2000.
20. S. Kaxiras and M. Martonosi, "Architectural Techniques for Low Power", Morgan & Claypool Publishers, 2008.
21. S. Kaxiras, Z. Hu, and M. Martonosi, "Cache Decay: Exploiting Generational Behavior to Reduce Cache Leakage Power", in Proceedings of 28th International Symposium on Computer Architecture (ISCA-28), 2001.

Chapter 11
REFLECT: Rendering FPGAs to Multi-core Embedded Computing

João M.P. Cardoso, Pedro C. Diniz, Zlatko Petrov, Koen Bertels, Michael Hübner, Hans van Someren, Fernando Gonçalves, José Gabriel F. de Coutinho, George A. Constantinides, Bryan Olivier, Wayne Luk, Juergen Becker, Georgi Kuzmanov, Florian Thoma, Lars Braun, Matthias Kühnle, Razvan Nane, Vlad Mihai Sima, Kamil Krátký, José Carlos Alves, and João Canas Ferreira

Abstract The relentless increase in capacity of Field-Programmable Gate-Arrays (FPGAs) has made them vehicles of choice for both prototypes and final products requiring on-chip multi-core, heterogeneous and reconfigurable systems. Multiple cores can be embedded as hard- or soft-macros, have customizable instruction sets, multiple distributed RAMs and/or configurable interconnections. Their flexibility allows them to achieve orders of magnitude better performance than conventional computing systems via customization. Programming these systems, however, is extremely cumbersome and error-prone and as a result their true potential is only achieved very often at unreasonably high design efforts. This project covers developing, implementing and evaluating a novel compilation and synthesis system approach for FPGA-based platforms. We rely on Aspect-Oriented Specifications to convey critical domain knowledge to a mapping engine while preserving the advantages of a high-level imperative programming paradigm in early software development as well as program and application portability. We leverage Aspect-Oriented specifications and a set of transformations to generate an intermediate representation suitable to hardware mapping. A programming language, LARA, will allow the exploration of alternative architectures and design patterns enabling the generation of flexible hardware cores that can be easily incorporated into larger multi-core designs. We will evaluate the effectiveness of the proposed approach using partner-provided codes from the domain of audio processing and real-time avionics. We expect the technology developed in REFLECT to be integrated by our industrial partners, in particular by ACE, a leading compilation tool supplier for embedded systems, and by Honeywell, a worldwide solution supplier of embedded high-performance systems.

J.M.P. Cardoso (✉)
Departamento de Engenharia Informática, Faculdade de Engenharia (FEUP), Universidade do Porto, Rua Dr. Roberto Frias, s/n, 4200–465 Porto, Portugal
e-mail: jmpc@acm.org

J.M.P. Cardoso and M. Hübner (eds.), *Reconfigurable Computing: From FPGAs to Hardware/Software Codesign*, DOI 10.1007/978-1-4614-0061-5_11,

11.1 Project Partners

1. Honeywell International s.r.o., HON, Czech Republic (Coordinator)
2. Instituto de Engenharia de Sistemas e Computadores Investigação e Desenvolvimento em Lisboa, INESC-ID, Portugal
3. Faculdade de Engenharia da Universidade do Porto, FEUP, Portugal
4. Technische Universiteit Delft, TUD, The Netherlands
5. Karlsruhe Institute of Technology, KIT, Germany
6. Imperial College of Science, Technology and Medicine, Imperial, UK
7. ACE Associated Compiler Experts b.v., ACE, The Netherlands
8. Coreworks – Projectos de Circuitos e Sistemas Electrónicos S.A., CW, Portugal

- Project Coordinator: Zlatko Petrov, Honeywell International s.r.o., HON, Czech Republic
- Project Scientific Coordinators: Pedro C. Diniz (INESC-ID), João M. P. Cardoso (FEUP)
- Start Date: 2010-01-01
- Expected End Date: 2012-12-31
- EU Program: 7th Framework Programme, FP7-ICT-2009-4, STREP Project No. 248976
- Global Budget: 3,703,577.00 €
- Global Funding by EU: 2,719,999.00 €
- Contact Author: João M. P. Cardoso, Email: jmpc@acm.org

11.2 Introduction

The increasing number of available transistors on a die has enabled the emergence of a new class of computing systems that can include multiple computing elements on a single chip. The functionality of those computing elements can be configured or tailored to specific functions, in some cases even at run-time. Contemporary Field-Programmable Gate-Arrays (FPGAs) devices and modern Multi-Core Processors are extreme examples of this trend in industry. Reconfiguration allows these systems to achieve orders of magnitude better performance than conventional processor systems as they can implement directly in hardware specific high-level operations either crystallized as custom instructions or as custom functional units. Reconfiguration techniques [1] are also seen as a way to mitigate, or even eliminate potential issues with reliability and intermittent faults in these large VLSI systems. As a result, these computing systems are now becoming commonplace in embedded and even in some cases in high-performance systems, that permeate everyday life systems from medical equipment to computer networks.

These advantages over traditional systems come at a cost. The large number of potential custom instructions or custom functional units coupled with the many choices of how to interconnect these functions makes the mapping of computations to these hardware/software architectures a highly non-trivial process. A typical application for these multi-core embedded systems has close to 100 K lines of C/C++ code sometimes permeated with snippets of assembly code for handling low-level functions.

As a result, the mapping of complex applications to these architectures is accomplished by a very labor intensive and error-prone manual process. Programmers must assume the role of hardware designers to synthesize or program various custom hardware units at a very low level, and also to understand how these units interact with software portions of application code. They must partition computations between a code that is executed on more traditional processor cores and a portion that is going to be synthesized in hardware with a consequent partitioning and mapping of data as it flows through the system. A complication of this mapping process is exacerbated by the fact that custom hardware units may internally exhibit different computation models (*e.g.*, data flow, concurrent synchronous processes) and architectural characteristics (*e.g.*, parallelism, customization), or that the various cores might support functional- or data-parallel concurrent execution paradigms.

The inherent complexity of this mapping process means long compilation and synthesis cycles fraught with errors, as programmers must manually perform many of the low-level translation steps from a high-level description of computations in application to multi-core hardware/software configurable fabric. As a consequence, and given the ever increasing pressures of tight time-to-market, it is practically impossible for programmers to explore a wide range of possible alternative designs. As a result programmers typically have to settle for "known" good designs with acceptable performance rather than exploring alternative designs with potentially large benefits in terms of performance, design area, consumed energy or any combination of the above.

It is an aim of REFLECT to develop methodologies and approaches to help designers achieve efficient FPGA-based heterogeneous multi-core computing systems. Our research and development agenda aims at developing a new foundation combining different areas of research: ***aspect-oriented specifications***, ***hardware compilation***, ***design patterns and hardware templates***. We strongly believe the complementarity and synergy of these concepts will allow us to deliver a prototype system that will enable designers to meet the challenges of mapping complex applications to future multi-core reconfigurable computing systems.

This project will develop, implement and evaluate a novel compilation and synthesis system approach for FPGA-based platforms. We rely on an Aspect-Oriented System Development (AOSD), with foundations on aspect-oriented programming approaches [2–5] to convey critical domain knowledge to mapping engines while preserving the advantages of a high-level imperative programming paradigm in early software development as well as program and application portability. We leverage aspect-oriented specifications using LARA (LAnguage for Reconfigurable Architectures), a new domain-specific aspect-oriented programming language, and a set of transformations to generate an intermediate representation using extensible mapping intermediate representations (LARA-IR). LARA specifications will allow a seamless exploration of alternative architectures and run-time adaptive strategies allowing generation of flexible hardware cores that can be easily incorporated into larger multi-core designs. By developing schemes to specify aspects of hardware templates, which are required for a compiler to be able to integrate them with other hardware templates forming more powerful computing systems, will also create foundations to leverage the generation of multi-core computing systems. The overall compiler design approach is complemented with techniques to decide among optimizations and design patterns based on previously best known practices. This will allow the compiler to explore very efficiently multiple design alternatives and thus achieving better solutions otherwise not possible within a reasonable compiling time.

In many applications, such as avionics, a model-based approach allows the "natural" specification of the design with inherent separation of certain aspects such as algorithm behavior, data-types, temporal coupling, parallelism, and fault tolerance. Thus an enhanced mapping of these applications to reconfigurable processing architectures, compared to current architectures, is only feasible if effective tools are available. We will use aspect-oriented specifications along with suitable intermediate representations to explore the extent of such improvements.

We are evaluating the effectiveness of the proposed approach using partner-provided codes and design models used in real-life applications. This evaluation includes the development of two demonstrators, respectively an avionics mission-critical embedded system evaluation and an audio encoder. Both demonstrators raise very realistic and demanding challenges that will highlight the power and impact of the base techniques and methodologies in the proposed REFLECT approach over traditional design and mapping methodologies.

An important part of the development and deployment of the proposed REFLECT technology for mission-critical embedded systems is the design assurance of safety, correct functional behavior, and real-time performance. We will develop fault-tolerance approaches such as redundancy, voting, and reconfiguration/checking to mitigate the effects of single-event-upsets and seamlessly integrate those in the design flow as aspect-oriented specifications. Real-time requirements for application/algorithm components will also be specified as aspects. These safety and real-time aspects are considered along with the core design aspects by the mapping engine allowing exploration of prioritization and trade-offs for a particular embedded system deployment.

This chapter is organized as follows. Section 11.2 presents an overview of the project and its main objectives. Section 11.3 illustrates the REFLECT design flow and the main tools being used. Section 11.4 presents the target applications that are being used to evaluate the REFLECT design flow and main ideas and R&D techniques. Section 11.5 describes briefly the use of aspects and design patterns in REFLECT including a number of examples. Section 11.6 presents some examples and results. Section 11.7 presents some related work. Finally, Sect. 11.8 concludes this chapter.

11.3 Main Overview and Objectives

REFLECT addresses the shortcomings of the process of mapping computations described in high-level programming languages to reconfigurable computing platforms, in particular, FPGAs, consisting of on-chip, heterogeneous, multi-core processing units. The target architecture we envision is depicted in Fig. 11.1 and consists of a general-purpose processor (GPP), *e.g.,* a Xilinx MicroBlaze or IBM PowerPC, tightly coupled with a reconfigurable hardware fabric where Configurable Custom Units (CCUs) exist according to application needs. Collectively, the CCUs define a reconfigurable computing system implementing various execution models in space and in time and can be programmable using parameterized and programmable hardware templates (*e.g.,* non von-Neumann IP *softcores*) or specialized hardware templates. The coupling and interface between the processor and the CCUs will be inspired on the Molen machine and programming paradigms [6]. We also envision high-end computing systems (akin to HPC systems) that are composed of several of these base reconfigurable systems interconnected using the traditional multiprocessor organization arrangements (*e.g.,* bus, hypercube or trees) and logically organized as distributed memory or shared memory heterogeneous multiprocessor systems. From a software-stack perspective the heterogeneous system is viewed as a co-processor device of a host system. The reconfigurable resources are not exposed to the operating system of the host system. Instead, there is a simple resident "monitor" system that is responsible for the communication of data and synchronization with the host system and/or I/O channels. It is out of the scope of the project to develop an operating system for this heterogeneous reconfigurable architecture.

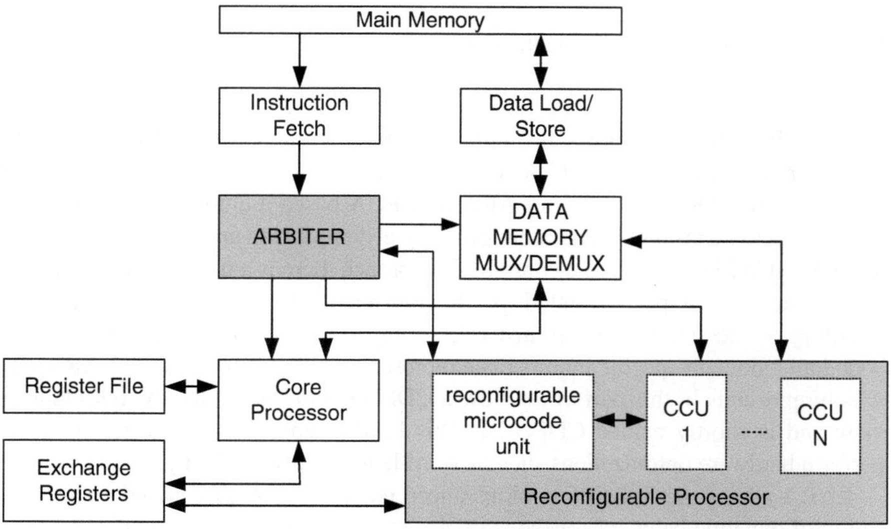

Fig. 11.1 Typical computing system addressed by REFLECT

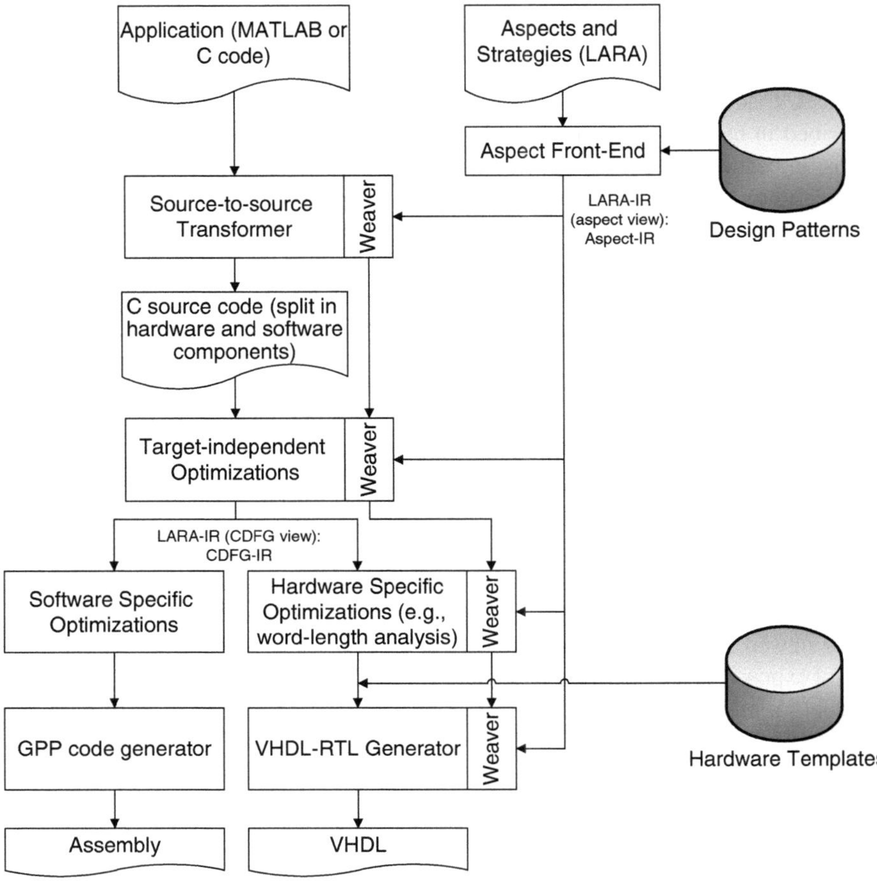

Fig. 11.2 Overall design flow for the REFLECT project

To address the challenges in mapping computations expressed in high-level programming languages (*e.g.,* C or stylized MATLAB/SciLab descriptions) using an Aspect-Oriented Specification in LARA for FPGA-based multi-core reconfigurable architectures, we propose a comprehensive novel compilation and mapping approach as depicted in Fig. 11.2. At the core of this approach there is a transformation engine that leverages the Aspect-Oriented specification and a set of techniques, such as term rewriting, successive refinement and refactoring to transform the input application code into a domain-specific intermediate representation language, named LARA-IR, which represents in this part of the flow a CDFG (control/data flow representation) view and is shortly named CDFG-IR. This LARA representation is then used to perform hardware optimizations such as word-length analysis [7–10].

LARA allows the description of advanced mapping strategies including specification of dynamic behavior for the mapping process, thus making reconfiguration a first-class abstraction in the mapping process rather than an afterthought. In its simplest form, LARA allows for specific high-level operators such as multiplications or additions to be mapped to specific hardware units, thus making the scheduling of the operation more explicit. More sophisticated mapping strategies include conditional mappings based on number of hardware operators or sequence of operations to be mapped to the same hardware unit. In addition, LARA will allow for data-dependent mapping decisions and reconfiguration. For instance, it will be possible to specify that for a specific data rate, the mapping shall take place in a specific set of hardware units to promote data streaming. Another critical mapping aspect is the co-location of data in internal storage and the functional units that will use that data.

The LARA intermediate representation, which thus includes explicit elements for reconfiguration and hardware-oriented directives such as data mapping and streaming information, is then translated into VHDL-RTL using either parameterized or programmable hardware templates. As a final result of the mapping process, we rely on existing vendor-specific synthesis tools (e.g., the Xilinx ISE tool chain for FPGAs) to generate concrete hardware implementations from the generated VHDL-RTL specifications.

A fundamental driver for the proposed approach and one of its distinguishing features is the extensive use of Aspect-Oriented specifications. Aspects allow the user to specify non-functional requirements, design patterns, and to expose domain-specific or algorithm-specific knowledge to the compiler non-intrusively and without compromising the semantics of the original specification. The knowledge conveyed by the aspects specification is used by the compiler and the architectural synthesis tools towards the implementation of highly specialized and domain-specific reconfigurable architectures according to the requirements.

In addition to performance, in its many facets, aspects also provide a handle into an important and often neglected issue in hardware design and implementation – verification. As mandated for aerospace and automotive systems requiring safety and high reliability, verification requires that designs shall correctly implement the functional behavior and timing implied by the application specification (in C or the corresponding data-flow models). We will leverage the ability of aspect-oriented specifications to allow the compiler not only to generate designs that comply with specific timing and rate requirements but also to develop comprehensive test-generation schemes for code coverage, design correctness and robustness, and timing.

Yet another important and unique feature of this approach is the inclusion of history-based, best-practices repository that will aid the design flow in dealing with very large designs. By understanding which sets of transformations and parameters lead to the best designs for application codes with specific aspects, this repository will substantially reduce the size of the design-search-space the compiler needs to cover in the pursuit of efficient, and correct designs that meet specific constraints. This repository can be built either with the help of designers for specific design patterns or using pattern extraction and matching techniques. When using a new architecture or a set of templates, the approach might be slow in deriving good designs and thus populating the space of best practices. With time, and with the use on many application codes and/or kernels, the system will use the knowledge of previous design mappings and deliver designs increasingly faster and of high quality.

The approach outlined for the REFLECT project will offer a series of advantages over current state-of-the-art, namely:

- Retains the obvious advantages of both a sequential input paradigm offered by an imperative programming language, and a data-flow modeling paradigm – for both which there is ample software basis as well as control algorithm models;
- Allows the specification in an extensible fashion of many conflicting goals via the aspect-oriented specification without compromising the clarity and understandability of the original code specification. This is crucial in program portability;
- Leverages the use of known architecture specific transformations allowing higher-levels of parallelism, fault-tolerance and energy savings via the design patterns and programmable templates without necessarily exposing their implementation complexity;
- Captures and retains the knowledge of best practices either manually or automatically using heuristics in building a repository of best design practices; This repository will allow the compiler to derive designs that meet specific constraints quickly by aggressively pruning the vast space of feasible designs.
- Enables the verification and generation of comprehensive tests for the resulting target implementation against a high-level programming notation to ensure the safety and reliability of the generated architecture and its implementation.

The overall objective of the proposed approach is to substantially improve the design time and also quality of the generated designs while assuring their correctness. While the technical approach presented in this chapter has been targeted specifically to FPGAs, it is extensible to other configurable and reconfigurable architectures by using a different set of design patterns and templates and a different set of code transformations. In this context, FPGAs and their extreme flexibility and minimal granularity allow the emulation of coarser granularity reconfigurable architectures. FPGAs are thus a natural vehicle for early validation of the mapping and synthesis concepts described here, which we believe will springboard into other more generic compilation and synthesis systems.

To put these objectives and techniques into perspective, in Table 11.1 we present the main issues in attacking the challenge of the REFLECT project, highlighting how the selected set of techniques and methods address the various high-level design issues in the context of the flow depicted in Fig. 11.2.

As can be seen, the REFLECT project brings together, in a synergistic fashion, techniques from a wide range of domains from programming language design and implementations (the Aspect-Oriented Programming) through compilation and synthesis, to the architecture description level, both in terms of execution models with the hardware architectural patterns to the more low-level hardware templates in the final stages of the mapping. The rigor of a compiler-based approach coupled with aspect-oriented specification will ultimately allow the development of correct and verifiable designs in a fraction of the time required with more traditional and labor-intensive approaches. Overseeing this process, we envision the application of history-based techniques (such as best-practices) to cope with the inherent complexity and need to shorten the design cycles for reconfigurable architectures.

Table 11.1 Main objectives and techniques to be researched

Main issues	Objective	Techniques	Support
Imperative model to computation models suited to multi-core FPGAs	To propose a transformation model from imperative programming constructs to the models suited for on-chip, multi-core, FPGA implementations	Successive refinement Term rewriting systems Refactoring Code transformations Computation and data distribution	Aspects support user knowledge as complementary program specifications
Domain Specific Language (DSL): LARA	To specify computations using FPGA resources and issues related to dynamically reconfigurable hardware	Adjustable synchronization schemes Temporal partitioning Dynamically reconfiguration specifications	Customizable soft-cores (with non Von Neumann models) Configuration controllers
Large design spaces to be explored	To reduce search space while meetings design constraints guaranteeing correctness and verifiability	Repository of best practices Design patterns Software/Hardware templates	Repository of hardware templates covering a set of application domains

In a wider context, this project takes a step further in the arena of application mapping and architecture design by bringing together techniques from diverse areas of computer science and computer engineering. With the increasing availability of transistor devices on a die and the emergence of multi-core and reconfigurable architectures, boosted by Moore's law, it is clear the more traditional techniques will eventually be limited by the inherent complexity of the current mapping algorithms and approaches. REFLECT is thus a step in the direction of overcoming these emerging challenges by combination of user-provided knowledge with flexible intermediate representation for increasingly dynamic target architectures.

Our approach of maintaining aspect modules as primary entities not embedded in the application code is important to preserve the code's readability and maintainability. This approach also promotes a clean application source as a single source for modifications, and enables the reusability of aspect modules. Multiple aspects modules and strategies will make the same input application specialized according to the target system organization (e.g., including hardware cores, interface between the GPP and hardware cores, memories connected to the FPGA, possible precisions). This concept will have as ultimate goal better adaptability of the tools to a specific target design.

Yet another aspect addressed in the framework is the definition and exploitation of a Design-Space-Exploration (DSE) engine. A key issue with this engine is the use of specific attributes such as data rates of input variables that will help the engines use design patterns to derive a sequence of representations that will meet (with high likelihood) desired applications requirements.

11.4 REFLECT Design Flow

We now describe in detail the basic concepts and components of the REFLECT design flow depicted in Fig. 11.2 and with major concepts shown in Fig. 11.3.

A. **Aspects**
We are using and extending concepts from aspect-oriented programming (AOP) [2] when considering mapping strategies. While these concepts are not new, their application and adaptation within the context of compilation and mapping for computations to reconfigurable architectures is, we believe, novel. A good and synthetic description of AOP is given in [3]: "AOP is based on the idea that computer systems are better programmed by separately specifying the various concerns (properties or areas of interest) of a system and description of their relationships, and then relying on mechanisms in the underlying AOP environment to weave or compose them together into a coherent program." We are adapting and extend the main ideas of AOP to accomplish an Aspect-Oriented Mapping, *i.e.,* aspects will be used to specify non-functional requirements (e.g., execution time, throughput, area, reliability concerns and fault-tolerance strategies), domain-specific knowledge, and design patterns, to help the compilation, synthesis and mapping process to achieve better design solutions. We envision the existence of short and long lifetime aspects during the compilation flow. Short time aspects will be alive from the input code until the generation of LARA-IR specifications for the CDFG view (i.e., CDFG-IR). Long lifetime aspects will survive up to the generation of the VHDL from the CDFG-IR specification. As such, AOP will ensure the application specifications last longer than their current implementations without obscuring the descriptions with details pertinent to today's architectures but that might be obsolete in tomorrow's architectures.

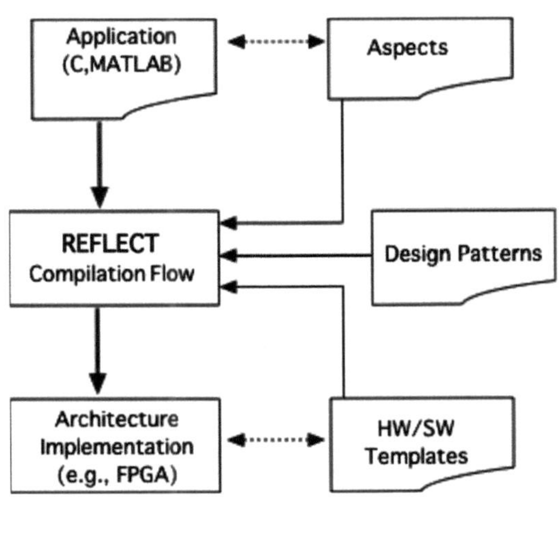

Fig. 11.3 REFLECT compilation flow concepts: aspect modules give the modularization and allow several implementations of the same input application

B. **Compiler Transformations and Hardware Execution Models**
In this component we are developing, refining and integrating in our compiler infrastructure source-level transformations that are specifically suited to multi-core reconfigurable architectures. These include many data-oriented transformations such as data partitioning and scalar replacement as well as floating-point specialization operator-level transformations. In addition we will develop execution "templates" geared towards pipelined, streaming and parallel computing execution models that are composed, in a parametric fashion (as much as possible) with the source-code transformation addressed in this component.

C. **Mapping (Transformations and Best Practices)**
In this component of the approach, named as the *transformation engine*, we are developing a series of mapping strategies that take into account specific features of the input codes and the target architecture to derive, in an automated fashion, good mapping strategies for each design. This component takes as input the application code and its aspects specification, along with a characterization of the target architecture parameters, and derives a sequence of transformations that need to be applied to map the input computation to the target architecture. During the first successive refinement steps source-to-source transformations are used. Then more transformations are done using intermediate representations. Internally, this engine engages in a design-space-exploration (DSE), generating LARA-IR specifications for design patterns to actually map and coordinate the execution of the mapped application onto the target architecture. In this DSE phase, the transformation engine will make use of a repository of best practices techniques to mitigate the complexity issues associated with the need to explore huge design spaces.

D. **LARA Language**
We are designing a simple, but powerful, aspect-oriented language named LARA. LARA will be able to describe aspects, strategies, hardware/software templates and design patterns to guide the compilation flow. This language, to be refined over the course of the project, will also guide a back-end capable of integrating the VHDL-RTL templates as well as taking advantage of architecture specific features such as the co-location of the RAMs and functional units for performance. In addition, we will also develop our own set of aspects and a weaver for code generation targeting the hardware templates mentioned above. A key novelty of LARA is the ability for designers to specify design patterns and characteristics of the target platform.

E. **Design Patterns**
We rely on design patterns [11] to derive effective target architectures, as well as to promote the reuse of best hardware design practices. We intend to use extended hardware templates with parameterization and internal dynamically reconfigurable features so that they can be easily adapted to changing scenarios and hence be used in the context of dynamic mappings.

A rich set of algorithm-level design patterns will be used to drive the implementations of the applications. These patterns and their variants can be captured using multiple aspects for mapping and generation. To meet the reliability goals of safety-critical functions, fault-tolerant design patterns such as TMR-voting, command/monitor redundancy, cross-lane signal consolidation will be employed as architectural-level design.

F. **Verification of Target Implementations by Test Generation**
We are investigating extensions to the use of commercially available toolsets for verifying the correctness of generated multi-core target configurations. This is essential for the long-term deployment of reconfigurable computing in safety-critical avionics and automotive applications. The approach will focus on the generation of a comprehensive set of test cases from the application models that are then executed on the target configuration implementation to ensure the following:

- The semantics of the functional behavior of library blocks is used to enumerate equivalence classes of tests and a search through the elements of the configuration to ascertain sequences of system inputs and expected outputs for comprehensive classes of functional and structural coverage.
- Enumeration of variants of these equivalence classes is generated for run-time reconfiguration options for reliability – making sure that each reconfiguration will produce safe and correct functional behavior.

We are leveraging the extensive experience of the industrial partners in this project to provide and adapt their best practices in this domain and take advantage of the aspects mechanism to automate the verification approach.

G. **Compilation Infrastructure and Synthesis Tools**
The proposed project will rely on the commercially available compiler infrastructure from the ACE Technologies partner. In addition to these tools, we will make use of popular FPGA vendor tools to generate complete hardware designs thus validating the entire design flow.

H. **Synthesis Templates in VHDL**
In addition to the use of commercially available tool sets, we are developing our own prototype tools allowing designers to specify synthesis and mapping aspects. We are working closely with our ACE project partner to interface with their internal compiler representation (CCMIR) so that the knowledge provided by our Aspect tool can be migrated into their CoSy™ compiler [19]. Understandably some of the partners will not adopt an entirely new infrastructure, but instead will establish an interface with their internal and proprietary tool sets.

11.5 Applications

We now describe a set of realistic applications provided by the REFLECT project industrial partners, to evaluate the potential of the core techniques and methodologies under development. The current applications include: a *3D Path Planning*, a *Stereo Navigation System*, an *MPEG 2 (layers I and II) audio encoder*, and a *G729 voice encoder*. They represent computationally intensive applications and are expressed as C implementations. They are amenable to performance, footprint, and cost improvements on configurable computing platforms – provided proper mapping, and generation technology is available. In this preliminary evaluation we describe specific metrics of improvement over current implementation approaches exercising the key features of our technology such as aspect mapping, design patterns, and translation to hardware. We now briefly describe two of the four applications referred.

Three-dimensional (3D) Path Planning: This kernel plans a 3D path between the current position of an autonomous robotic vehicle and a predetermined goal position. The algorithm uses a map of the environment with obstacles and uses Laplace's equation for collision avoidance by constructing a path $r(t)$, through the domain by assigning a potential value of $v(r) = 0$ for r on any boundaries or obstacle, and a potential of $v(r) = -1$ for r on the goal region. The Laplace's equation is solved in the interior of the 3D region leaving a global minimum of $v(r) = -1$ for r on the goal region, and global maxima of $v(r) = 0$ for r on any boundaries or obstacle. A path from any initial point $r(0)$, to the goal, is constructed by following the negative gradient of the potential v.

A numerical approach to solve the Laplace's equation can be obtained by gridding the domain, and iteratively setting the potential at each interior point, equal to the average of its nearest neighbors. By varying the grid size (halving or doubling cell length at each step) from the coarsest that still leaves paths between obstacles, to the finest that is required for smooth paths, the iteration can be made to converge in a time proportional to the number of cells in the finest grid. The solution on coarse grids is computationally inexpensive, and is used to initialize the solution on finer grids (see e.g. [12, 13]). As an output, the algorithm estimates a vehicle's position and velocity.

The C implementation is structured as a sequence of three, main tasks at a very top level, namely obstacle map producer, collision avoidance, based on a multi-grid Laplacian solver and a speed controller. The solver, the computationally more expensive task is internally structured as a sequence of four tasks, namely grid initialization, interpolation (with approximately 14% of the execution time), iteration (with approx. 80% of the execution time), and updating.

MPEG-1/2 Audio Encoder: We consider the code for MPEG-2 Layers I and II audio encoding with more than two channels (up to 5.1 channels) MPEG [14, 15]. Internally the C implementation is structured into three tasks, namely, input filtering, quantization and encoding.

The input audio stream passes through a filter bank that divides the input into multiple frequency sub-bands. The input also passes through a psychoacoustic model that determines the ratio of the signal energy to the masking threshold for each sub-band. The bit or noise allocation block uses the signal-to-mask ratios to decide how to assign the total number of code bits available for the quantization of the sub-band signals to minimize the audibility of the quantization noise. Finally, the last block takes the representation of the quantized sub-band samples and formats the data and side information into a coded bitstream. In this application, the function consuming most execution time is the *filter_subband* (approx. 31%) followed by *II_f_f_t* (approx. 14%).

11.5.1 Mapping Opportunities

The applications currently considered are real-time applications with tight execution time constraints that may require different implementation approaches according to the characteristics of the target architecture. As such they have the potential to take advantage of the configurability of FPGA-based architectures in aspects such as parallelization, both fine-grained and coarse-grained, as well as specializations and word-length optimizations.

These applications allow the evaluation of the REFLECT's design flow over all its dimensions, from transformations and optimizations to meet non-functional requirements, to the flexibility on targeting FPGA-based systems with different characteristics.

The applications provide a vast number of mapping opportunities and the possibility to achieve hybrid patterns of fault-tolerance – such as triple module redundancy (TMR), self-checking pairs, etc. Furthermore, opportunities such as data-streaming, task-level pipelining, data-flow parallelism, data replication and distribution are present in all the applications currently considered.

11.6 Aspects and Strategies

There are several concerns that can be captured as aspect-oriented modules. Examples include the addition of code to monitor, debug, and visualize data. In addition, by specifying certain concerns, aspect and design pattern modules can be used to specialize the implementation of an application code when targeting heterogeneous hardware and/or software architectures, thus deriving customized solutions. Typically, we are interested in different implementations for the same input code, so that the code itself does not become too architecture-specific.

In this context we distinguish three (3) main abstractions in the REFLECT compilation/synthesis flow, namely:

- **Application aspects**: These Aspects allow developers to specify application characteristics such as precision representation, input data rates or even reliability requirements for the execution of specific code sections. These features act as "requirements" of the design solution sought and cannot be easily expressed using common programming languages like C. In REFLECT high-level aspects are usually associated to the specification of non-functional requirements such as "throughput of 30 frames per second" and "precision error in the output data less than 1E-3". These aspects are internally decomposed into a number of low-level aspects that, when applied in sequence, guide the design flow to generate an implementation which meets the input requirements. Some low-level aspects used and the ordering to be applied can be specified by the user using strategies[1] or defined by a Design Space Exploration (DSE) approach. Strategies are defined as particular specifications of the aspects to apply, including a particular ordering. Ultimately, strategies can be seen as rules that enforce the design flow to apply a specific design pattern.

- **Design patterns**[2]: These design patterns act as a collection of transformations or "actions" with pre-conditions and post-conditions to be used to transform the code in search for specific implementations that meet the desired goals as specified by the applications aspects. These transformations are defined over the program representation symbols, such as variables, procedures over which we have defined a series of attributes. The pre-conditions and post-conditions of transformations are defined in terms of these attributes or newly introduced attributes, so that the compilation system can reason about the legality of specific sequence of transformations. For example, an execution time requirement for a specific code section might require the concurrent execution of a specific function. This in turn will require the use of a design pattern or transformations (applying strategies) that performs loop unrolling and data partitioning, so that data are available to all the concurrently executing units.
- **Hardware/software templates**: These templates, including a mix of hardware and software implementations, define the "lower" layers of the mapping hierarchy as they define the characteristics of the target resources. These templates are the components used to implement a solution given the design patterns used. The hardware versions of a parameterized FIFO or streaming buffer can be considered hardware templates. On the other hand, the software versions of the same components can be considered software templates. Most hardware/software templates are characterized in terms of resource usage and number of clock cycles in a specific custom design (e.g., as in FPGA).

Overall, the developer defines as a first approach the application aspects related to the code at hand, relying on a wealth of existing design patterns and hardware/software templates together with DSE support to find a suitable set of transformations or design patterns that can lead to a specific feasible implementation. The proposed design flow or compilation/synthesis framework lays the foundation for the development of mapping strategies, which can and should be complemented with the inclusion of cost evaluation functions to define "goodness" criteria among multiple feasible designs.

---

[1] The term is used herein in a more generic way than in [16].

[2] In practice, design patterns try to fulfill high level requirements that require a global strategy to be accomplished. The selection of the strategies is based on heuristics extracted from good design practices.

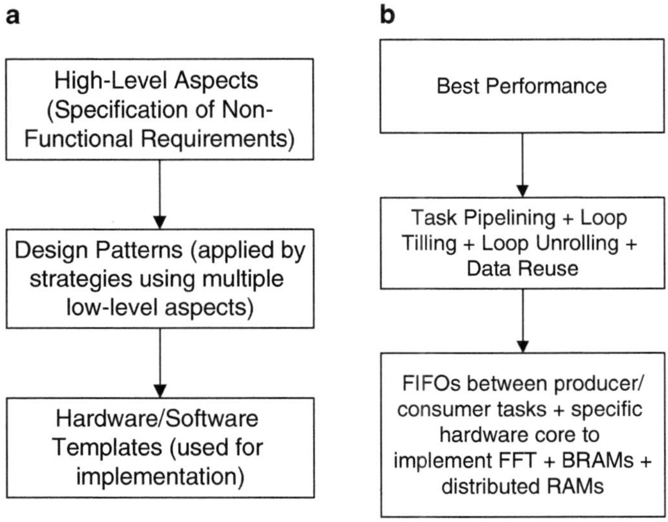

Fig. 11.4 REFLECT's hierarchy of concepts: (**a**) the three concepts; (**b**) an example of the concepts

The REFLECT's compilation flow depicted in Fig. 11.2, with major concepts depicted in Fig. 11.3, will benefit from aspects to produce better FPGA implementations. The approach, however, is also applicable to other contemporary reconfigurable and non-reconfigurable computing architectures. Figure 11.4 shows and exemplifies the three levels of concepts in REFLECT.

We envision the use of aspects, strategies, and design pattern modules to focus on four types of features:

- **Specializing**: Specialization of an input code to be more suitable for the particular target system (e.g., specializing data types, numeric precision, and input/output data rates);
- **Mapping and guiding**: Specification of mapping actions to guide the tools in some decisions (e.g., mapping array variables to memories, specifying FIFOs to communicate data between cores).
- **Monitoring**: Specification of which implementation features, such as current value of a variable or the number of items written to a specific data structure, provide insight for the refinement of other implementation-related aspects.
- **Retargeting**: Specification of certain characteristics of the target system in order to make the tools adaptable and aware of those characteristics (i.e., retargetable).

An important component of the aspect-oriented programming model is the notion of a *weaver*. A weaver is a compilation component that receives as input the code of an intermediate representation of the application along with the aspect modules, and produces a new version of the code or of the intermediate representation for the application as result of applying the descriptions (rules) in the aspect modules.

The aspect modules usually have *select-apply* type operations,[3] being *select* the operator where *join points*[4] are specified and *apply* the action to be done over that *join points*. An example of *select* and *apply* operations include "find invocations of functions" and "test if array arguments have size greater than 0", respectively. In this case, the weaver will insert additional code at each function invocation site testing if array arguments have size greater than zero.

The REFLECT's design flow relies on a *weaver* per design flow stage. Each weaver is responsible for the aspect modules at the appropriate level/stage in the design flow. The main stages of the REFLECT's design flow use partner's specific tools leveraging specific optimizations which are being extended to enhance the REFLECT's design flow. As an example, Harmonic [18] is used as a source-to-source transformation tool and it is responsible for analyzing and giving hints about code compliance[5]; partitioning the input applications in software/hardware components; inserting communication primitives, and providing support for code execution monitoring. Aspects related to these tasks are identified by the Aspect Front-End tool and input to Harmonic in an intermediate representation (Aspect-IR). Harmonic also performs cost estimations for a given platform to assist in the software/hardware partitioning of the input application code. When partitioning between hardware and software, Harmonic extends the software components on the host processor with primitives to communicate data and synchronize the overall execution of the hardware/software system.

Strategies, defined as sequences of application of aspects, are described in LARA using constructs based on scripting languages. These strategies enhance the Design Space Exploration (DSE) via the practice of try-and-feedback schemes, implementation of the design patterns and their strategies, and alternative flows for host simulation and target compilation. The C code output from Harmonic is input to a CoSy™-based compiler is then responsible for a number of compiler optimizations. These compiler optimizations may be guided by strategies specified by the user or resultant from the DSE. The CoSy™ compiler directly invokes the subsequent design flow stages, including the weavers to implement some aspects, further target-required optimizations and transformations, and word-length optimizations.

These design flow stages communicate through CDFG-IR, a common intermediate representation implemented internally using CoSy™'s CCMIR [20] and extended with annotations.[6] This representation is common among the design flow stages integrated as depicted in Fig. 11.2. Then, the compiler is responsible for the generation of the VHDL-RTL code (hardware components) and RISC code (software component). A further design flow for the software components may include the generation of a low-level C representation of the part of the application that runs in

---

[3] AspectJ [17], a well-known AOP approach for Java, uses pointcut and advice as operators.

[4] They can refer to the entire application, a specific function, all or a specific loop, or even a specific location in the code for which an annotated label in the C code is used.

[5] E.g., the C-to-VHDL generator used in the back-end of CoSy may not support all C constructs.

[6] It includes annotations of customized bit-widths, statement latencies and scheduling, etc.

the microprocessor and feed it through a specialized compiler and linker (such as `mb-gcc`, or `ppc-gcc`).

The CDFG-IR carries all information between the compilation stages: the transformed and gradually specialized and mapped representation of the application, and all kinds of attributes, not only simple attributes (such as memory spaces of variables) and structured (such as loop-nest information and dependences), but also those that support aspects. At some point in the design flow, the intended partitioning is reflected in the CDFG-IR by creating one partition per target architecture, thus separating the design flows for each component.

The optimizations performed by CoSy™ will be visible at the CDFG-IR representation. The CDFG-IR is then used by the hardware optimizer. This stage uses sophisticated methods for word-length analysis to optimize the word-lengths and representations of the variables and computations. This tool outputs a new version of the CDFG-IR that is then used by the hardware generation stage to generate the optimized VHDL code of the hardware core. Lastly, we use a tool based on the Dwarv workbench [21] to generate VHDL for the hardware kernels using the same CDFG-IR and the same options for arranging the order of transformations as described above. In particular, it applies transformations required to translate a computation to a structural model more suitable for FPGAs, implement the weavers meant for this part of the design flow, and carry out the design space exploration (DSE) for generating high-quality VHDL descriptions of hardware modules.

We consider aspect lifetime as the span in the compilation flow during which an aspect can be used for weaving. Low-level aspects will be used by strategies to allow a specific design pattern application, whereas high-level aspects are decomposed in strategies that specify a sequence of low-level aspects to be applied. The application of the low-level aspect features is based on priorities and the ordering specified by the user. Possible conflicts among aspect modules and tools need to be reported, giving a chance to the user to resolve those conflicts. The tools in the design flow will report the applicability of each aspect module when applying a given strategy.[7]

11.7 Code Examples and Design Results

We now present examples of the specification the application of transformations and mapping aspects to realistic code kernels derived from the set of application we are using in the evaluation of the REFLECT design flow. We begin with a description of the codes and aspects followed by the corresponding implementation results targeting a FPGA-based architecture.

[7] During the design flow a join point might be removed by a previous tool's action or might be different from the original when a previous tool transforms the code.

```
common.h: ...
#define FAR /*far*/ ...
#define      SBLIMIT      32 ...
#define      HAN_SIZE     512 ...
encode.c: ...
void filter_subband(double FAR z[HAN_SIZE], double FAR s[SBLIMIT] ) {
  double y[64];
  int i,j; ...
  typedef double MM[SBLIMIT][64]; // array of constants (bank filters) loaded from a file
  ...
  for (i=0; i<64; i++)
     for (j=0, y[i] = 0; j<8; j++) y[i] += z[i+64*j];
  for (i=0; i<SBLIMIT;i++)
     for (j=0, s[i]= 0; j<64; j++) s[i] += (*m)[i][j] * y[j];
} ...
```

Fig. 11.5 Original C source code for function *filter_subband* from the MPEG Encoder application

```
aspectdef maxmizePerformance
  select A: *.function{name="filter_subband"}
  apply to A: map to hardware(1)
  apply to A: strategy1
  constraints:
    function{name="filter_subband"}.arg{name="s"}.noise_power <= 1e-3;
    function{name="filter_subband"}.arg{name="z"}.range = [-40..120];
  end
end
```

Fig. 11.6 An aspect module invoking a strategy with the goal to maximize performance and including constraints for word-length optimizations

11.7.1 Code Example and Mapping Aspects and Strategies

Figure 11.5 presents a C source code implementation of the Polyphase Filter Bank component of an MPEG-2 Audio Encoder (layers I and II). This filter component is responsible for a large fraction of the execution time of the MPEG Encoder (about 31.5% when profiling with a number of input *pcm* sources on a Pentium4 processor). The function receives 512 audio samples and outputs 32 equal-width frequency sub-bands.

An aspect module can be used to specify a concern such as "implement it trying to achieve the minimum latency". To achieve this goal, different strategies (using different design patterns) are pursued as is the case of using loop unrolling and data reuse. The aspect module in Fig. 11.6 considers the execution of a specific strategy ("strategy1"). The constraints defined in the aspect module (see the *range* and *power_error* attributes in Fig. 11.6) will be used by the word-length optimization included in the hardware optimizer stage.

```
aspectdef localArrays2DistMemories
begin
  select A: function{*}.var{*}
  apply to A: map $var to DistRAM
  condition: $var.is_array && $var.scope == "local" && $var.size <= 2048
end
```

a

```
aspectdef localArrays2BlockRAM
begin
  select A: function{*}.var{*}
  apply to A: map $var to BlockRAM
  condition: $var.is_array && $var.scope == "local" && $var.size <= 2048
end
```

b

Fig. 11.7 Aspect modules for mapping array variables to local on-chip memories: (**a**) to distributed memories; (**b**) to block memories

Figure 11.7 shows two aspects that can be used by two different strategies. One strategy guides the hardware compiler to map local array variables smaller than a certain size to distributed memories (Fig. 11.7a) and the other to block RAMs (Fig. 11.7b).

We now describe possible optimizations and design patterns that can be specified by strategies using LARA.[8]

A possible optimization relies on code motion and partial scalar replacement to reduce the number of accesses to arrays *y* and *s* thereby reducing by 36.56% the total number of loads and by 96.25% the total number of stores. The array *m* stores the table of coefficients used by the filter. This array can be mapped to a local memory (e.g., BRAM in the case of Xilinx FPGAs). When mapped to a BRAM its values can be part of the FPGA bit-stream eliminating any overhead for the BRAM initialization. This is a very efficient implementation as in a software variant the values need to be stored in the memory addresses corresponding to the elements of the array in the first time the function is executed.

Figure 11.8 depicts an implementation of the function taking advantage of the fact that both *y* and *m* can be mapped to local memories with two read and two write ports (e.g., distributed RAMs or BRAMs present in Xilinx FPGAs). In this case, the implementation can take advantage of two simultaneous memory accesses (reads or writes). The outermost loop in the code is unrolled twice and then *jammed.* The innermost loop of the second loop block is unrolled twice. This allows the implementation to carry out two simultaneous write operations to array *y* and two simultaneous read operations from array *m*.[9]

---

[8] LARA is currently being designed and the version presented here uses and extends some of the concepts proposed in [22].

[9] Strength reduction can also be applied to optimize the calculation of array indexing of z, i.e., 64*j used in z[i+64*j] and in z[i+1+64*j], lines 11 and 12 in Fig. 9. The expression 64*j can be translated to j<< 6 which can be implemented in hardware with the 26 least significant bits of j concatenated with six zeros.

```
void filter_subband(double z[512], double s[32]) {

  double y[64], s_aux, y_aux1, y_aux2, t1, t2;
  int i,j;
  static const double m[32][64] = {...}; // 2048

  for (i=0;i<64; i+=2) { // unroll 2x outer loop and jam
    y_aux1 = 0.0;
    y_aux2 = 0.0;
    for (j=0, j<8;j++) {
      y_aux1 += z[i+64*j];
      y_aux2 += z[i+1+64*j];
    }
    y[i] = y_aux1;
    y[i+1] = y_aux2;
  }
  for (i=0;i<32;i++) { // unroll 2x inner loop
    s_aux= 0.0;
    for (j=0, j<64; j+=2) {
      t1 = m[i][j] * y[j];
      t2 = m[i][j+1] * y[j+1];
      s_aux += t1+t2;
    }
    s[i] = s_aux;
  }
}
```

Fig. 11.8 Code of the function *filter_subband* after partial unrolling and jam

This code example illustrates various opportunities for different implementations. A different possibility is to map the array variable *y* to a rotating register with 64 stages. The first loop block is responsible for creating the 64 values and then the 64 values are used 32 times. This is ensured by the rotating register, as for every 64 values the rotating register is in the initial state (i.e., with the same values in the same positions as after its creation in the first loop block).

The two sets of nested loops in the *filter_subband* function correspond to typical producer/consumer behavior. In this case, we can use task-level pipelining to allow an implementation that executes concurrently both sets of loops and the values of the original array variable y are now communicated by a streaming buffer and using blocking versions of *peek*, *get*, and *put* streaming buffer access primitives. In this implementation, for each calculation of *y* in the first set of loops (i.e., for each iteration of the outermost loop in the first set of loops) the innermost loop of the second set of loops is able to perform the 64 iterations.

```
aspectdef tofixed
begin
  select A: function{name="filter_subband"}.var{name="s", "y", "z", "m"}
  apply to A: $var.type = "signed fixed(64,36)"
end
```

Fig. 11.9 An aspect module for changing the types of specific variables in function *filter_subband* to a fixed-point representation

Data type conversions are especially important in the presence of architectures that allow custom implementations. These conversions can be specified by the user using aspect modules. For instance, Fig. 11.9 presents an aspect to instruct the compiler to change the types of a set of variables to a fixed-point representation.

A second and more advanced option is the integration of word-length optimizations in the compiler. In this case, aspect modules may specify non-functional requirements (see the constraints defined in the aspect module of Fig. 11.6) that drive the word-length optimizations to specific conversions that continue meeting those requirements. One type of word-length optimization is scaling analysis, which converts floating-point descriptions into descriptions using fixed-point representations. The word-length optimization will be also responsible for inferring the sufficient word-lengths for the loop control variables. In this case, different variables to control each loop will be used and each one will have a specific word-length according to the initiation and limit values. Thus, for the *filter_subband* example we will obtain: first nested loops: 7 bits for *i*, and 4 bits for *j*; second nested loops: 6 bits for *i*, and 7 bits for *j*.

Other concerns specified in the strategy used for this example are only applied at the hardware generation stage. Examples of such concerns are the mapping of certain array variables to Xilinx BRAMs or to Xilinx Distributed RAMs. Aspect definitions may also affect the hardware generation stage by defining the specific behavior to access a hardware component (e.g., IP core, memory).

Yet another alternative implementation can be obtained by a strategy derived using an aspect defining the primitives to access data-streaming buffers as illustrated in Fig. 11.10.

11.7.2 Implementation Results

We conducted a set of implementation experiments with partial automation of the REFLECT compilation and synthesis tool flow targeting a Xilinx XC4VFX100 FPGA accommodating custom-computing units in the Molen Reconfigurable architecture. In this section we report results for the *filter_subband* function from the *MPEG-1/2 encoder* application and from the *gridIterate* function from the *3D-Path*

```
aspectdef MapToStreamingBuffer
begin
  // map array variable to a streaming buffer when there are only
  // reads or only writes to the array variable
   select A: function{*}.arg{*}.array{*}
   apply A: map to StreamingBuffer($array.name, input)
   condition: $array.writes == 0
   apply A: map to StreamingBuffer($array.name,output)
   condition: $array.reads == 0
end
```

Fig. 11.10 An aspect module to map specific array variables to streaming buffers

Table 11.2 *MPEG-1/2 encoder* and *3D Path Planning* applications: implementations considered

Implementation	Data types	Description
fsubb-base-d	Double	Array m as argument
fsubb-base-f	Float	Array m as argument
fsubb-sr	Float	Scalar replacement and code motion
fsubb-unroll+jam	Float	Scalar replacement, code motion, and using unroll and jam
fsubb-fixed	Fixed-point	Array m as argument
fsubb-hw1	Float	Using a hardware template to perform multiply and add of single-precision (32 bits) data types
gridIt-base-f	Float	Original, but using unidimensional arrays
gridIt-fixed	Fixed-point	Original, but using unidimensional arrays

Planning application described in Sect. 11.4. The mapping and place-and-route results were obtained using the Xilinx ISE v9 tool.

To illustrate the use of Aspects and Design Patterns we obtained different implementations for the two functions as described in Table 11.2.

Figure 11.11 shows the number of slices and the execution time for the different implementations of the *filter_subband* function, respectively. The implementations use Xilinx DSP48s blocks. The first implementation of the function (*fsubb-base-d*) uses 16 DSP48s blocks and the others only four. The maximum clock frequencies reported by ISE after placement-and-routing are similar to the one reported for the first implementation being the highest differences with the *fsubb-sr* (1.14×) and *fsubb-fixed* (1.12×). As observed in Fig. 11.11, there are substantial differences among the implementations. As expected, the implementation using fixed-point representations (*fsubb-fixed*) is the one requiring fewer slices.

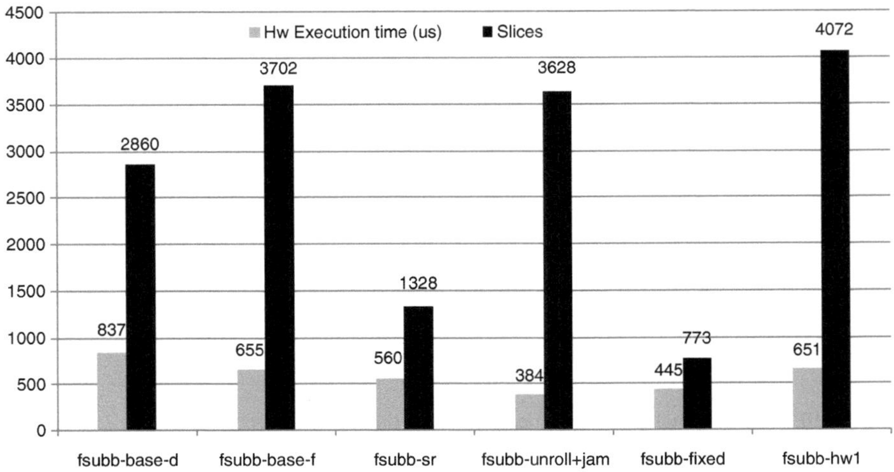

Fig. 11.11 Used slices and execution time for different filter-subband function implementations

Table 11.3 Results for the *gridIterate* function used in the 3D Path Planning application

Implementation	Slices	DSP48s	Max. clock freq. (MHz)	Latency (#ccs)	Hw execution time (us)
gridIt-base-f	2,757	4	130	7,311,750	56,191
gridIt-fixed	4,768	0	125	5,539,000	44,245

As observed in Fig. 11.11 the *fsubb-unroll + jam* implementation variant achieves a speedup of 2.18× over the first implementation. Other noteworthy implementations exhibit speedup of 1.88× and 1.49× for *fsubb-sr* and for *fsubb-fixed*, respectively. These results confirm the importance of applying different design patterns as for a particular solution we may not be interested in using the resources needed for the *fsubb-unroll + jam* implementation and thus may prefer the use of *fsubb-sr* or the use of *fsubb-fixed*. These particular implementations may be selected according to energy consumption figures, FPGA area available for a particular CCU, and/or execution time deadlines.

The implementation referred as *fsubb-fixed* is included here to show the current support of the tool chain to hardware templates supplied by the user or already existent in the library.

Table 11.3 shows results when considering the *gridIterate* function from the 3D Path Planning application. The first implementation (*gridIt-base-f*) uses single-precision floating-point data types (*float*) and the second implementation (*gridIt-fixed*) uses fixed-point data types. A speedup of 1.27× was achieved by using fixed-point representations. Note, however, that the fixed-point representations used 64 bits instead of the 32 bits of the single precision floating-point representation and this has increased the number of used slices. The *gridIt-fixed* implementation does not need to use multipliers as the involved multiplication is implemented with a shift and an addition.

11.8 Related Work

Many previous efforts have addressed the problem of mapping of computations described in high-level programming languages to FPGA-based systems (see, e.g., [23, 24]). The approach proposed in REFLECT is naturally influenced by experiences of the project members (e.g., [25–35]), in particular the hArtes [36] project.

In an effort to make reconfigurable architectures accessible to a wider range of programmers as well as to take advantage of a large software basis currently developed and tested for imperative programming languages, various efforts, both in academia and in industry, aimed at the automatic translation of such applications codes directly to hardware. Examples of these efforts include the Handel-C language [37], a derivate of the C language which exposes hardware details to designers such as parallelism, resource usage and word-lengths for variables and expressions, Mitrionics' [38], a derivate of the C language targeting the Mitrion Virtual Processor, and Nallatech's Dime-C [39], a subset of the ANSI-C language annotated with hardware pragmas indicating specific DimeTalk components.

Other efforts relevant to our approach are the ones combining parallelizing compiler technology with high-level synthesis technology. An example is DEFACTO – a Design Environment for Adaptive Computing TechnOlogy [25] automatically maps computations written in high-level imperative programming languages such as C to multi-FPGA-based systems.

Still, other efforts have focused very specific domains by mapping a very limited subset of programming constructs (e.g., Trident [40]). Approaches, such as the Streams-C [41], use a CSP-oriented execution model with a C-like syntax, in an attempt to ease the learning curve of their language and systems. Overall, one of the problems of current high-level synthesis tools and hardware compilers from C (e.g., ROCCC [42], Impulse-C [43]) is the lack of programmer's control over hardware generation besides the use of simple constraints and a selection of very limited optimization options. Example of features extending programming languages in order to permit some designer's control over the generated hardware has been shown with Handel-C [37] in which authors have used a combination of cycle-accurate and behavioral descriptions [35]. The Mentor Graphics' Catapult-C [44] allows extensive designer input in the form of design constraints (using a GUI) to guide the applied optimizations and mapping process.

REFLECT extends the reach of these approaches by the use of aspect-oriented programming. This approach, rather than the localized use of *pragmas* or annotations (e.g., as in [45]), will allow programmers to retain the advantages of an imperative programming paradigm and installed software basis, while conveying to a compilation tool the domain knowledge about important application requirements, code characteristics, and design patterns. To the best of our knowledge, our approach is the first one to aggressively use the powerful features of aspect-oriented design to specify architecture and domain specific knowledge and different implementation strategies for the same algorithm. We envision the use of aspects to help produce optimized architectures beyond the state-of-the-art. To the best of our knowledge the use of aspect-oriented programming in the context of hardware compilation and synthesis and with the reach and breath proposed in the REFLECT project is novel.

The most relevant EU funded projects are the hArtes (see Chap. 5) and the MORPHEUS (see Chap. 4). The hArtes (Holistic Approach to Reconfigurable Real Time Embedded Systems) project aims the development of hardware platform comprising embedded processors, digital signal processing and reconfigurable hardware. The hArtes project targets the entire design cycle for a heterogeneous computing platform and using a more conventional approaches in the hardware generation for its algorithm exploration and hardware-software system co-design tool chain. Unlike hArtes, the REFLECT project does not exclusively aim at designing and development of a heterogeneous hardware platform. Instead REFLECT focuses on a hardware generation approach using design patterns and aspect-oriented strategies for guiding and controlling processes such as transformation, optimization, and mapping.

The MORPHEUS (Exploitation of reconfiguration for increased run-time flexibility and self-adaptive capabilities in future SoCs) project aims at the development of a heterogeneous reconfigurable architecture including an embedded processor and coarse- and fine-grained reconfigurable IP. MORPHEUS also aims at developing an integrated design flow supporting fast exploration of hardware and software alternatives. It envisions a multi-application environment and the use of a real-time operating system that will support such execution environment. Existing system-level tools provide the partitioning and mapping of the different tasks, and are coupled to cope with the implementation requirements of different architectures of the SoC. In contrast, REFLECT exploits the use of aspect oriented methods enabling the use of FPGA architectures using hardware templates code generation techniques with higher abstraction levels.

The approach outlined for the REFLECT project differs from these efforts in several aspects. Firstly, REFLECT takes the pragmatic approach that the input application specification has to be based on the imperative programming paradigm given the enormous investment in this paradigm in current software basis. Rather than taking the hard route of automatic extraction of knowledge from traditional sequential specifications, in REFLECT we use the concepts of Aspect-Oriented specifications to add complementary domain-specific knowledge to the input code. This approach, to the best of our knowledge is novel. Secondly, our approach also combines the information provided by aspect-oriented specifications with high-level program transformations and software engineering methodologies (*e.g.,* refactoring and successive refinement) for generating hardware and application- and domain-specific architectures. Thirdly, we will develop an intermediate reconfigurable-aware representation that allows the compiler to internally specify dynamic mapping schemes capable of taking advantage of the flexibility of our parameterizable hardware and architecture templates. Finally, and as with other projects, we will evaluate the potential for the techniques developed in the context of realistic and real-life applications kernels in a series of staged demonstrators.

11.9 Conclusion

This chapter briefly described the main objectives of the REFLECT project and presented preliminary results of the application of its underlying concepts, namely, the use of aspects-oriented techniques for mapping sophisticated computations onto FPGA-based architectures. This project aims at developing, in a unified framework design, mapping techniques that will, we believe, greatly improve the quality and design time of complex embedded and reconfigurable systems. Furthermore, and given the diversity of the techniques brought together into REFLECT, we expect some, if not all of them, to lay the foundation of newer methodologies and techniques whose lifetimes, we believe, will surpass the duration of this project.

The techniques and the approach will be evaluated using a prototype compiler and synthesis flow that includes existing academic and industrial tools. Code transformations, optimizations, and design methodology for reconfigurable computing systems, will be integrated into the CoSy™ compiler framework. Hardware/Software compilation and synthesis templates will serve as the basis for a wide range of designs of interest to our industrial partners. The work being carried-on will result in a novel compilation and synthesis system that uses application and domain-specific knowledge to map computations to the emerging reconfigurable computing architectures.

References

1. K. Paulsson, M. Hübner, and J. Becker, "Strategies to On- Line Failure Recovery in Self-Adaptive Systems based on Dynamic and Partial Reconfiguration," in *First NASA/ESA Conf. on Adaptive Hardware and Systems (AHS'2006)*, 15–18 June 2006, pp. 288–291.
2. G. Kiczales, "Aspect-Oriented Programming," in *ACM Computing Surveys (CSUR)*, special issue: position statements on strategic directions in computing research, 1996. 28(4es).
3. T. Elrad, R. Filman, A. Bader, "Aspect-Oriented Programming," in *Communications of the ACM*, Vol. 44, Issue 10, Oct. 2001, pp. 29–32.
4. G. Kiczales, J. Lamping, A. Mendhekar, C. Maeda, C. Lopes, J.-M. Loingtier, and J. Irwin, "Aspect Oriented Programming," in *Proc. of the European Conference on Object-Oriented Programming (ECOOP'97)*, Finland. Springer-Verlag LNCS 1241. June 1997.
5. J. Irwin, J.-M. Loingtier, J. Gilbert, G. Kiczales, J. Lamping, A. Mendhekar, and T. Shpeisman, "Aspect-Oriented Programming of Sparse Matrix Code," in *Proc. Int'l Scientific Computing in Object-Oriented Parallel Environments (ISCOPE'97)*, Springer-Verlag, LNCS 1343, 1997, pp. 249–256.
6. S. Vassiliadis, S. Wong, G. Gaydadjiev, K. Bertels, G. Kuzmanov, and E. Panainte, "The Molen Polymorphic Processor," in *IEEE Trans. on Computers*, Nov. 2004, 53(11):1363–1375.
7. D. Boland, and G. Constantinides, "Automated Precision Analysis: A Polynomial Algebraic Approach" in *Proc. IEEE Int'l Symp. on Field-Programmable Custom Computing Machines (FCCM'10)*, 2–4 May, 2010, pp. 157–164.
8. J. Clark, G. Constantinides, and P. Cheung, "Word-length selection for power minimization via nonlinear optimization," in *ACM Trans. Design Autom. Electron. Syst. (TODAES)*, 14(3), May 2009.

9. D. Lee, J. Villasenor, "A Bit-Width Optimization Methodology for Polynomial-Based Function Evaluation," in *IEEE Trans. on Computers*, April, 2007, pp. 567–571.
10. W. Osborne, R. Cheung, J. Coutinho, W. Luk, and O. Mencer, "Automatic Accuracy-Guaranteed Bit-Width Optimization for Fixed and Floating-Point Systems," in *Proc. IEEE Int'l Conf. on Field Programmable Logic and Applications (FPL'07)*, 27–29 Aug. 2007, pp. 617–620.
11. A. DeHon, J. Adams, M. DeLorimier, N. Kapre, Y. Matsuda, H. Naeimi, M. Vanier, and M. Wrighton, "Design Patterns for Reconfigurable Computing," in *Proc. IEEE Symp. on Field-Programmable Custom Computing Machines (FCCM'04)*, April 20–23, 2004, pp. 13–23.
12. Z. Li, and T. Bui, "Robot Path Planning Using Fluid Model," in *Journal of Intelligent and Robotic Systems*, v. 21, 1998, pp. 29–50.
13. K. Valavanis, T. Hebert, R. Kolluru, and N. Tsourveloudis, "Mobile Robot Navigation in 2-D Dynamic Environments Using an Electrostatic Potential Field," in *IEEE Trans. Sys. and Cybernetics*, 30(2), March 2000, pp. 187–196.
14. ISO/IEC 11172–3 "Information technology – Coding of moving pictures and associated audio for digital storage media at up to about 1.5 Mbit/s – Part 3: Audio".
15. ISO/IEC 13818–3 "Information technology – Generic coding of moving pictures and associated audio information – Part 3: Audio".
16. R. Lämmel, E. Visser, and J. Visser, "Strategic programming meets adaptive programming," In *Proc. 2nd Int'l Conf. on Aspect-Oriented Software Development (AOSD'03)*, Boston, Mass., March 17–21, 2003. ACM, New York, NY, USA, pp. 168–177.
17. J. Gradecki, and N. Lesiecki, *Mastering AspectJ: Aspect-Oriented Programming in Java*, Wiley, 2003.
18. W. Luk, J. Coutinho, T. Todman, Y. Lam, W. Osborne, K. Susanto, Q. Liu, and W. Wong, "A High-Level Compilation Toolchain for Heterogeneous Systems," in *Proc. IEEE Int'l SOC Conference (SOCC'09)*, Sept. 2009, pp. 9–18.
19. ACE CoSy compiler development system, http://www.ace.nl/compiler/cosy.html.
20. ACE – Associated Compiler Experts bv., "CoSy CCMIR Definition", Ref. CoSy-8002-ccmir, 2008.
21. Y. Yankova, K. Bertels, S. Vassiliadis, R. Meeuws, and A. Virginia, "Automated HDL Generation: Comparative Evaluation," In *Proc. Int'l Symp. on Circuits and Systems (ISCAS'07)*, May 2007, pp. 2750–2753.
22. J. Cardoso, P. Diniz, M. Monteiro, J. Fernandes, and J. Saraiva, "A Domain-Specific Aspect Language for Transforming MATLAB Programs," in *Domain-Specific Aspect Language Workshop (DSAL'2010)*, part of *AOSD'2010*, Rennes & Saint Malo, France, March 15–19, 2010.
23. J. Cardoso, and P. Diniz, *Compilation Techniques for Reconfigurable Architectures*, Springer, Oct. 2008.
24. J. Cardoso, P. Diniz, and M. Weinhardt, "Compiling for Reconfigurable Computing: A Survey," in *ACM Computing Surveys (CSUR)*, Vol. 42, No. 4, Article 13, June 2010, pp. 1–65.
25. P. Diniz, M. Hall, J. Park, B. So and H. Ziegler, "Automatic Mapping of C to FPGAs with the DEFACTO Compilation and Synthesis Systems," in *Elsevier Journal on Microprocessors and Microsystems*, Vol. 29, Issues 2–3, 1 April 2005, pp. 51–62.
26. J. Cardoso, and H. Neto, "Compilation for FPGA-Based Reconfigurable Hardware," in *IEEE Design & Test of Computers Magazine*, March/April, 2003, vol. 20, no.2, pp. 65–75.
27. M. Weinhardt, W. Luk, "Pipeline vectorization," in *IEEE Trans. on CAD of Integrated Circuits and Systems*, 20(2), 2001, pp. 234–248.
28. Q. Liu, G. Constantinides, K. Masselos, and P. Cheung, "Combining Data Reuse with Data-Level Parallelization for FPGA Targeted Hardware Compilation: A Geometric Programming Framework", in *IEEE Trans. on Computer-Aided Design*, Vol. 28, Issue 3, March 2009, pp. 305–315
29. D. Lee, A. Abdul Gaffar, O. Mencer, and W. Luk, "Optimizing hardware function evaluation," in *IEEE Transactions on Computers*, vol. 54, no. 12, Dec. 2005, pp. 1520–1531.

30. Y. Yankova, K. Bertels, S. Vassiliadis, R. Meeuws, and A. Virginia, "Automated HDL Generation: Comparative Evaluation," In *Proc. Int'l Symp. on Circuits and Systems (ISCAS2007)*, May 2007.
31. E. Panainte, K. Bertels, and S. Vassiliadis, "The Molen Compiler for Reconfigurable Processors," in *ACM Trans. in Embedded Computing Systems (TECS)*, 6(1), Article 6, Feb. 2007.
32. E. Panainte, K. Bertels, and S. Vassiliadis, "Interprocedural Compiler Optimization for Partial Run-Time Reconfiguration," in *Journal of VLSI Signal Processing*, 43(2), June 2006, pp. 161–172.
33. J. Cardoso, "On Combining Temporal Partitioning and Sharing of Functional Units in Compilation for Reconfigurable Architectures," in *IEEE Trans. on Computers*, Vol. 52, No. 10, Oct. 2003, pp. 1362–1375.
34. Q. Liu, T. Todman, J. Coutinho, W. Luk, and G. Constantinides, "Optimising designs by combining model-based and pattern-based transformations", in *Proc. 19th Int'l Conf. on Field Programmable Logic and Applications (FPL'09)*, Aug. 31-Sept. 2, 2009, pp. 308–313.
35. J. Coutinho, J. Jiang, and W. Luk, "Interleaving behavioral and cycle-accurate descriptions for reconfigurable hardware compilation," in *Proc. IEEE Symp. on Field Programmable Custom Computing Machines (FCCM'05)*, 18–20 April, 2005, pp. 245–254.
36. K. Bertels, V. Sima, Y. Yankova, G. Kuzmanov, W. Luk, J. Coutinho, F. Ferrandi, C. Pilato, M. Lattuada, D. Sciuto, A. Michelotti, "HArtes: Hardware-Software Codesign for Heterogeneous Multicore Platforms," in *IEEE Micro*, 30(5): 2010, pp. 88–97.
37. M. Bowen, *Handel-C Language Ref. Manual*, Embedded Solutions Ltd., 2.1 ed., 1998.
38. Mitrionics AB Inc., *The Mitrion Processor, Product Overview*, Sweden, 2005. http://www.mitrion.com (accessed on Dec. 2010).
39. Nallatech, http://www.nallatech.com (accessed on Dec. 2010).
40. J. Tripp, M. Gokhale, K. Peterson, "Trident: From High-Level Language to Hardware Circuitry," in *IEEE Computer*, March 2007, vol. 40, no 3, pp. 28–37.
41. M. Gokhale, J. Stone, J. Arnold, and M. Kalinowski, "Stream-Oriented FPGA Computing in the Streams-C High Level Language," in *Proc. IEEE Symp. on FPGAs for Custom Computing Machines (FCCM'00)*, April 2000, pp. 126–135.
42. Z. Guo, W. Najjar, and A. Buyukkurt, "Efficient Hardware Code Generation for FPGAs," in *ACM Trans. on Architecture and Compiler Optimizations (TACO)* Vol. 5, No. 1, Article 6, May 2008.
43. Impulse Accelerated Technologies, Inc., http://www.impulseaccelerated.com/ (Dec. 2010).
44. Catapult C Synthesis Overview, http://www.mentor.com/esl/catapult/ (Dec. 2010).
45. C. Huang, S. Ravi, A. Raghunathan, and N. Jha, "Synthesis of heterogeneous distributed architectures for memory-intensive applications," in *Proc. Int'l Conf. Computer-Aided Design (ICCAD'03)*, Nov. 2003, pp. 46–53.

Chapter 12
Conclusion

João M.P. Cardoso and Michael Hübner

Although the maturity level already achieved, Reconfigurable Computing is far from being completely exploited. Becoming increasingly important also makes reconfigurable computing technology very attractive and interesting. The interests come from diverse areas of computing, from embedded to high-performance computing, and from various domains of applications. Recognizing the importance of reconfigurable computing, the EU (European Union) funded, through its framework programs, a number of research projects during recent years.

This book presented a selection of the most interesting FP6 and FP7 EU funded projects with strong focus on reconfigurable computing. The book started with two chapters presenting the role of reconfigurable computing as a mainstream technology and discussing some of the major challenges in order to widespread the reconfigurable computing technology and approach. Following those two chapters, eight chapters presented eight prominent EU funded projects about reconfigurable computing.

Reconfigurable technology underwent phases where scientists from industry and academia thought to have an all-round solution for a variety of applications. Many projects started and had delivered a tremendous amount of new results and extraordinary experiences. Very fast, industry and academia found out, that a pure implementation of systems with reconfigurable hardware, may not be the solution to meet all requirements. Pure reconfigurable hardware solutions may not be able to achieve some of the required characteristics of the system, being performance and/or the power consumption two of the most relevant. This led then to a novel era of multicore hardware, where different kind of processing cores, like state-of-the-art processors, coarse- and fine-grained reconfigurable hardware components were combined on a

J.M.P. Cardoso (✉)
Departamento de Engenharia Informática, Faculdade de Engenharia (FEUP),
Universidade do Porto, Rua Dr. Roberto Frias, s/n, 4200–465 Porto, Portugal
e-mail: jmpc@acm.org

J.M.P. Cardoso and M. Hübner (eds.), *Reconfigurable Computing: From FPGAs to Hardware/Software Codesign*, DOI 10.1007/978-1-4614-0061-5_12,

single chip. Most of the EU projects presented in this book recognized the importance to use reconfigurable computing as a vehicle to enhance contemporary computing systems, by adding customized accelerator units, for instance, and as an important component in heterogeneous multicore computing systems. Nowadays, the trend of multiprocessor systems follows partly the results of these EU projects.

The discussion if homogeneous or heterogeneous multiprocessors, including reconfigurable hardware, provide a beneficial hardware for future applications is still ongoing. The debate is currently active in every major conference related to computer science and computer engineering, but it is being also addressed in new EU projects. The future will show what solutions finally combine the technological advantages and restrictions in terms of massive parallelism and dependability. Also, novel nano-structures will play a major role in future hardware architectures, but even this technological avenue seems to highlight reconfigurable computing as a prominent technology as it is able to dynamically adapt hardware structures according to computing demands and reliability issues.

This wave of large and interesting projects, partially funded by the EU, and being most of the representative ones included in this book, is undoubtedly going to shape ongoing and future research and the use of reconfigurable computing technology. All the projects have had the participation of industry partners (most of them are companies from Europe) and proved the importance of reconfigurable computing technology.

This book was edited with the intention to assist people interested in reconfigurable computing, on having an overview about highly relevant EU projects in the subject, and on acquiring knowledge about recent achievements in reconfigurable hardware. The authors, as well as the editors, envisioned with this book to encourage more scientists to research in this challenging domain and to propose novel technologies, methods and tools.

It is interesting to see how the projects described in this book have already addressed or are addressing the eight challenges in reconfigurable computing introduced in Chap. 3. We leave such analysis and the conclusions to the readers.

Trends show that reconfigurable hardware can be effectively used in application tailored accelerators for general purpose processors. We believe that the role of reconfigurable computing will be wider than that and it is our role (i.e., the role of the reconfigurable computing community) to make this happen!

Index

J.M.P. Cardoso and M. Hübner (eds.), *Reconfigurable Computing: From FPGAs to Hardware/Software Codesign*, DOI 10.1007/978-1-4614-0061-5,

MIX
Papier aus verantwortungsvollen Quellen
Paper from responsible sources
FSC® C105338

If you have any concerns about our products,
you can contact us on
ProductSafety@springernature.com

In case Publisher is established outside the EU,
the EU authorized representative is:
Springer Nature Customer Service Center GmbH
Europaplatz 3, 69115 Heidelberg, Germany

Printed by Libri Plureos GmbH
in Hamburg, Germany